# 162 Topics in Current Chemistry

# Advances in the Theory of Benzenoid Hydrocarbons II

Editor: Ivan Gutman

With contributions by
J. Brunvoll, B. N. Cyvin, S. J. Cyvin, I. Gutman

With 64 Figures and 58 Tables

Springer-Verlag
Berlin Heidelberg GmbH

This series presents critical reviews of the present position and future trends in modern chemical research. It is addressed to all research and industrial chemists who wish to keep abreast of advances in their subject.

As a rule, contributions are specially commissioned. The editors and publishers will, however, always pleased to be receive suggestions and supplementary information. Papers are accepted for "Topics in Current Chemistry" in English.

ISBN 978-3-662-14963-8     ISBN 978-3-540-46609-3 (eBook)
DOI 10.1007/978-3-540-46609-3

Library of Congress Catalog Card Number 74-644622

© Springer-Verlag Berlin Heidelberg 1992
Originally published by Springer-Verlag Berlin Heidelberg New York in 1992
Softcover reprint of the hardcover 1st edition 1992

The use of general descriptive names, registered namen, trademarks, etc. in this publication does not imply, even in the absence of a specific statement, that such names are exempt from the relevant protective laws and regulations and therefore for general use.

Typesetting: Th. Müntzer, Bad Langensalza;

51/3020-5 4 3 2 1 0 — Printed on acid-freepaper

# Guest Editor

Prof. Dr. *Ivan Gutman*
University of Kragujevac, Faculty of Science, P.O. Box 60,
YU-34000 Kragujevac, Yugoslavia

# Editorial Board

# Preface

In 1988 the editor of this volume of Topics in Current Chemistry together with Sven J. Cyvin of Trondheim wrote a book devoted to the theory of benzenoid molecules [Gutman I, Cyvin SJ (1989) Introduction to the theory of benzenoid hydrocarbons, Springer, Berlin Heidelberg New York]. Due to the introductory nature of that book, a number of topics in which active research is currently taking place had either to be omitted or presented in a succinct and somewhat oversimplified manner. In order to compensate for this, the same authors edited an issue of Topics in Current Chemistry (Vol. 153) entitled "Advances in the theory of benzenoid hydrocarbons" in which a large number of active researchers reported on the most recent achievements in the field. The aim of present volume is also to complement the above-mentioned book. Here, however, emphasis is given to those directions of research in which Cyvin and Gutman (separately) gave their most numerous contributions. Their own works are, of course, outlined together and in connection with the related results obtained by many other contemporary scientists.

Chapter 1 summarizes the first twenty years of Gutman's investigations of "topological" properties of benzenoid hydrocarbons. Chapter 2 is devoted to the classical, fifty-year-old problem of the structure-dependency of total $\pi$-electron energy. Chapters 3 and 4 provide a complete survey of the efforts of Cyvin's group (as well as of several other research teams) on the enumeration and classification of benzenoid systems and benzenoid molecules.

Kragujevac, January 1991                                     Ivan Gutman

# Preface

# Table of Contents

**Topological Properties of Benzenoid Systems**
I. Gutman . . . . . . . . . . . . . . . . . . . . . 1

**Total π-Electron Energy of Benzenoid Hydrocarbons**
I. Gutman . . . . . . . . . . . . . . . . . . . . . 29

**Enumeration of Benzenoid Systems and Other Polyhexes**
B. N. Cyvin, J. Brunvoll, and S. J. Cyvin . . . . . . . . . 65

**Benzenoid Chemical Isomers and Their Enumeration**
J. Brunvoll, B. N. Cyvin, and S. J. Cyvin . . . . . . . . . 181

**Author Index Volumes 151 – 162** . . . . . . . . . . . 223

# Table of Contents of Volume 153

Benzenoid Hydrocarbons in Space: The Evidence and
Implications
L. J. Allamandola

The Distortive Tendencies of Delocalized $\pi$ Electronic Systems. Benzene,
Cyclobutadiene and Related Heteroannulenes
P. C. Hiberty

The Spin-Coupled Valence Bond Description of Benzenoid Aromatic
Molecules
D. L. Cooper, J. Gerratt, M. Raimondi

Semiempirical Valence Bond Views for Benzenoid
Hydrocarbons
D. J. Klein

Scaling Properties of Topological Invariants
J. Cioslowski

Molecular Topology and Chemical Reactivity of Polynuclear
Benzenoid Hydrocarbons
M. Zander

A Periodic Table for Benzenoid Hydrocarbons
J. R. Dias

Calculating the Numbers of Perfect Matchings and of Spanning Trees,
Pauling's Orders, the Characteristic Polynomial, and the Eigenvectors of a
Benzenoid System
P. John, H. Sachs

The Existence of Kekulé Structures in a Benzenoid System
F. J. Zhang, X. F. Guo, R. S. Chen

Peak-Valley Path Method on Benzenoid and Coronoid Systems
W. C. He, W. J. He

Rapid Ways to Recognize Kekuléan Benzenoid Systems
R. Q. Sheng

Methods of Enumerating Kekulé Structures, Exemplified by Applications
to Rectangle-Shaped Benzenoids
R. S. Chen, S. J. Cyvin, B. N. Cyvin, J. Brunvoll, D. J. Klein

Clar's Aromatic Sextet and Sextet Polynomial
H. Hosoya

Caterpillar (Gutman) Trees in Chemical Graph Theory
S. El-Basil

# Topological Properties of Benzenoid Systems

**Ivan Gutman**

Faculty of Science, University of Kragujevac, P.O. Box 60, YU-34000 Kragujevac, Yugoslavia

## Table of Contents

**1 Introduction** . . . . . . . . . . . . . . . . . . . . . . . . . . . . 3
  1.1 Reminiscences . . . . . . . . . . . . . . . . . . . . . . . . . . 3
  1.2 Some Terminological Remarks . . . . . . . . . . . . . . . 4
  1.3 Scope . . . . . . . . . . . . . . . . . . . . . . . . . . . . . . 5

**2 Structural Features** . . . . . . . . . . . . . . . . . . . . . . . . 5
  2.1 Inner Dual, Excised Internal Structure, Branching Graph . . . . . 8

**3 Spectral Properties** . . . . . . . . . . . . . . . . . . . . . . . . 9
  3.1 Eigenvalues . . . . . . . . . . . . . . . . . . . . . . . . . . 10
  3.2 Characteristic Polynomial . . . . . . . . . . . . . . . . . . 11
  3.3 Spectral Moments . . . . . . . . . . . . . . . . . . . . . . . 12

**4 Kekulé and Clar Structures** . . . . . . . . . . . . . . . . . . . 13
  4.1 Kekuléan and Non-Kekuléan Benzenoid Molecules . . . . . . . 13
  4.2 Bounds for the Kekulé Structure Count . . . . . . . . . . . 15
  4.3 A Simple Invariant of the Kekulé Structures . . . . . . . . 16
  4.4 Sextet Polynomial . . . . . . . . . . . . . . . . . . . . . . . 18
  4.5 Corals . . . . . . . . . . . . . . . . . . . . . . . . . . . . . 21

**5 Topological Indices** . . . . . . . . . . . . . . . . . . . . . . . 23
  5.1 Wiener Index . . . . . . . . . . . . . . . . . . . . . . . . . . 23

**6 Bibliographic Note** . . . . . . . . . . . . . . . . . . . . . . . . 25

**7 Addendums** . . . . . . . . . . . . . . . . . . . . . . . . . . . . 25

**8 References** . . . . . . . . . . . . . . . . . . . . . . . . . . . . 26

Topics in Current Chemistry, Vol. 162
© Springer-Verlag Berlin Heidelberg 1992

Ivan Gutman

The article reports investigations of the "topological" properties of benzenoid molecules which the author has performed in the last 20 years. Emphasis is given on recent developments and other scientists' contributions to these researches. Topics covered in recent books and reviews are avoided. The article outlines spectral properties, some aspects of the study of Kekulé and Clar structures, the Wiener index as well as a number of graphs derived from benzenoid systems (inner dual, excised internal structure, Clar graph, Gutman tree, coral and its dual).

# 1 Introduction

## 1.1 Reminiscences

As a young assistant at the "Ruđer Bošković" Institute in Zagreb, Yugoslavia the present author discovered in the library a fascinating paper by Dewar and Longuet-Higgins [1], reporting the remarkable formula

$$\det A = (-1)^{n/2} K^2 . \tag{1}$$

This happened somewhere around the end of 1970 or in early 1971. Already then, and ever since, the author was deeply impressed by the beauty and power of the Dewar — Longuet-Higgins formula and a great part of his long-lasting interests and activities in the theory of benzenoid systems can be related to Eq. (1).

Formula (1) has, at least, three noteworthy features. First, its left-hand side is the determinant of the adjacency matrix A, hence a genuine algebraic object. Its right-hand side has a purely combinatorial interpretation: K is the number of Kekulé valence formulas (in the language of chemistry) or the number of perfect matchings (in the language of mathematics). Thus Eq. (1) connects two seemingly unrelated fields of mathematics — linear algebra and combinatorics. (Recall that in Eq. (1) n stands for the order of the matrix A i.e. the number of carbon atoms of the respective benzenoid molecule i.e. the number of vertices of the respective molecular graph.) Second, A is related with the Hamiltonian operator in the Hückel tight-binding molecular orbital approach whereas K is a typical quantity appearing in valence-bond and resonance-theoretical considerations. Thus Eq. (1) connects two seemingly unrelated fields of quantum chemistry — molecular-orbital theory and resonance theory. The original intention of Dewar and Longuet-Higgins [1] seems just to be the revealing of this kind of interrelation between chemical theories. Third, Eq. (1) is not obeyed by all polycyclic conjugated hydrocarbons. Conjugated systems possessing rings whose sizes are different than six often fail to satisfy the Dewar — Longuet-Higgins formula. In other words, by means of Eq. (1) a special class of polycyclic conjugated hydrocarbons is shown to play a distinguished role in theoretical chemistry. These are the conjugated systems possessing exclusively condensed six-membered rings, traditionally referred to as benzenoid systems/benzenoid hydrocarbons.

Meditating about the "message" contained in Eq. (1) one necessarily arrives at the following two questions.

1. What is so peculiar in the structure (= "topology") of benzenoid systems that they form an outstanding and well-separated class of polycyclic conjugated molecules?

2. Are there, in addition to the Dewar — Longuet-Higgins formula, other far-reaching mathematical regularities obeyed by (and only by) benzenoids?

Attempts to find answers to the above questions resulted in numerous (not always successful) studies of benzenoid systems. In 1974 the present author published an article [2] entitled *"Some Topological Studies of Benzenoid Systems"*

Ivan Gutman

in which he communicated some of his early findings and observations. This was later considered as Part 1 of the series *"Topological Properties of Benzenoid Systems"* which nowadays embraces some 80 papers (c.f. Sect. 6). The aim of this article is to give a survey of the main directions of these investigations and to comment on them from the point of view of the most recent achievements in the field. Fortuitously, this article is being written on the twentieth anniversary of the author's encounter with the Dewar — Longuet-Higgins formula, that is twenty years after he started his journey through the magic kingdom of benzenoid molecules.

## 1.2 Some Terminological Remarks

The terminology and notation employed in the present article follows as much as it is possible that of the book *"Introduction to the Theory of Benzenoid Hydrocarbons"* [3]. There, a precise definition of a benzenoid hydrocarbon/benzen-

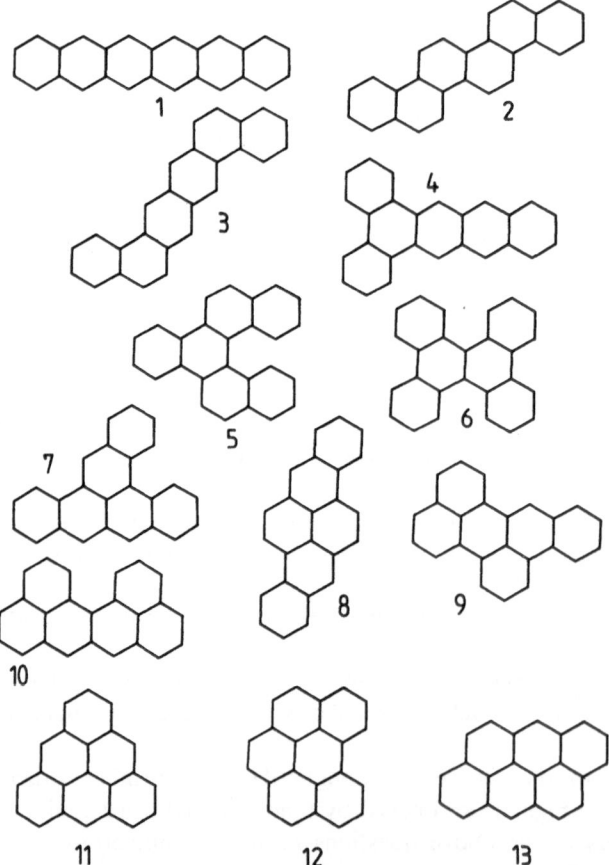

**Fig. 1.** Some benzenoid systems with six hexagons

4

oid system/benzenoid graph can be found. In Fig. 1 are depicted 13 benzenoid systems with six hexagons (out of a total of 81 hexacyclic benzenoids). From the inspection of these examples it should be perfectly clear what benzenoid systems are and how they are constructed from congruent regular hexagons.

The exclusion of nonplanar helicenic and hollow coronoid species from the class of benzenoids was maybe not fully justified from a chemist's point of view, but there were good and convincing mathematical reasons for this; anyway we use the term "benzenoid" in the same sense as in the book [3]. On the other hand, we find that it serves no purpose to strictly distinguish between benzenoid hydrocarbons (chemical objects) and benzenoid systems (mathematical objects), since this distinction is always obvious from the context. We note in passing that what we call "benzenoid system" is the same as "hexagonal system" or "hexagonal animal" in the mathematical literature.

Throughout many years the words "topology" and "topological property" were used by numerous theoretical chemists (also including the present author) with a meaning completely different to those in mathematics. This caused a considerable amount of confusion. In most cases the chemists' "topology" is synonymous to "structure" when under "structure" we understand the connectedness of the atoms in the molecule, represented by classical structural formulas. A clear and satisfactory analysis of chemical and mathematical "topologies" as well as their mutual relations can be found in a recent treatise by Merrifield and Simmons [4].

## 1.3 Scope

The present article is meant to review the author's investigations of the "topological" properties of benzenoid systems. Not all such investigations could be outlined. First of all, the work on total π-electron energy and related matter is covered by another article in the same issue. Further, basic facts and notions from the theory of benzenoid molecules were presented in the book [3] and will be repeated here only to a very limited extent. The reader's attention is also called to the first volume [5] of "*Advances in the Theory of Benzenoid Hydrocarbons*" where additional studies of the "topological properties" of benzenoid molecules can be found; they are often complementary to the present article.

The investigations outlined in this article are grouped into four sections: (a) questions concerned with the structure of benzenoid systems, (b) spectral properties and graph polynomials, (c) works related to Kekulé and Clar structures and (d) topological indices. It will become clear, however, that all these researches are intimately interrelated and that there exist quite a few unexpected connections between them.

## 2 Structural Features

In spite of numerous efforts, we are still far from being able to offer any satisfactory answer to question 1 posed in Sect. 1.1. One obvious starting point in this direction is to try to characterize those (classes of) graphs for which Eq. (1) holds. For instance, Eq. (1) applies to all acyclic graphs as well as to graphs obtained by attaching acyclic branches to benzenoid graphs. Further, the sizes of the rings need not be 6, they can be any even integer not divisible by four (i.e. 10, 14, 18, ...). Thus, the notion of a benzenoid system could be generalized, without violating the Dewar — Longuet-Higgins formula. Results of this kind were first offered in [6] and somewhat more recently (without being aware of [6]) in the works of He and He [7, 8] and Sheng [9]. It seems, however, that such generalizations are just a mathematical set-up and have little chemical relevance. Generalized benzenoid systems have hardly any chemical counterparts whereas normal benzenoid systems usually represent well-characterized molecular species [3].

A more profound structural property of benzenoid systems is the following

*Theorem 1.* Let C be a cycle of a benzenoid system. The size of C is necessarily an even integer. If the size of C is divisible by four, then in the interior of C there is an odd number of vertices. Otherwise, in the interior of C there are either no vertices or their number is even.

This result has long been known (e.g. see [6]), but its complete proof was offered quite recently [10]. One of its proper consequences is that all conjugated circuits in all benzenoid systems have sizes not divisible by four (see pp. 85–87 in [3]). Another consequence of Theorem 1 is

*Corollary 1.1.* Catacondensed benzenoid systems do not possess cycles whose sizes are divisible by four.

In other words, a cycle in a catacondensed benzenoid is of the size 6 or 10 or 14 or 18 . . . This is because catacondensed systems (by definition [3]) possess no internal vertices. On the other hand, according to Theorem 1 the existence of a cycle whose size is divisible by four implies the existence of at least one internal vertex.

Recall that in Fig. 1 catacondensed are the systems 1, 2, 3, 4, 5 and 6.

Since a catacondensed system possesses no internal vertex its perimeter embraces all the vertices. Consequently, the perimeter of a catacondensed system is a Hamiltonian cycle. In other words, all catacondensed benzenoid systems are Hamiltonian.

In pericondensed benzenoid systems a Hamiltonian cycle may, but need not exist. The problem of the existence of a Hamiltonian cycle is solved by the following result [11].

Let e be a set of edges of a benzenoid system B, such that $B^* = B - e$ is also a bezenoid system. Then the transformation $B \Rightarrow B^*$ is called an e-transformation.

*Theorem 2.* A (pericondensed) benzenoid system B possesses a Hamiltonian cycle if and only if there exists an e-transformation, $B \Rightarrow B^*$, such that $B^*$ is catacondensed.

The benzenoid system 14 is Hamiltonian because of the e-transformations:

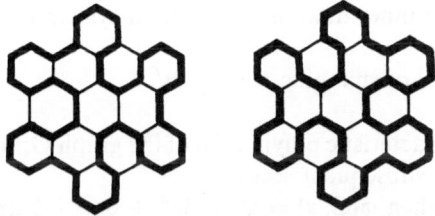

$$e = \{1, 3, 5\} \qquad 14 \qquad e = \{2, 4, 6\}$$

Consequently, 14 has two distinct Hamiltonian cycles:

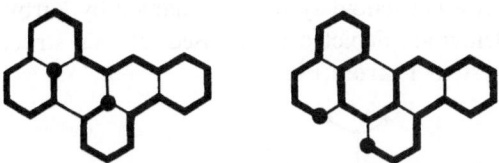

It has been claimed [11] that the Hamiltonian cycle in a Hamiltonian benzenoid system in unique. The above example shows that this is not always the case.

The pericondensed systems 7, 8, 9, 10, 11 and 13 from Fig. 1 are not Hamiltonian whereas 12 is.

Finding the necessary and sufficient conditions for the existence of a Hamiltonian path (= path which embraces all the vertices) of a benzenoid system seems to be a much more difficult task. Kirby [12, 13] calls benzenoid systems possessing a Hamiltonian path "traceable". Of course, Hamiltonian benzenoids are necessarily traceable, but the reverse is not true. There exist many traceable non-Hamiltonian benzenoids, e.g. the systems 7, 8, 9, and 13 in Fig. 1. Two Hamiltonian paths of the system 9 are indicated in the diagrams below:

In spite of recent efforts [12, 13], a complete characterization of traceable benzenoids has not yet been achieved.

7

## 2.1 Inner Dual, Excised Internal Structure, Branching Graph

A large number of graphs has been associated with benzenoid systems. Here we mention three of them, neither of which characterizing the benzenoid system up to isomorphism.

The vertices of the *inner dual* (ID) represent the hexagons of a benzenoid system. Vertices corresponding to adjacent hexagons are adjacent. For example, ID(1) is the inner dual of the systems 1, 2 or 3 (see Fig. 1) whereas ID(9) is the inner dual of the system ) 9:

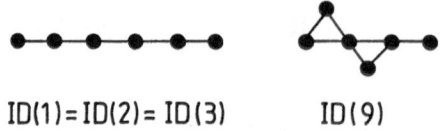

ID(1) = ID(2) = ID(3)          ID(9)

A remarkable property of the inner dual is summarized below.

*Theorem 3* [14, 15]. If ID is the inner dual of a benzenoid system B, then

$$\varphi(ID, 6) = \text{number of spanning trees of B}.$$

Here $\varphi(G, x)$ stands for the characteristic polynomial of the graph G; the definition of $\varphi(G, x)$ can be found in the subsequent section.

For example, if $G = ID(1)$ then $\varphi(G, x) = x^6 - 5x^4 + 6x^2 - 1$ and therefore both 1, 2 and 3 (from Fig. 1) have $6^6 - 5 \cdot 6^4 + 6 \cdot 6^2 - 1 = 40391$ spanning trees. Four spanning trees of the system 1 are depicted below; the fourth example is a Hamiltonian path.

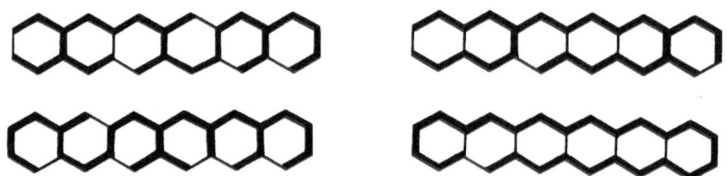

The method for counting spanning trees, described in Theorem 3, was recently further elaborated by John and Sachs [16, 17].

The subgraph spanned by the internal vertices of a benzenoid system was named by Dias [18–20] the *excised internal structure* (EIS). The subgraph spanned by the three-valent vertices of a benzenoid system was named by Kirby [12, 13] the *branching graph* (BG). Below are depicted the excised internal structure and the branching graph of the system 11 from Fig. 1:

EIS(11)                    BG(11)

The EIS-concept was much used for the enumeration and classification of benzenoid systems [18–21]. It has been demonstrated recently [22] that EIS contains information about the existence/nonexistence of Kekulé structures in the respective benzenoid molecule. An application of EIS is found in Theorem 13.

Several applications of the BG-concept were put forward by Kirby [12, 13]. It has been shown recently [23] that the number of 2-factors of a benzenoid system equals the number of 1-factors (=perfect matchings) of the branching graph. This latter result is of considerable relevance in Clar's aromatic sextet theory.

# 3 Spectral Properties

If a benzenoid hydrocarbon is represented by a molecular graph G in the usual manner [3, 21, 24] and if the vertices of G are labeled by 1, 2, ..., n (in an arbitrary order), then the adjacency matrix A of the graph G is defined via its matrix elements as

$$A_{rs} = \begin{cases} 1 & \text{if the vertices r and s are adjacent} \\ 0 & \text{otherwise} . \end{cases}$$

The eigenvalues of A will be denoted by $x_1, x_2, ..., x_n$. They form the spectrum of the graph G.

The characteristic polynomial of G is just the characteristic polynomial of the adjacency matrix. It will be denoted by $\varphi(G)$. Then

$$\varphi(G) = \varphi(G, x) = \det(x I - A) \tag{2}$$

where I stands for the unit matrix of order n. According to a well-known result of linear algebra,

$$\varphi(G, x) = \prod_{i=1}^{n} (x - x_1) . \tag{3}$$

The spectral theory of graphs is well elaborated, both in the case of general graphs [25] and graphs of interest in chemistry [24]. In this section we are concerned with the graph spectra which are specific for benzenoid systems.

One such regularity has already been mentioned, namely the Dewar – Longuet-Higgins formula, Eq. (1). Bearing in mind Eqs. (2) and (3) as well as the pairing theorem

$$x_1 + x_{n-1+1} = 0; \quad i = 1, 2, ..., n$$

the Dewar – Longuet-Higgins formula can be rewritten as

$$\prod_{i=1}^{n/2} x_i = K . \tag{4}$$

9

The identity (4) assumes that n is even. If n is odd then (in a trivial manner) both the product of graph eigenvalues and K are equal to zero.

As already discussed in detail, formulas (1) or (4) represent a very important (spectral) property of benzenoid systems. Although this result has been known since 1952, not much progress in the spectral theory of benzenoid systems has been made in the meantime.

We first mention an interesting result by Heilbronner [26] relating the inverse of the adjacency matrix with the Pauling bond order:

$$(A^{-1})_{rs} = K\{B - r - s\}/K\{B\} .\tag{5}$$

Here B is the respective benzenoid system and r and s are its two vertices. Formula (5) holds only if r and s are adjacent, a detail overlooked in [26] but eventually corrected [6]. Twenty years after the publication of the paper [26] formula (5) was rediscovered by Kiang and Chen [27], who also committed the same mistake as in [26].

A recent result on the spectra of benzenoid systems is outlined in the "Addendum".

## 3.1 Eigenvalues

The fact that certain numbers frequently occur among the eigenvalues of benzenoid systems was observed a long time ago. Hall [28] proposed an explaination of such regularities based on symmetry-relations. It should be noted, however, that such common eigenvalues are found in both symmetric and non-symmetric benzenoids.

It is well known [24, 25] that the eigenvalues of benzenoid systems belong to the interval $(-3, +3)$ and are symmetric with respect to $x = 0$. Thus the only integer eigenvalues which may occur are $-2$, $-1$, $0$, $+1$ and $+2$.

From the Dewar — Longuet-Higgins formula, it follows that a zero eigenvalue is contained in the spectrum of a benzenoid system if and only if $K = 0$. We examine this question in more detail in Sect. 4.1 where the conditions for the existence of Kekulé structures are discussed.

Eigenvalues $+1$ and $-1$ occur in very many benzenoid systems, often with considerably high multiplicities [29, 30]. This fact has attracted the attention of several authors. Dias [31] seems to be the first to have systematically examined this question. He conjectured that the divisibility of n by four implies the existence of $+1$ eigenvalues. This conjecture was shown to be false [32], but it nevertheless stimulated a large amount of work on the elucidation of the structural requirements for the existence of $+1$ eigenvalues [21, 32–34]. In spite of all progress achieved we still do not possess a complete structural characterization (in terms of necessary and sufficient conditions) for the existence of $+1$ in the spectrum of a benzenoid system.

Curiously enough, $+2$ occurs in the spectrum of only a limited number of benzenoids. Almost nothing is known about structural factors stipulating this eigenvalue.

## 3.2 Characteristic Polynomial

Since benzenoid graphs are bipartite, their characteristic polynomials can be written in the form

$$\varphi(B, x) = \sum_{k=0}^{[n/2]} (-1)^k \, b(B, k) \, x^{n-2k}$$

where $b(B, k) \geq 0$ for all values of k.

Special methods for the calculation of the characteristic polynomials of benzenoid graphs have recently been designed by Sachs and John [16, 17]. Their method is especially efficient in the case of catacondensed systems.

Denote by m(B, k) the number of k-matchings ($=$ number of ways in which k mutually independent edges are selected in B); by definition, $m(B, 0) = 1$ and $m(B, 1) = m =$ number of edges of B.

The following relations are established between the numbers $b(B, k)$ and $m(B, k)$.

*Theorem 4* [35]. For any benzenoid system B and for any value of k, $0 \leq k \leq [n/2]$,

$$m(B, k) \leq b(B, k) \leq m(B, k)^2 \,.$$

*Corollary* 4.1. The coefficient $b(B, k)$ is equal to zero if and only if $m(B, k)$ is equal to zero.

*Theorem 5.* If $k = 0, 1$ or 2 then $b(B, k) = m(B, k)$. If $k = 3$ then $b(B, k) = m(B, k) + 2 h$ where h is the number of hexagons of B. If n is even and $k = n/2$ then $b(B, k) = m(B, k)^2$ and $m(B, k) = K =$ number of perfect matchings of B.

The quantities Z and Z* defined via

$$Z = \sum_{k=0}^{m} m(G, k) ; \qquad Z^* = \sum_{k=0}^{m} b(G, k)$$

are called the *Hosoya index* and the *modified Hosoya index* [36, 37], respectively, of the molecular graph G. From Theorems 4 and 5 we immediately see that in the case of benzenoid systems Z* is strictly greater than Z. In [38] the relation between the Z and Z* indices of benzenoid molecules was examined and a good linear correlation between log Z and log Z* established.

Explicit combinatorial expressions are known for the first few coefficients of $\varphi(B, x)$ [2, 21, 33, 39, 40]. We will skip these results because in the subsequent paragraph the spectral moments are discussed at due length. Using the Newton identities [24] it is easy to compute the coefficients of the characteristic polynomial from spectral moments and vice versa.

Ivan Gutman

## 3.3 Spectral Moments

The k-th spectral moment of a graph is defined as

$$M_k = \sum_{i=1}^{n} (x_i)^k .$$

Spectral moments of molecular graphs find various applications both in theoretical chemistry of conjugated molecules and in physical chemistry of solid state. In all such applications it is necessary to know their dependence on molecular structure. Several recent works are devoted to the solution of this problem, especially in the case of benzenoid systems [39, 41–45].

In addition to the long-known results for $M_0$, $M_2$ and $M_4$ (i.e. b(B, 0), b(B, 1) and b(B, 2)) [2], Dias [39] and Hall [42] discovered the actual form of the dependence of $M_6$ on the structure of a benzenoid system. Dias [39] also reported a formula for b(B, 4), valid for catacondensed systems only. Formulas for $M_8$ and $M_{10}$ of arbitrary benzenoids and for $M_{12}$ of catacondensed systems were obtained quite recently [45]. The first few of these expressions read:

$$M_0 = n$$

$$M_2 = 2m$$

$$M_4 = 18m - 12n$$

$$M_6 = 158m - 144n + 48 + 6b$$

$$M_8 = 1330m - 1364n + 704 + 80B + 168C + 256F + 16h_0 + 8h_1$$

where n and m denote the numbers of vertices and edges, respectively, $b = B + 2C + 3F$ is the number of bay regions, B, C and F count the (simple) bays, coves and fjords, respectively, whereas $h_0$ and $h_1$ count the hexagons with no and with one vertex of degree two, respectively.

The formulas for $M_{10}$ and $M_{12}$ are similar, but much more complicated [45].

The spectral moments of benzenoid systems can always be represented in the form [45]

$$M_k = \alpha_k n + \beta_k m + \gamma_k + r_k$$

where $\alpha_k$, $\beta_k$ and $\gamma_k$ are constants (depending solely on k) whereas $r_k$ is the k-th residual which is much smaller than $\alpha_k n + \beta_k m + \gamma_k$. In all the cases examined $r_k$ was found [45] to be divisible by k. It would be interesting to see whether this curious regularity is a generally valid result.

# 4 Kekulé and Clar Structures

In the book [3] the role of both Kekulé and Clar structures in various (contemporary) chemical theories as well as their relevance for practical chemistry were outlined in detail. A recent book [46] by Cyvin and the present author is devoted to the enumeration of Kekulé structures of benzenoid molecules. In addition to this, the first volume of *"Advances in the Theory of Benzenoid Hydrocarbons"* contains several review articles [47–51] dealing with topics of relevance for our considerations. In order to avoid repetition and overlapping we will just briefly mention the work on the elaboration and application of the John — Sachs theorem for the enumeration of Kekulé structures [52–55], the search for concealed non-Kekuléan benzenoid systems [2, 56–59], examination of fully benzenoid (=all-benzenoid) systems [60–62, 135] as well as the enumeration of Kekulé structures in long and random benzenoid chains [63–65].

## 4.1 Kekuléan and Non-Kekuléan Benzenoid Molecules

The early history of the search for non-Kekuléan benzenoid systems (=systems for which no Kekulé structural formula can be written = systems for which K = 0) is described elsewhere (see pp. 62–66 in [3]). Some time was needed for theoretical chemists to recognize that for large benzenoids it is not quite simple to decide whether K > 0 (Kekuléans) or K = 0 (non-Kekuléans).

From the Dewar — Longuet-Higgins formula, Eqs. (1) and (4), it is immediately seen that the above problem is equivalent to the question whether there exist zero eigenvalues in the spectrum of a benzenoid graph. Indeed, in computer-aided searches, constructions and classifications of benzenoid systems, the easiest and most efficient way to recognize non-Kekuléan species is just to compute det A. At this point it should be mentioned that Hall [66] recently proposed a new easy method for rapid calculation of det A of a benzenoid system.

After a large number of erroneous attempts (see [67, 68] and the references quoted therein), the first complete structural characterization of Kekuléan/non-Kekuléan benzenoids was achieved in 1985 independently by Zhang, Chen and Guo [69] and Kostochka [70]. Their results differ only in minor details.

The vertices of a benzenoid system can be colored by two colors, say black and white, so that first neighbors have different colors [3]. Since every double bond in a Kekulé structure lies between a black and a white vertex, every Kekuléan benzenoid system must have equal numbers of black and white vertices. (Recall that the K = 0 benzenoids having equal numbers of black and white vertices are called concealed non-Kekuléan systems [3].)

Consider a benzenoid system B whose vertices are colored in the above described manner. An edge-cut of B is a collection $e_1, e_2, ..., e_t$ of edges of B, such that

(a) by deleting the edges $e_1, e_2, ..., e_t$ from B, it decomposes into two parts $F_1$ and $F_2$;

(b) for each edge $e_i$, i = 1, 2, ..., t, its black end-vertex belongs to $F_1$ (and therefore its white end-vertex belongs to $F_2$);

(c) each pair of edges $e_i, e_{i+1}$, i = 1, 2, ..., t − 1, belongs to the same hexagon of B while $e_1$ and $e_t$ belong to the perimeter.

*Theorem 6* [69, 70]. A benzenoid system B is Kekuléan if, and only if, it has equal numbers of black and white vertices, and if for all edge-cuts of B, the fragment $F_1$ does not have more white vertices than black vertices.

Below is depicted a benzenoid system 15 (with colored vertices) and two of its edge-cuts. In these edge-cuts only the vertices belonging to the fragment $F_1$ are colored. In the first edge-cut $F_1$ has more black than white vertices. In the second edge-cut $F_1$ has more white than black vertices. This latter cut reveals that 15 is (concealed) non-Kekuléan.

It has been pointed out by G. G. Hall and Dias [68] that dissection approaches (like the one described in Theorem 6) are just special cases of previously known general graph-theoretical results by P. Hall (1935) and Tutte (1947).

Zhang and coworkers [71–73] and Sheng [74] have further sharpened the method described in Theorem 6. Their efforts are reviewed in [47, 49]. Sheng [49] also put forward an algorithmic approach for rapid (paper-and-pencil) recognition of Kekuléan benzenoids. Other such tests are also quite numerous in the recent chemical literature [22, 68, 75–80, 136]. In particular, in [22] and [68] the

Kekuléan/non-Kekuléan nature of a benzenoid system is deduced from the properties of its excised internal structure.

We wish to conclude this paragraph with a quotation from the author's first paper [2] on benzenoid systems: *"There is no simple recipe to decide by inspection of the molecular graph whether $K = 0$ or not. In other words, the necessary and sufficient conditions for the existence of Kekulé structures seem to be rather complicated"*. Fifteen years later one may optimistically state that this hard problem of the topological theory of benzenoid hydrocarbons is completely settled.

## 4.2 Bounds for the Kekulé Structure Count

Finding upper and lower bounds for the K-value of benzenoid systems is intimately related to the identification of benzenoids with extremal (minimum and maximum) number of Kekulé structures. Clearly, such benzenoids are expected to possess unusual chemical properties (e.g. to be highly reactive or exceptionally stable).

For catacondensed systems with h hexagons the following bounds have been obtained [81]

$$h + 1 \leq K \leq 2^{h-1} + 1 .$$

Later Cyvin and Chen [82, 83] improved the above upper bound as:

$$K \leq \begin{cases} 3^{h/2} & \text{if } h \text{ is even} \\ 2 \cdot 3^{(h-1)/2} & \text{if } h \text{ is odd} . \end{cases}$$

Recently John [84] offered an even better upper bound, viz.

$$K \leq a(h)$$

where $a(1) = 2$, $a(2) = 3$, $a(3) = 5$ and

$$a(h) = a(h - 2) + 3a(h - 3) \quad \text{for} \quad h \geq 4 . \tag{6}$$

It can be shown that [84]

$$a(h) = C_1 \alpha^h + [C_2 \cos (xh) - C_3 \sin (xh)] \beta^h$$

with

$$C_1 = 1.37301 \ldots$$
$$C_2 = 0.37301 \ldots$$
$$C_3 = 0.67296 \ldots$$
$$\alpha = 1.671699 \ldots$$
$$\beta = 1.33962 \ldots$$
$$x = 0.8970 \ldots \text{ (radians)} .$$

The only available exact result of this kind, applicable to all benzenoid systems is a complicated, but by no means a sharp upper bound [85]:

$$K \le [2m/n + R(n/2 - 1)^{1/2}]^{1/2} [2m/n - R(n/2 - 1)^{-1/2}]^{(n-2)/4}$$

where

$$R = \frac{1}{n}(18mn - 12n^2 - 4m^2)^{1/2}$$

and where n and m stand for the numbers of vertices and edges, respectively.

Empirical experience suggests [86] that among benzenoids with a fixed number of hexagons the maximum K-value is achieved for catacondensed systems. If so, then we arrive at bounds which until now have eluded a rigorous proof:

*Conjecture.* For a Kekuléan benzenoid system with h hexagons,

$$h + 1 \le K \le a(h)$$

where a(h) is given by Eq. (6).

## 4.3 A Simple Invariant of the Kekulé Structures

If a benzenoid system is drawn so that some of its edges are vertical, then the double bonds in such a system (in a Kekulé structure) can have one of the following three orientations:

A      B      C

Denote by x, y and z the numbers of double bonds of type A, B and C, respectively. For example, in the Kekulé structure below $x = 7$, $y = 3$ and $z = 3$.

$k_1$

Evidently, $x + y + z = n/2$.

The following property of the Kekulé structures seems to have been overlooked for many decades. It was put forward as late as in 1986:

*Theorem 7* [87]. All Kekulé structures of a benzenoid molecule have the same triplet (x, y, z).

In Fig. 2 are depicted the seven Kekulé structures of benzoanthracene. The reader can easily verify that each of them has x = 2, y = 3, z = 4.

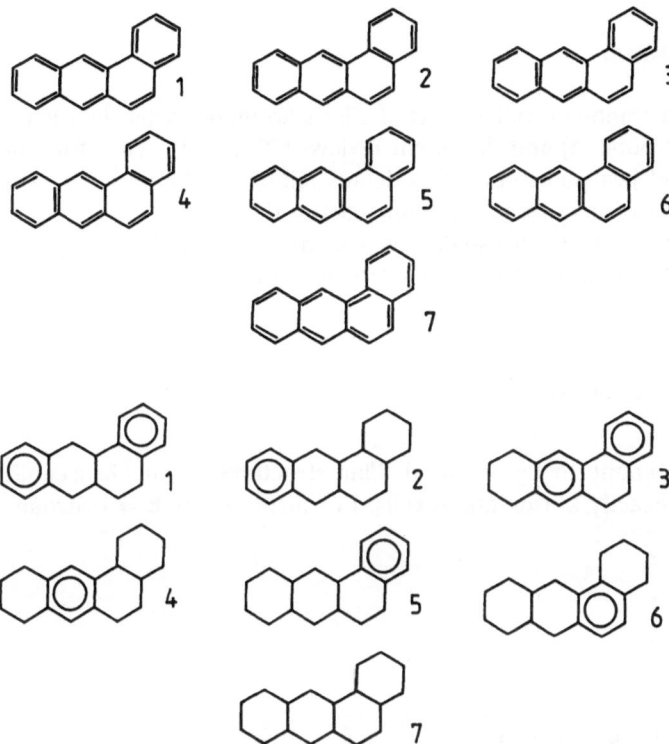

**Fig. 2.** Kekulé structures and generalized Clar structures of benzo[a]anthracene

Not all triplets of positive integers correspond to benzenoid systems. The difficult problem of the characterization of such triplets was solved by Zhang and Guo [88, 137].

*Theorem 8* [88]. A triplet (x, y, z), x ≤ y ≤ z, corresponds to a benzenoid system if, and only if, one of the conditions (a), (b), (c) or (d) is satisfied.

(a) x = 1,    y = z
(b) x = 2   and
     [y = 2, 2 < z ≤ 4]   or   [y = 3, 3 < z ≤ 6]   or   [y = 4, z ≠ 9, 11]   or
     [y = 5, z ≠ 12]   or   [y ≥ 6]

17

(c) x = 3  and

    [y = 3, z ≠ 11]  or  [y ≥ 4]

(d) x ≥ 4

*Theorem 9* [88]. A triplet (x, y, z), x ≤ y ≤ z, corresponds to a catacondensed benzenoid system if and only if x + y + z is odd and x + y ≥ z + 1.

Bearing in mind that the Kekulé structures have found a variety of applications in chemistry [3] it was hoped that also the invariant (x, y, z) will be of some use. The few attempts made until now [87] have been far from successful.

## 4.4 Sextet Polynomial

Various algebraic and combinatorial aspects of Clar's aromatic sextet theory are outlined in the recent book [3] and the recent reviews [50, 51, 89]. Therefore, in this paragraph we will just point out a few details related to the author's own research and mention the most recent developments in the filed.

The crucial impetus for these investigations was given by the short paper of Hosoya and Yamaguchi [90] in which they introduced the numbers s(B, k) and the sextet polynomial

$$\sigma(B, x) = \sum_{k=0}^{s} s(B, k)\, x^k .$$

The numbers s(B, k) count the generalized Clar structures of the benzenoid system B, containing exactly k aromatic sextets. For instance, for **B** = benzoan-thracene we have (see Fig. 2):

    s(B, 0) = 1

    s(B, 1) = 4

    s(B, 2) = 2

    s(B, k) = 0  for  k ≥ 3

and consequently,

$$\sigma(B, x) = 1 + 4x + 2x^2 .$$

This example is also an illustration of the peculiar fact that the number of generalized Clar structures coincides with the number of Kekulé structures, i.e.

$$\sigma(B, 1) = \sum_{k=0}^{s} s(B, k) = K\{B\} . \tag{7}$$

The identity (7) is just a consequence of a one-to-one correspondence between generalized Clar and Kekulé structures. In particular, the i-th Clar structure in

Fig. 2 is constructed so that in the i-th Kekulé structure each triplet of double bonds arranged as A was replaced by a circle and all other double bonds were abandoned.

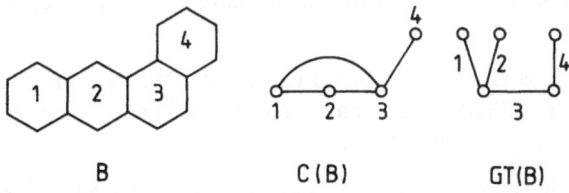

**A**

The first exact result obtained along these lines was

*Theorem 10* [91]. Eq. (7) holds for all catacondensed benzenoid systems.

Eq. (7) does not hold for all pericondensed benzenoids and its range of validity was established in a series of investigations [90, 92–96]. The work of Ohkami [96] can be considered as a complete solution of this problem.

*Theorem 11* [94]. Eq. (7) holds for a pericondensed benzenoid system B if and only if one of the conditions (a), (b) or (c) is satisfied.
(a) B is not coronene.
(b) B does not contain a coronene subunit.
(c) If B contains a coronene subunit *Cor*, then for each such subunit, B-*Cor* does not possess Kekulé structures.

The numbers s(B, k) can be interpreted in the following straightforward manner [97].

Define the *Clar graph* C(B) of the benzenoid system B as a graph whose vertices correspond to the hexagons of B. Two vertices of C(B) are adjacent if, and only if, it is not possible to simultaneously arrange aromatic sextets (=circles in the generalized Clar formulas) in the respective hexagons of B.

Below is depicted the Clar graph of benzoanthracene:

**B**　　　　**C(B)**　　　　**GT(B)**

Denote by n(G, k) the number of ways in which k independent (=mutually nonadjacent) vertices can be selected in a graph G. By definition, n(G, 0) = 1 and n(G, 1) = n = number of vertices of G.

*Theorem 12* [97]. The identity s(B, k) = n(C(B), k) holds for all benzenoid systems B and for all values of k, except k = 1.

Since the independence numbers, i.e. the numbers n(G, k), are easily determined by means of pertinent graph-theoretical algorithms (see pp. 111–112 in [3]),

19

Theorem 12 provides a straightforward and general method for the calculation of the sextet polynomial [97, 98].

An obvious question arising from Theorem 12 is when the identity $s(B, k) = n(C(B), k)$ holds also for $k = 1$. Since $n(C(B), 1)$ is just the number of hexagons of B, we are interested in those benzenoid systems in which all hexagons are resonant. This problem was first examined in [11, 99] and eventually solved by Zhang and Chen [100, 101] and independently by John [102]:

*Theorem 13.* The identity $s(B, 1) = n(C(B), 1)$ holds if, and only if, the excised internal structure of the benzenoid system possesses a perfect matching.

*Corollary 13.1.* The identity $s(B, 1) = n(C(B), 1)$ holds for all catacondensed benzenoids.

Another result of this kind is reported in the recent work by He and He [103]. A benzenoid systems is said to be "normal" [3] if none of its edges corresponds to a fixed double or a fixed single bond.

*Theorem 14.* The identity $s(B, 1) = n(C(B), 1)$ holds if, and only if, B is a normal benzenoid system.

*Corollary 14.1.* A benzenoid system B is normal if, and only if, (a) B is catacondensed or (b) the excised internal structure of B possesses a perfect matching.

*Corollary 14.2.* Any normal benzenoid with h hexagons (h > 1) can be generated by adding a hexagon to a normal benzenoid with $h - 1$ hexagons.

The statement formulated here as Corollary 14.2 was first conjectured by Cyvin and the present author [104]. It played a significant role in the construction and enumeration of normal benzenoids [3, 104]. Its formal proof is given in [105].

The line graph L(G) of a graph G is defined as follows. The vertices of L(G) correspond to the edges of G and two vertices of L(G) are adjacent if the corresponding edges of G have a common vertex.

It turns out that in some cases the Clar graph is a line graph. This fact was conceived even before the discovery of the Clar-graph concept (of course in a somewhat different form). We may thus reformulate the result originally obtained in 1977 as follows:

*Theorem 15* [106]. If B is an unbranched catacondensed benzenoid system then there exist a graph GT(B), such that $C(B) = L(GT(B))$. The graph GT is connected and acyclic (i.e. it is a tree).

*Corollary 15.1.* If B is an unbranched catacondensed benzenoid system then $s(B, k) = m(GT(B), k)$ holds for all values of k. Here $m(G, k)$ denotes the number of k-matchings of a graph G (cf Sect. 3.2).

*Corollary 15.2.* The characteristic polynomial of GT(B) is of the form

$$\varphi(GT(B), x) = \sum_{k=0}^{s} (-1)^k s(B, k) x^{h-1-2k}$$

where h is the number of hexagons of B.

*Corollary 15.3.* If B is an unbranched catacondensed benzenoid system then all the zeros of its sextet polynomial are real and negative numbers.

The graph GT was eventually named, the *"Gutman tree"* [51, 89, 107–109] and its properties were extensively studied [51, 89, 106–112].

A further method for calculating the sextet polynomial of an unbranched catacondensed benzenoid molecule, not based on the Gutman-tree concept, was reported in [113].

## 4.5 Corals

The study of generalized Clar structures and their relations to the Kekulé structures led to the introduction of a new concept [92] which was eventually named *"coral"* [114].

Three double bonds in a six-membered ring can have two distinct arrangements, A and B:

A          B

Let $k_1, k_2, \ldots, k_k$ be the Kekulé structures of a benzenoid molecule. Define a mapping f which transforms a Kekulé structure $k_i$ into another Kekulé structure $k_j$ so that all A-type arrangements of double bonds in $k_i$ are changed into arrangements of type B whereas all other double bonds are left unchanged. This will be denoted as $f(k_i) = k_j$.

For example, $f(k_1) = k_2$ and $f(k_2) = k_3$. Furthermore, $f(k_3) = k_3$ because $k_3$ does not have any A-type arrangements of double bonds.

$k_1$          $k_2$          $k_3$

The Kekulé structures of benzoanthracene (see Fig. 2) are mapped in the following manner: $f(1) = 4, f(2) = 4, f(3) = 4, f(4) = 6, f(5) = 6, f(6) = 7, f(7) = 7$.

21

It is clear that f maps the set $\{k_1, k_2, ..., k_k\}$ onto itself. Such a mapping can be visualized by means of a diagram. For instance, COR is the diagramatical representation of the mapping of the Kekulé structures of benzoanthracene.

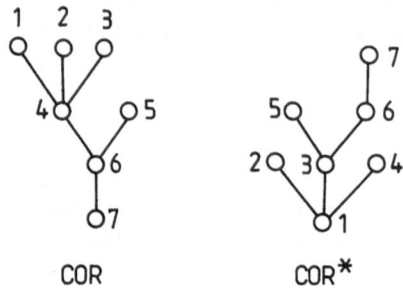

Note that COR is not a graph because here the vertex lying above is mapped onto the vertex lying below. Objects of this type are called Hasse diagrams, but because in our case they possess certain special properties (see below) we proposed for them the name "corals".

The Kekulé structure $k_0$ having the property $f(k_0) = k_0$ is called a *root*.

*Theorem 16* [92]. Every Kekuléan benzenoid system has a root Kekulé structure and this structure is unique.

*Theorem 17* [115]. The Hasse diagram corresponding to the mapping f is connected and possesses no cycles.

By $f^p(k)$ we denote the Kekulé structure which is obtained from the Kekulé structure k by repeating the mapping f p times. Then Chen's Theorem 17 means that for any k and some sufficiently large p, $f^p(k) = k_0$. Furthermore, for any $p > 0$, the equation $f^p(k) = k$ has only one solution, namely $k = k_0$.

For the construction of a coral it was substantial that A-type arrangements of double bonds were transformed into B-type arrangements. On the other hand, there is no a priori reason for preferring the transformation $A \Rightarrow B$ over the transformation $B \Rightarrow A$. The mapping f* defined in the same way as f, but using the transformation $B \Rightarrow A$ is said to be the dual of f. The coral induced by f* is called the *dual coral*.

For instance, the dual mapping of the Kekulé structures of benzoanthracene (Fig. 2) reads: $f^*(1) = 1$, $f^*(2) = 1$, $f^*(3) = 1$, $f^*(4) = 1$, $f^*(5) = 3$, $f^*(6) = 3$, $f^*(7) = 6$. The respective dual coral, COR*, is depicted above. Even from this example we see that the coral and its dual may have quite dissimilar structures. Nevertheless, the coral and its dual agree in a number of structural details. First of all, Theorems 16 and 17 remain valid if f is interchanged by f*.

A Kekulé structure which cannot be obtained from another Kekulé structure by means of the mapping f is called a *bud*. The number of buds is the *width* of the respective coral. The maximum distance between a bud and the root is the *height* of the respective coral.

*Theorem 18* [114]. The coral and its dual have equal widths. The coral and its dual have equal heights.

*Theorem 19* [114]. The root of a coral is a bud in the dual coral and vice versa.

Hence both the width and the height of the coral are independent of the (otherwise arbitrary) choice A $\Rightarrow$ B or B $\Rightarrow$ A. Attempts have been made [114, 116] to correlate these parameters with various physico-chemical properties of benzenoid hydrocarbons. Only a limited success was achieved, especially in the case of resonance energy [116].

Many problems in the theory of corals are still open. For instance, we do not know how (if) the coral can be reconstructed from its dual. Particular emphasis should be given on those properties which are the same for both the coral and its dual. Theorem 18 may represent just the first step in this direction.


# 5 Topological Indices

The name "topological index" is used in theoretical chemistry for what mathematicians prefer to call "graph invariant". Thus any quantity $I = I(G)$ which somehow can be determined from the structure of the graph G and which has the property $I(G) = I(H)$ whenever the graphs G and H are isomorphic, can be considered as a "topological index". In theoretical chemistry, of course, the choice of topological indices is restricted by the requirement that there should be some reasonable connection (preferably a correlation) between I and some physico-chemical properties of the respective molecule [117]. Nevertheless, the number of topological indices, currently used in chemical graph theory is legion [118].

We have already examined a few graph invariants which, according to the above definition, could be included among the topological indices of benzenoid molecules. These are the number of Kekulé structures, the eigenvalues, spectral moments, (coefficients of) the characteristic and sextet polynomials, to mention just some. The total π-electron energy is surveyed elsewhere in this volume.

In addition to these, only a limited number of other topological indices of benzenoid molecules have been studied. With a few not too important exceptions, generally valid mathematical results were obtained only for one of them — namely for the Wiener index. Therefore the remaining part of this section is devoted to the Wiener index of benzenoid systems. (Further graph invariants worth mentioning in connection with benzenoids, especially unbranched catacondensed systems, are the Hosoya index [119–121], the Merrifield — Simmons index [122, 123], the modified Hosoya index [38] and the polynomials associated with them.)


## 5.1 Wiener Index

Let G be a connected graph and let its vertices be labeled by $1, 2, ..., n$. The distance $d(r, s)$ between the vertices r and s is the length of the shortest path which connects r and s. The Wiener index W is then the sum of all distances in

Ivan Gutman

the graph G:

$$W = W(G) = \sum_{r<s} d(r, s) = \tfrac{1}{2} \sum_{r=1}^{n} \sum_{s=1}^{n} d(r, s).$$

The Wiener index has been extensively studied in the chemical literature (for reviews see [118, 124]) and therefore it is not surprising that it was also examined in the case of benzenoids.

Recurrence relations have been established for the Wiener index of catacondensed benzenoid systems, both unbranched [125] and branched [126]. Based on these relations it could be shown that for unbranched catacondensed systems $U_h$ with h hexagons [127],

$$W(H_h) \leq W(U_h) \leq W(L_h)$$

where $H_h$ and $L_h$ are the helicene and the linear polyacene, respectively, with h hexagons and

$$W(H_h) = \tfrac{1}{3} (8h^3 + 72h^2 - 26h + 27)$$
$$W(L_h) = \tfrac{1}{3} (16h^3 + 36h^2 + 26h + 3).$$

Thus $W(U_h)$ is bounded from both below and above by cubic polynomials in the variable h. One could therefore anticipate that also the expected value $W_h$ of the Wiener number of a random benzenoid chain containing h hexagons is a cubic polynomial in h. This, however, is not the case [128]. The expected value is given by $W_1 = 27$, $W_2 = 109$, $W_3 = 271 + 8q$ and

$$W_h = 4h^3 + 16h^2 + 6h + 1 + \tfrac{4}{3} q(h^3 - 3h^2 + 2h)$$
$$- \tfrac{4}{3} (p_1 - p_2)^2 F(h, q)$$

for $h \geq 4$, where

$$F(h, q) = \sum_{k=1}^{h-3} k(k + 1) (k + 2) q^{h-3-k}$$

and where $p_1$, $p_2$ and q are, respectively, the probabilities for annelation of the types $\alpha$, $\beta$ and $\gamma$, $p_1 + p_2 + q = 1$.

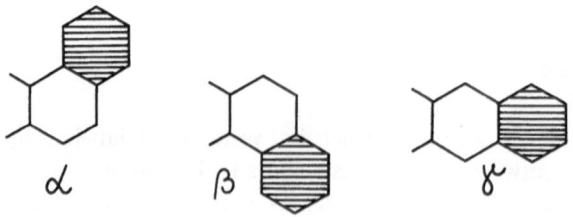

For catacondensed benzenoids a curious and fully unexpected regularity was first empirically observed [129] and then rigorously proved [127, 130].

*Theorem 20.* If $B_1$ and $B_2$ are catacondensed benzenoid systems with equal numbers of hexagons, then $W(B_1) \equiv W(B_2)$ (mod 8).

*Corollary 20.1.* If B is a catacondensed benzenoid system with h hexagons and if h is of the form $4k + j$, $j = 0, 1, 2$ or 3, then $W(B) \equiv 2j + 1$ (mod 8).

It is really surprising that the Wiener indices have such number-theoretical properties. This kind of modular behavior was observed never before for any of the numerous topological indices studied in chemical graph theory. Results analogous to those given in Theorem 20 were later found for other classes of (non-benzenoid) graphs [130, 131], but their extension to pericondensed benzenoids was never accomplished.

It is difficult to imagine any benefit which an experimental chemist could have from theoretical results like Theorem 20. Although its discovery was a serious achievement, this theorem may serve as a drastic example that not all products of the "topological" investigations in chemistry have chemical relevance. Obscurities of this kind have occurred and certainly will occur in future researches. Nevertheless, we are positive that the totality of our efforts in the last twenty years has increased the chemists' understanding of the physical and chemical nature of benzenoid hydrocarbons.

# 6 Bibliographic Note

This article is Part 78 of the series *"Topological Properties of Benzenoid Systems"*. Parts 75, 76 and 77 are the references [128], [123] and [132]. For review of earlier work see Parts 17 [133], 21 [134] and 59 [67]. Other parts of the series, quoted in this article are 1 [2], 2 [106], 3 [91], 9 [97], 14 [35], 16 [81], 19 [99], 24 [98], 25 [11], 29 [38], 32 [111], 37 [110], 40 [75], 40a [77], 44 [76], 46 [87], 47 [125], 49 [127], 50 [52], 51 [120], 52 [129], 53 [53], 57 [62], 58 [10], 60 [113], 65 [114], 70 [122] and 71 [116].

# 7 Addendum

After the completion of the text of this article a remarkable discovery has been made in the spectral theory of benzenoid molecules. In the 1980s serveral authors tried to find isospectral benzenoid systems (i.e. benzenoids having equal spectra, cf Sect. 3). These efforts were, however, not successful. Finally, Cioslowski conjectured [138] that such systems do not exist at all. After that Babić succeeded to design a method by which arbitrarily many isospectral benzenoids can be constructed [139]. One of Babić's isospectral pairs are the following benzenoid systems, each possessing 9 hexagons and 33 vertices.

It has been recently shown [140] that the above example is the smallest possible and that it is unique. All isospectral benzenoids constructed by Babić's method have odd numbers of vertices and, consequently, have no Kekulé structures. In the present moment (December 1991) isospectral Kekuléan benzenoids are not known.

# 8 References

1. Dewar MJS, Longuet-Higgins HC (1952) Proc Roy Soc London A214: 482
2. Gutman I (1974) Croat Chem Acta 46: 209
3. Gutman I, Cyvin SJ (1989) Introduction to the theory of benzenoid hydrocarbons, Springer, Berlin Heidelberg New York
4. Merrifield RE, Simmons HE (1989) Topological methods in chemistry, Wiley, New York
5. Gutman I, Cyvin SJ (eds) (1990) Advances in the theory of benzenoid hydrocarbons, (Topics in Current Chemistry 153), Springer, Berlin Heidelberg New York
6. Cvetković D, Gutman I, Trinajstić N (1974) J Chem Phys 61: 2700
7. He WC, He WJ (1986) Theor Chim Acta 70: 447
8. He WJ, He WC (1989) Theor Chim Acta 75: 389
9. Sheng RQ (1989) Math Chem 24: 201
10. Gutman I, Cyvin SJ (1989) J Mol Struct (Theochem) 184: 159
11. Gutman I (1983) In: Cvetković D, Gutman I, Pisanski T, Tošić R (Eds) Graph theory, Univ Novi Sad, Novi Sad, pp 151–160
12. Kirby EC (1990) J Math Chem 4: 31
13. Kirby EC (1990) J Chem Soc Faraday Trans 86: 447
14. Cvetković D, Gutman I (1981) Publ Inst Math (Beograd) 29: 49
15. Gutman I, Mallion RB, Essam JW (1983) Molec Phys 50: 859
16. John P, Sachs H (1990) J Chem Soc Faraday Trans 86: 1033
17. John P, Sachs H (1990) Topics Curr Chem 153: 145
18. Dias JR (1984) J Chem Inf Comput Sci 24: 124
19. Dias JR (1984) Canad J Chem 62: 2914
20. Dias JR (1989) Z Naturforsch 44a: 765
21. Dias JR (1987) Handbook of polycyclic hydrocarbons Part A Benzenoid hydrocarbons, Elsevier, Amsterdam
22. Gutman I, Dias JR (1990) In: Bodendiek R (ed) Contemporary methods in graph theory, BI-Wissenschaftsverlag, Mannheim, pp 249–259
23. Gutman I, Kirby EC (1991) Math Chem 26: 111
24. Gutman I, Polansky OE (1986) Mathematical concepts in organic chemistry, Springer, Berlin Heidelberg New York
25. Cvetković D, Doob M, Sachs H (1980) Spectra of graphs — theory and application, Academic Press, New York
26. Heilbronner E (1962) Helv Chim Acta 45: 1722
27. Kiang YS, Chen ET (1983) Pure Appl Chem 55: 283
28. Hall GG (1977) Molec Phys 33: 551
29. Wild U, Keller J, Günthard HH (1969) Theor Chim Acta 14: 383
30. Wild U (1980) Theor Chim Acta 54: 245
31. Dias JR (1985) Nouv J Chim 9: 125
32. Gutman I, Kruszewski J (1985) Nouv J Chim 9: 669

33. Dias JR (1987) J Mol Struct (Theochem) 149: 213
34. Jiang YS, Chen GG (1990) Theor Chim Acta 76: 437
35. Gutman I (1983) J Chem Soc Faraday Trans II 79: 337
36. Hosoya H (1971) Bull Chem Soc Japan 44: 2332
37. Hosoya H, Hosoi K, Gutman I (1975) Theor Chim Acta 38: 37
38. Gutman I, Shalabi A (1984) Z. Naturforsch 39a: 797
39. Dias JR (1985) Theor Chim Acta 68: 107
40. Dias JR (1987) J Chem Educ 64: 213
41. Cioslowski J (1985) Z Naturforsch 40a: 1167
42. Hall GG (1986) Theor Chim Acta 70: 323
43. Kiang YS, Tang AC (1986) Int J Quantum Chem 29: 229
44. Jiang YS, Zhang HX (1989) Theor Chim Acta 75: 279
45. Marković S, Gutman I (1991) J Mol Struct (Theochem) 235: 81
46. Cyvin SJ, Gutman I (1988) Kekulé structures in benzenoid hydrocarbons (Lecture Notes in Chemistry 46), Springer, Berlin Heidelberg New York
47. Zhang FJ, Guo XF, Chen RS (1990) Topics Curr Chem 153: 181
48. He WC, He WJ (1990) Topics Curr Chem 153: 195
49. Sheng RQ (1990) Topics Curr Chem 153: 211
50. Hosoya H (1990) Topics Curr Chem 153: 273
51. El-Basil S (1990) Topics Curr Chem 153: 291
52. Gutman I, Cyvin SJ (1987) Chem Phys Letters 136: 137
53. Gutman I, Su LX, Cyvin SJ (1987) J Serb Chem Soc 52: 263
54. Cyvin SJ, Cyvin BN, Brunvoll J, Gutman I (1987) Z Naturforsch. 42a: 722
55. Bodroža O, Gutman I, Cyvin SJ, Tošić R (1988) J Math Chem 2: 287
56. Brunvoll J, Cyvin SJ, Cyvin BN, Gutman I, He WJ, He WC (1987) Math Chem 22: 105
57. He WC, He WJ, Cyvin BN, Cyvin SJ, Brunvoll J (1988) Math Chem 23: 201
58. Guo XF, Zhang FJ (1989) Math Chem 24: 85
59. Cyvin SJ, Cyvin BN, Brunvoll J (1990) J Math Chem 4: 47
60. Polansky OE, Gutman I (1980) Math Chem 8: 269
61. Cyvin BN, Brunvoll J, Cyvin SJ, Gutman I (1988) Math Chem 23: 163
62. Gutman I, Cyvin SJ (1988) Math Chem 23: 175
63. Gutman I, Cyvin SJ (1988) Chem Phys Letters 147: 121
64. Gutman I (1988) Graph Theory Notes NY 16: 26
65. Brunvoll J, Gutman I, Cyvin SJ (1989) Z Phys Chem (Leipzig) 270: 982
66. Hall GG (1988) Chem Phys Letters 145: 168
67. Gutman I, Cyvin SJ (1988) J Serb Chem Soc 53: 391
68. Hall GG, Dias JR (1989) J Math Chem 3: 233
69. Zhang FJ, Chen RS, Guo XF (1985) Graphs Comb 1: 383
70. Kostochka AV (1985) Proc 30 Int Wiss Koll TH Ilmenau F: 49
71. Zhang FJ, Chen RS (1987) Natur J 10: 163
72. Zhang FJ, Guo XF (1988) Math Chem 23: 229
73. Zhang FJ, Chen RS (1989) Acta Math Appl Sinica 1: 1
74. Sheng RQ (1989) Math Chem 24: 207
75. Gutman I, Cyvin SJ (1986) J Mol Struct (Theochem) 138: 325
76. Cyvin SJ, Gutman I (1987) J Mol Struct (Theochem) 150: 157
77. Cyvin SJ, Gutman I (1988) J Mol Struct (Theochem) 164: 183
78. Sheng RQ, Cyvin SJ, Gutman I (1989) J Mol Struct (Theochem) 187: 285
79. Sheng RQ (1987) Chem Phys Letters 142: 196
80. Guo XF, Zhang FJ (1990) J Math Chem 5: 157
81. Gutman I (1982) Math Chem 13: 173
82. Cyvin SJ (1986) Math Chem 20: 165
83. Chen RS, Cyvin SJ (1987) Math Chem 22: 175
84. John P (1990) private communication
85. Gutman I, Cioslowski J (1987) Publ Inst Math (Beograd) 42: 21
86. Balaban AT, Brunvoll J, Cyvin BN, Cyvin SJ (1988) Tetrahedron 44: 221
87. Zhang FJ, Chen RS, Guo XF, Gutman I (1986) J Serb Chem Soc 51: 537

88. Zhang FJ, Guo XF (1987) Math Chem 22: 181
89. El-Basil (1987) J Math Chem 1: 153
90. Hosoya H, Yamaguchi T (1975) Tetrahedron Letters: 4659
91. Gutman I, Hosoya H, Yamaguchi T, Motoyama A, Kuboi N (1977) Bull Soc Chim Beograd 42: 503
92. Ohkami N, Motoyama A, Yamaguchi T, Hosoya H, Gutman I (1981) Tetrahedron 37: 1113
93. Ohkami N, Hosoya H (1983) Theor Chim Acta 64: 153
94. Zhang FJ, Chen RS (1986) Math Chem 19: 179
95. He WJ, He WC (1986) Theor Chim Acta 70: 43
96. Ohkami N (1990) J Math Chem 5: 23
97. Gutman I (1982) Z Naturforsch 37a: 69
98. Gutman I, El-Basil S (1984) Z Naturforsch 39a: 276
99. Gutman I (1983) Wiss Z TH Ilmenau 29: 57
100. Zhang FJ, Chen RS (1989) Math Chem 24: 323
101. Zhang FJ, Chen RS (1991) Dicrete Appl Math 30: 63
102. John P (1989) Z Phys Chem (Leipzig) 270: 1023
103. He WC, He WJ (1990) Math Chem 25: 225
104. Cyvin SJ, Gutman I (1986) Z Naturforsch 41a: 1079
105. He WC, He WJ (1990) Math Chem 25: 237
106. Gutman I (1977) Theor Chim Acta 45: 309
107. El-Basil S (1984) Theor Chim Acta 65: 191
108. El-Basil S (1984) Theor Chim Acta 65: 199
109. El-Basil S (1986) J Chem Soc Faraday Trans II 82: 299
110. Gutman I, El-Basil S (1985) Z Naturforsch. 40a: 923
111. Gutman I, El-Basil S (1985) J Serb Chem Soc 50: 25
112. El-Basil S, Randić M (1988) J Chem Soc Faraday Trans II 84: 1875
113. Gutman I, Cyvin SJ (1989) Bull Chem Soc Japan 62: 1250
114. Gutman I, Teodorović AV, Kolaković N (1989) Z Naturforsch 44a: 1097
115. Chen ZB (1985) Chem Phys Letters 115: 291
116. Gutman I, Teodorović AV, Kolaković N (1990) J Serb Chem Soc 55: 363
117. Rouvray DH (1986) Sci Amer 255: 36
118. Balaban AT, Motoc I, Bonchev D, Mekenyan O (1983) Topics Curr Chem 114: 21
119. Hosoya H, Ohkami N (1986) J Comput Chem 4: 585
120. Gutman I (1988) Z Naturforsch 43a: 939
121. Randić M, Hosoya H, Polansky OE (1989) J Comput Chem 10: 683
122. Gutman I, Kolaković N (1990) Bull Serb Acad Sci 102: 39
123. Gutman I (1991) Rev Roum Chim, in press
124. Rouvray DH (1987) J Comput Chem 8: 478
125. Gutman I, Polansky OE (1986) Math Chem 20: 115
126. Polansky OE, Randić M, Hosoya H (1989) Math Chem 24: 3
127. Gutman I (1987) Chem Phys Letters 136: 134
128. Gutman I, Kennedy JW, Quintas LV (1990) Chem Phys Letters 173: 403
129. Gutman I, Marković S, Luković D, Radivojević V, Rančić S (1987) Coll Sci Papers Fac Sci Kragujevac 8: 15
130. Gutman I (1988) Publ Inst Math (Beograd) 43: 3
131. Gutman I, Rouvray DH (1990) Comput Chem 14: 29
132. Gutman I, Utvić D, Mukherjee AK (1991) J Serb Chem Soc 56: 59
133. Gutman I (1982) Bull Soc Chim Beograd 47: 453
134. Gutman I (1983) Croat Chem. Acta 56: 365
135. Gutman I, Babić D (1992) J Mol Struct (Theochem), in press
136. Sheng RQ (1991) Math Chem 26: 205
137. Zhang FJ, Chen RS, Guo XF, Gutman I (1991) Math Chem 26: 229
138. Cioslowski J (1991) J Math Chem 6: 111
139. Babić D (1992) J Math Chem, in press
140. Gutman I, Marković S, Grović V (1991) J Serb Chem Soc 56: 553

# Total π-Electron Energy of Benzenoid Hydrocarbons

Ivan Gutman

Faculty of Science, University of Kragujevac, P.O. Box 60, YU-34000 Kragujevac, Yugoslavia

## Table of Contents

1 Introduction . . . . . . . . . . . . . . . . . . . . . . . . . . . 31
  1.1 Basic Definitions . . . . . . . . . . . . . . . . . . . . . . . 31
  1.2 Total π-Electron Energy . . . . . . . . . . . . . . . . . . . . 32
  1.3 A Note on the Hückel Molecular Orbital Theory . . . . . . . . 33

2 Total π-Electron Energy and the Thermodynamic Stability
  of Benzenoid Hydrocarbons . . . . . . . . . . . . . . . . . . . . 34

3 Elements of the Theory of Cyclic Conjugation . . . . . . . . . . . 35

4 Identities for E, TRE, ef and efa . . . . . . . . . . . . . . . . 40

5 Bounds for Total π-Electron Energy . . . . . . . . . . . . . . . . 42
  5.1 The McClelland Inequality . . . . . . . . . . . . . . . . . . . 43
  5.2 Other Inequalities . . . . . . . . . . . . . . . . . . . . . . 45

6 Approximate Formulas for Total π-Electron Energy . . . . . . . . . 47
  6.1 The McClelland Formula . . . . . . . . . . . . . . . . . . . . 47
  6.2 Other Formulas of (n, m)-Type . . . . . . . . . . . . . . . . . 48
  6.3 Formulas of (n, m, K)-Type . . . . . . . . . . . . . . . . . . 51

7 Dependence of Total π-Electron Energy on Molecular Structure . . . 53
  7.1 The (n, m)-Dependence . . . . . . . . . . . . . . . . . . . . . 53
  7.2 The K-Dependence . . . . . . . . . . . . . . . . . . . . . . . 54
  7.3 Beyond the (n, m, K)-Dependence . . . . . . . . . . . . . . . . 57

8 Cyclic Conjugation in Benzenoid Hydrocarbons . . . . . . . . . . . 58

9 Bibliographic Note . . . . . . . . . . . . . . . . . . . . . . . . 60

10 References . . . . . . . . . . . . . . . . . . . . . . . . . . . 60

Topics in Current Chemistry, Vol. 162
© Springer-Verlag Berlin Heidelberg 1992

Ivan Gutman

The theory of the HMO total $\pi$-electron energy ($E$) of benzenoid hydrocarbons is surveyed with particular emphasis on the research of its dependence on molecular structure. Identities, bounds and approximate formulas for $E$ are considered. The dependence of $E$ on the size of the molecule and on the number of Kekulé structures is discussed in detail. The effect of cycles on $E$, and six-membered rings in particular, is considered within the framework of the theory of cyclic conjugation.

# 1 Introduction

In this article we will outline the investigations concerned with the total π-electron energy of benzenoid hydrocarbons and its dependence on molecular structure. This topic was one of the main themes examined within the project „*Topological Properties of Benzenoid Systems*" (c.f. Sect. 9). We have excluded it from the survey [1] and decided to present it separately only because of a relative large number of results known in this area and because of the lack of any previous review.

The elucidation of the dependence of various chemical and physical properties of substances on molecular structure can be considered as one of the main goals of theoretical chemistry. Although an immense knowledge has accumulated in this field, a fairly limited number of direct, causal and quantitative (or at least semiquantitative) structure-property relations have been discovered so far. The main reason for this is the enormous complexity of the quantum-chemical calculations, by means of which the contemporary theoretical chemists try to describe and predict the behaviour of molecules. During such calculations the insight into the actual connection between the input (e.g. molecular structure) and output (e.g. certain molecular properties) is usually completely lost.

The total π-electron energy (as calculated within the HMO model, see below) seems to be a favourable exception. Its mathematical form is relatively simple and therefore we still have a chance to look for direct relations with molecular structure. Its mathematical form, however, is not too simple and therefore the relations with molecular structure are far from being trivial and can be revealed only by means of a proficient analysis. As a consequence of this the structure-dependency of total π-electron energy has continuously attracted the attention of theoretical chemists for more than 50 years. (For some works of historical importance see [2–11]; more recent research will be mentioned in the subsequent parts of this article.)

## 1.1 Basic Definitions

The definition of a benzenoid hydrocarbon/benzenoid system/benzenoid graph as well as a sufficient number of examples can be found in the preceding article [1] and elsewhere [12]. We shall not reintroduce the notation and terminology described in [1], except that for the readers convenience we list the most frequently employed symbols. Let BH be a benzenoid hydrocarbon and let B stand for the corresponding benzenoid system/benzenoid graph. Then:

$n$ = number of carbon atoms of BH = number of vertices of B
$m$ = number of carbon-carbon bonds of BH = number of edges of B
$h$ = number of six-membered rings of BH = number of hexagons of B
$h$ = $m - n + 1$
$K$ = number of Kekulé structures of BH = number of perfect matchings of B
$A$ = $[A_{rs}]$ = adjacency matrix of B

$A_{rs} = 1$ if the vertices r and s are adjacent and $A_{rs} = 0$ otherwise

$x_1, x_2, ..., x_n$ = eigenvalues of A i.e. eigenvalues of B

$x_1 \geq x_2 \geq ... \geq x_n$

$\varphi(B, x) = \det(xI - A) $ = characteristic polynomial of B

$b(B, k)$ = k-th coefficient of the characteristic polynomial

$$\varphi(B, x) = \sum_{k=0}^{[n/2]} (-1)^k \, b(B, k) \, x^{n-2k}$$

$$M_k = \sum_{i=1}^{n} (x_i)^k = \text{k-th spectral moment of B}.$$

For further details on graph-theoretical notions important in the "topological" studies of conjugated molecules and benzenoid hydrocarbons in particular see [13].

## 1.2 Total π-Electron Energy

Within the framework of the simple tight-binding Hückel molecular orbital (HMO) approximation (see the next paragraph) the total π-electron energy of a conjugated molecule is given by

$$E_\pi = \sum_{i=1}^{n} g_i E_i \tag{1}$$

where $E_i$ is the energy of the i-th MO and $g_i$ is the respective occupation number. For a conjugated hydrocarbon in its ground electronic state, $g_1 = g_2 = ... = g_{n/2} = 2$ and $g_{n/2+1} = ... = g_n = 0$. Furthermore,

$$E_i = \alpha + \beta x_i \tag{2}$$

where $x_i$ is the i-th eigenvalue of the molecular graph and where $\alpha$ and $\beta$ are certain semiempirical parameters, assumed to have the same values for all conjugated hydrocarbons. From Eqs. (1) and (2) follows

$$E_\pi = n\alpha + E\beta \tag{3}$$

with

$$E = 2 \sum_{i=1}^{n/2} x_i. \tag{4}$$

The only non-trivial term in Eq. (3) is the quantity E which is usually identified with the HMO total π-electron energy. This can be achieved by formally setting $\alpha = 0$ and $\beta = 1$ and then we speak about total π-electron energy expressed in the units of the resonance integral $\beta$.

Anyway, what we are studying in the present article is the quantity E defined via Eq. (4). It will be referred to as the total π-electron energy. For alternant hydrocarbons and thus for all benzenoid molecules Eq. (4) is readily transformed into a more symmetric form [6, 7], namely

$$E = \sum_{i=1}^{n} |x_i| .$$ (5)

Formula (5) is the starting point for almost all considerations on the structure-dependency of E.

The general theory of the HMO total π-electron energy as well as its chemical applications are outlined in full detail in the book [13]. There exists one more review on E [14], but in the Serbo-Croatian language. Whenever E is applied in chemistry one should bear in mind that the resonance integral β is a negative-valued parameter.

The right-hand side of Eq. (5) is defined in the case of an arbitrary graph. Accordingly, by means of Eq. (5) we can define a novel graph-spectral quantity, named the *energy of a graph*. Some basic properties of this graph energy are surveyed in [15].

## 1.3 A Note on the Hückel Molecular Orbital Theory

As it is well known (see, for example, [16, 17]), the Hückel molecular orbital theory is based on a Hamiltonian operator $\mathscr{H}$ defined by means of the matrix elements

$$\langle p_r| \mathscr{H} |p_r\rangle = \alpha_r$$

$$\langle p_r| \mathscr{H} |p_s\rangle = \beta_{rs} \qquad r \neq s$$

where $|p_r\rangle$, $r = 1, 2, ..., n$ is an orthogonal basis usually interpreted so that $|p_r\rangle$ is a p-orbital centered at the nucleus of the r-th atom. For conjugated hydrocarbons (and thus for benzenoid systems) these matrix elements are approximated so that $\alpha_r = \alpha$ for all atoms r, that $\beta_{rs} = \beta$ whenever there exists a chemical bond between the atoms r and s, and that $\beta_{rs} = 0$ if the atoms r and s are chemically not bonded. These assumptions lead straightforwardly to Eqs. (1) and (2) [18].

It would be completely outdated to search for some physical justification of the above approximations. In the early thirties, when the HMO model was invented, such drastic simplifications were inevitable because of the lack of computing machines. Since then quantum chemistry has made extraordinary advances in both theory and technology. Nowadays the HMO model can be considered only as a historical episode in the development of quantum chemistry.

As a consequence of this, the HMO model should no longer be used for calculations of π-electron properties of particular conjugated compounds. For this purpose there exist much more accurate and reliable quantum-theoretical techniques. (We, of course, do not deny the potentials of the HMO theory in chemical education.)

There is, however, one aspect where the HMO model can still beat the more advanced quantum-chemical approaches. Because of its extreme simplicity, the HMO Hamiltonian is related to the structure of the respective conjugated molecule in an obvious manner and is fully determined by the connectedness of its carbon-atom skeleton. Therefore it is possible (although not easy) to find direct connections between the results of HMO calculations and molecular structure. For more sophisticated quantum-chemical methods the finding of such connections seems to be far from feasible.

In the case of the HMO model the success of the search for structure-property relations is much enhanced by the possibility of applying the powerful mathematical apparatus of graph-spectral theory [18, 19]. There is hardly any application of graph (spectral) theory in more sophisticated molecular orbital models.

It is a remarkable and not yet satisfactorily explained phenomenon that in spite of the apparent shortcomings and suspicious quantum-physical basis of the HMO theory, in some cases its results are in good quantitative agreement with experimental data. In particular, this applies to the total $\pi$-electron energy, especially in the case of benzenoid hydrocarbons. This fortunate feature of the HMO theory is discussed in more detail in the subsequent section.

# 2 Total $\pi$-Electron Energy and the Thermodynamic Stability of Benzenoid Hydrocarbons

It is an often repeated claim that the HMO $\pi$-electron energies are in good agreement with experimental enthalpies of the respective conjugated compounds. This statement could easily be tested provided experimental enthalpies were available. Unfortunately, they are known only for a limited number of conjugated hydrocarbons. In particular, heats of formation are tabulated for only 24 benzenoid hydrocarbons [20].

In Fig. 1 we show the correlation between E and experimental heats of formation for the (complete) set of $C_{22}H_{14}$ benzenoid isomers. For comparison we also present some recent data for the same set of compounds, obtained by a semiempirical MNDO method [21] and by the MMX/PI version of molecular mechanics calculations [22]. The only conclusion we wish to draw from Fig. 1 is that HMO theory is capable of reproducing the experimental enthalpies of benzenoid hydrocarbons with an accuracy which is not much worse than that of the much more sophisticated (and highly parametrized) molecular orbital and molecular mechanics approaches.

Much theoretical effort was made to explain this somewhat surprising success of the HMO total $\pi$-electron energy. Here we briefly mention the arguments supporting the opinion that E accounts for (at least a part of) the electron interaction [23–26] and even the electron correlation effects [27]. In particular, Ichikawa and Ebisawa [26] reported an almost perfect linear correlation between E and the kinetic energy of the $\pi$-electrons, as calculated by means of STO-3G ab initio methods.

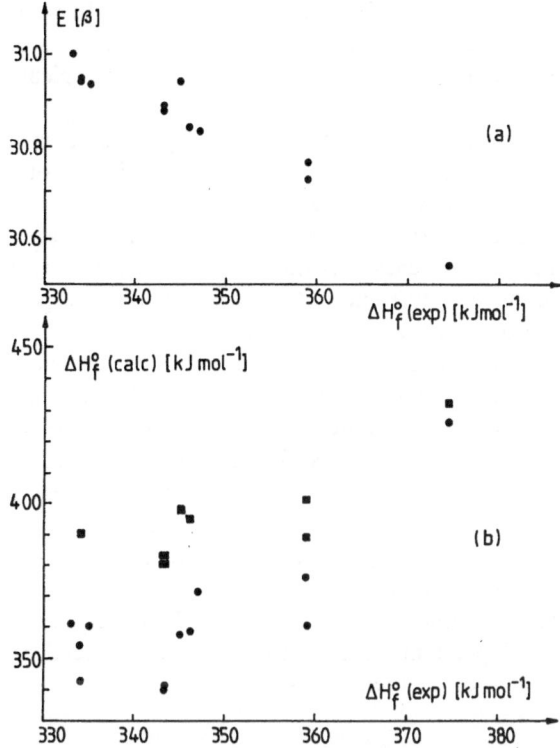

**Fig. 1. a** HMO total π-electron energies vs. experimental standard gas-phase heats of formation [20] for the twelve benzenoid isomers $C_{22}H_{14}$; correlation coefficient $-0.966$; $\beta = -97.0$ kJ mol$^{-1}$; **b** Calculated vs experimental standard gas-phase heats of formation for the same molecules; *squares*: MNDO results [21], correlation coefficient 0.797; *dots* MMX/PI results [22], correlation coefficient 0.780

Especially interesting seems to us the work of Schaad and Hess [28] where it was shown that E is a good measure not only of the energy of the π-electrons, but also of the energy of the σ-electron carbon-carbon framework. According to [28], heats of atomization computed by the HMO method are accurate to 0.1%, implying that E is accurate up to $\pm 0.005\beta$ units. By the way, Schaad and Hess [28] recommend the value $\beta = -137.00$ kJ mol$^{-1}$ for thermochemical purposes.

# 3 Elements of the Theory of Cyclic Conjugation

Among organic chemists it is nowadays commonly accepted that the unusually high stability (so called "aromaticity") of some conjugated systems as well as the exceptionally low stability (so called "antiaromaticity") of some others is somehow related to the presence of cycles in their carbon-atom skeleton. The traditional view is that these effects come from π-electrons. However, Hiberty and coworkers gave recently very convincing arguments in favour of the σ-electron origin of

Ivan Gutman

aromaticity/antiaromaticity (see [29] and the references quoted therein). In what follows we remain at the π-electron-viewpoint, mainly because it is the basis of literally everything what has been done in this field my means of "topological" methods.

The idea that the effect of cyclic conjugation can be measured by the difference between the total π-electron energy of the conjugated molecule and the total π-electron energy of a pertinently chosen acyclic reference structure seems to have been first put forward in 1963 by Breslow and Moháchi [30] and soon thereafter by Dewar [31–33]. Crucial in these approaches was the finding of a convenient "total π-electron energy of the acyclic reference structure". In this respect graph-theoretical considerations proved to be of major importance.

Since the theory of cyclic conjugation is mainly concerned with non-benzenoid polycyclic conjugated systems (containing rings of sizes other than six), we shall skip its complete presentation and focus our attention to only those details which are needed in the study of benzenoid hydrocarbons. The readers interested in furhter aspects of this theory should consult the papers [34–40] and the references quoted therein.

Let G be a molecular graph and $Z_1, Z_2, ..., Z_t$ its cycles. For example, in Fig. 2 are depicted the t = 10 cycles of benzoanthracene.

Denote by $G - Z_i$ the subgraph obtained by deleting from G the vertices of $Z_i$. Two cycles of G are said to be independent if they possess no common vertices. If $Z_i$ and $Z_j$ are independent cycles of G then the subgraph $G - Z_i - Z_j$ is

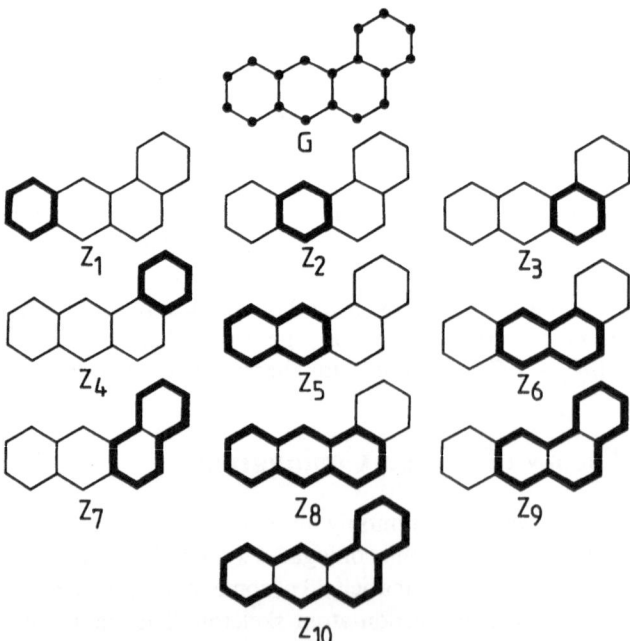

**Fig. 2.** The molecular graph G of benzo[a]anthracene and the ten cycles contained in it

defined as $(G - Z_i) - Z_j$. The subgraphs $G - Z_i - Z_j - Z_k$ etc. are defined analogously.

In Fig. 2 only five pairs of cycles are independent, namely $Z_1, Z_3; Z_1, Z_4; Z_1, Z_7;$ $Z_2, Z_4$ and $Z_4, Z_5$. No three cycles of benzoanthracene are mutually independent.

The polynomial

$$\alpha(G) = \alpha(G, x) = \sum_{k=0}^{m} (-1)^k m(G, k) x^{n-2k}$$

is called the matching polynomial of the graph G. Here $m(G, k)$ stands for the number of k-matchings of G i.e. the number of selections of k independent edges in G [1]. For details of the theory of the matching polynomial see [41–43].

Now, the characteristic polynomial $\varphi(G) = \varphi(G, x)$ of a graph G conforms to the identity

$$\varphi(G) = \alpha(G) - 2\sum_i \alpha(G - Z_i) + 4\sum_{i,j} \alpha(G - Z_i - Z_j)$$

$$- 8 \sum_{i,j,k} \alpha(G - Z_i - Z_j - Z_k) + \dots . \qquad (6)$$

The second, third etc. summations on the right-hand side of Eq. (6) run over all pairs, triplets etc. of independent cycles of G. For instance, if G = benzoanthracene then we have (see Fig. 2):

$$\varphi(G) = \alpha(G) - \sum_{i=1}^{10} \alpha(G - Z_i) + 4 [\alpha(G - Z_1 - Z_3)$$

$$+ \alpha(G - Z_1 - Z_4) + \alpha(G - Z_1 - Z_7) + \alpha(G - Z_2 - Z_4)$$

$$+ \alpha(G - Z_4 - Z_5)] .$$

Equation (6) can be viewed as the explicit expression for the dependence of the characteristic polynomial of the graph G on the cycles $Z_1, Z_2, \dots, Z_t$ contained in this graph. Since, on the other hand, the graph eigenvalues $x_1, x_2, \dots, x_n$ are related with the characteristic polynomial via

$$\varphi(G, x) = \prod_{i=1}^{n} (x - x_i) \qquad (7)$$

and since the total π-electron energy is computed from the eigenvalues via Eq. (5), we arrive at a method for expressing the effect of the cycles on the E-value of a conjugated molecule.

Combining Eqs. (5)–(7) we see that the value of E is influenced by all the cycles $Z_1, Z_2, \dots, Z_t$. Suppose now that we ignore all the terms on the right-hand side of Eq. (6), depending on $Z_1, Z_2, \dots, Z_t$. Then only the matching polynomial will remain. Presenting $\alpha(G, x)$ in the form

$$\alpha(G, x) = \prod_{i=1}^{n} (x - y_i)$$

it follows that the total-$\pi$-electron-energy-like quantity $E^R$ defined as

$$E^R = \sum_{i=1}^{n} |y_i| \tag{8}$$

accounts for the effects of all structural details on E, except of the cycles. Consequently, $E^R$ can be viewed as the total $\pi$-electron energy of the acyclic reference structure. Furthermore, the difference

$$TRE = E - E^R \tag{9}$$

can be interpreted as the *effect of all cycles on the total $\pi$-electron energy of the conjugated molecule whose molecular graph is G.*

The abbreviation TRE comes from "topological resonance energy", as this measure of the overall cyclic conjugation in a polycyclic conjugated molecule was originally named [35, 44]. The TRE-concept was introduced in 1976 independently by Aihara [34] and by Milun, Trinajstić and the present author [35, 44, 45]. Aromatic conjugated molecules have positive whereas antiaromatic molecules have negative TRE-values.

The TRE-concept fails to correctly describe the cyclic conjugation in many important classes of conjugated molecules [46–51]. Among them are also non-Kekuléan benzenoid hydrocarbons [50, 51]. On the other hand, in the case of Kekuléan benzenoids the predictions made on the basis of TRE are usually in reasonable agreement with experimental findings [50, 52, 53].

We mention in passing that the zeros of the matching polynomial, namely the numbers $y_1, y_2, ..., y_n$, are always real-valued [42, 54].

Using a fully analogous reasoning as in the case of TRE, we see that the effect of an individual cycle $Z = Z_k$ on E can be measured by means of a quantity $ef = ef(Z) = ef(G, Z)$, defined as

$$ef(Z) = E - E^R(Z) \tag{10}$$

where

$$E^R(Z) = \sum_{i=1}^{n} |y_i(Z)|$$

and where $y_i(Z)$, $i = 1, 2, ..., n$ are the zeros of the polynomial $\varphi(G, Z, x)$. This latter polynomial is obtained by ignoring all the terms containing the cycle $Z_k$ on the right-hand side of Eq. (6). It can be shown that [55]

$$\varphi(G, Z, x) = \varphi(G, x) + 2\varphi(G - Z, x) \tag{11}$$

which makes the calculation of ef quite easy [36, 55].

By means of ef we can estimate the energy-effect of the interactions of π-electrons along a particular cycle in a polycyclic conjugated molecule. Thus ef measures the effect of cyclic conjugation caused by an individual cycle on the thermodynamic stability of the respective molecule. We could also say that ef measures the local aromaticity. For the purpose of the present considerations it is important that ef(G, Z) is the *effect of an individual cycle Z on the total π-electron energy of the conjugated molecule whose molecular graph is G* [56].

It has been shown that [57]

$$\text{TRE} \approx \sum_{k=1}^{t} \text{ef}(Z_k) \, .$$

The above approximation is found to be especially well-satisfied in the case of benzenoid molecules.

Aihara [58] proposed another similar recipe for expressing the effect of an individual cycle on the conjugation in a polycyclic molecule. Instead of ignoring the cycle $Z = Z_k$, he ignored on the right-hand side of Eq. (6) all cycles except $Z_k$. The resulting polynomial is denoted by $\alpha(G, Z, x)$ and it is easy to see that

$$\alpha(G, Z, x) = \alpha(G, x) - 2\alpha(G - Z, x) \, . \tag{12}$$

Then Aihara's measure of cyclic conjugation is

$$\text{efa}(Z) = E^A(Z) - E^R \tag{13}$$

where $E^R$ is given by Eq. (8),

$$E^A(Z) = \sum_{i=1}^{n} |x_i^A(Z)|$$

and where $x_1^A(Z), x_2^A(Z), \dots, x_n^A(Z)$ are the zeros of the polynomial $\alpha(G, Z, x)$.

In connection with the definition of ef(G, Z) Herndon [59] expressed the objection that the zeros $y_1(Z), y_2(Z), \dots, y_n(Z)$ of the polynomial $\varphi(G, Z, x)$ are not always real-valued numbers. The difficulties caused by this were later resolved [60]. The zeros $y_1^A(Z), y_2^A(Z), \dots, y_n^A(Z)$ of the polynomial $\alpha(G, Z, x)$, needed for the calculation of efa(G, Z), were conjectured to be real-valued for all G and for all Z [58, 61]. Only very limited progress in proving the reality of these zeros has been achieved so far [61–63].

The basic conceptual difference between ef(G, Z) and efa(G, Z) is that ef can whereas efa cannot be conceived as the effect of the cycle Z on the E-value of the molecular graph G. Furthermore, whereas the polynomial $\varphi(G, Z, x)$ is easily computed by means of the identity (11), the computation of $\alpha(G, Z, x)$ from Eq. (12) is difficult, especially in the case of large polycyclic molecular graphs. As a consequence of this, ef-values are known for very many conjugated systems (including benzenoids [39, 40, 64]). The number of available efa-values is very small [58]. Anyway, a systematic comparative study of ef and efa has never been undertaken.

In Sect. 8 the ef-values of benzenoid hydrocarbons are discussed at due length.


# 4 Identities for E, TRE, ef and efa

In this section we report the Coulson-type integral formulas for total $\pi$-electron energy and their modifications applicable for TRE, ef and efa. We consider here only the special cases when the respective conjugated systems are benzenoid hydrocarbons.

The integral formulas for total $\pi$-electron energy were invented by Coulson [2, 3, 6]. Later, they were extensively elaborated (see [13], pp. 139–147). Their real usefulness in structure-property analysis became evident only after they were combined with the results of graph spectral theory [13, 18, 65, 66].

Let B be the molecular graph of a benzenoid hydrocarbon. Then [66, 67]

$$E(B) = (2/\pi) \int_0^\infty x^{-2} \ln \sum_{k=0}^{[n/2]} b(B, k)\, x^{2k}\, dx$$

where $b(B, k), k = 0, 1, \ldots, [n/2]$ are the coefficients of the characteristic polynomial of B (see Sect. 1.1). Recall that $b(B, 0) = 1$, $b(B, 1) = m$, $b(B, n/2) = K^2$ and $b(B, k) \geq 0$ for all values of k [1].

Let $B_1$ and $B_2$ be molecular graphs of two isomeric benzenoid hydrocarbons. Then [6]

$$E(B_1) - E(B_2) = (2/\pi) \int_0^\infty \ln \frac{\varphi(B_1, ix)}{\varphi(B_2, ix)}\, dx$$

$$= (2/\pi) \int_0^\infty \ln \frac{\sum_{k=0}^{[n/2]} b(B_1, k)\, x^{n-2k}}{\sum_{k=0}^{[n/2]} b(B_2, k)\, x^{n-2k}}\, dx \qquad (14)$$

where $i = \sqrt{-1}$.

Formulas of the above type were studied in [68] were it was established that the main contribution to the right-hand side integrals comes from near-zero values of x. On the other hand, when x approaches to zero, then the integrand

$\ln \left[ \varphi(B_1, ix)/\varphi(B_2, ix) \right]$ tends to $2 \ln \left[ K\{B_1\}/K\{B_2\} \right]$. Nevertheless, the assertion [68] that the energy difference of benzenoid isomers is proportional to the difference between the logarithms of their Kekulé structure counts was not confirmed by later investigations (see Sect. 7.2).

Slight modifications of Eq. (14) lead us to Coulson-type integral representations of TRE, ef and efa, as defined by Eqs. (9), (10) and (13), respectively. These integral formulas read:

$$\text{TRE} = (2/\pi) \int_0^\infty \ln \frac{\varphi(B, ix)}{\alpha(B, ix)}\, dx$$

$$= (2/\pi) \int_0^\infty \ln \frac{\sum\limits_{k=0}^{[n/2]} b(B, k)\, x^{n-2k}}{\sum\limits_{k=0}^{[n/2]} m(B, k)\, x^{n-2k}}\, dx \tag{15}$$

$$\text{ef}(B, Z) = -(2/\pi) \int_0^\infty \ln \left| 1 + 2\, \frac{\varphi(B - Z, ix)}{\alpha(B, ix)} \right|\, dx$$

$$= -(2/\pi) \int_0^\infty \ln \left| 1 + (-1)^{z/2}\, 2\, \frac{\sum\limits_{k=0}^{[(n-z)/2]} b(B - Z, k)\, x^{n-z-2k}}{\sum\limits_{k=0}^{[n/2]} b(B, k)\, x^{n-2k}} \right|\, dx\,. \tag{16}$$

$$\text{efa}(B, Z) = (2/\pi) \int_0^\infty \ln \left[ 1 - 2\, \frac{\alpha(B - Z, ix)}{\alpha(B, ix)} \right]\, dx$$

$$= (2/\pi) \int_0^\infty \ln \left[ 1 - (-1)^{z/2}\, 2\, \frac{\sum\limits_{k=0}^{[(n-z)/2]} m(B - Z, k)\, x^{n-z-2k}}{\sum\limits_{k=0}^{[n/2]} m(B, k)\, x^{n-2k}} \right]\, dx\,. \tag{17}$$

In Eqs. (16) and (17) z denotes the size (= number of vertices) of the cycle Z.
The following generally valid regularities were deduced from Eqs. (15)–(17).

*Theorem 1* [69]. For any catacondensed benzenoid system, TRE > 0.

*Theorem 2* [51]. For any benzenoid systems, TRE > 0.

Ivan Gutman

*Theorem 3* [56]. If **B** is a benzenoid systems and Z its cycle, such that the size of Z is divisible by four, then ef(B, Z) < 0. (Note that in this case **B** must be pericondensed [1].)

*Theorem 4* [70]. If **B** is a benzenoid systems and Z its cycle, such that the size of Z is divisible by four, then efa(B, Z) < 0. If the size of Z is not divisible by four, then efa(B, Z) > 0.

*Corollary 4.1.* If **B** is a catacondensed benzenoid system, then efa(B, Z) > 0 for all cycles Z of **B**.

Comparing Theorems 3 and 4 we immediately observe that the former does not contain a statement about the sign of ef(Z) when Z is a (4k + 2)-membered cycle. There exist non-benzenoid alternat hydrocarbons in which (4k + 2)-membered cycles have negative ef-values [36]. Therefore it is somewhat risky to formulate the following

*Conjecture.* If **B** is a benzenoid system and Z its cycle, such that the size of Z is not divisible by four, then ef(B, Z) > 0. In particular, if **B** is a catacondensed benzenoid system, then ef(B, Z) > 0 for all cycles Z of **B**.

Numerous numerical examples support this conjecture [39, 40, 64], but one should note that practically all benzenoid systems examined possessed Kekulé structures.

The above conjecture is true if **B** is a catacondensed benzenoid system and Z is its perimeter. Clearly, the size of Z is then equal to n = 4h + 2 whereas $\varphi(B - Z, x) \equiv 1$. Then Eq. (16) reduces to

$$ef(B, Z) = -(2/\pi) \int_0^\infty \ln\left[1 - \frac{2}{\sum_{k=0}^{n/2} b(B, k) x^{n-2k}}\right] dx . \tag{18}$$

Since $b(B, n/2) = (h + 1)^2$, it follows that for all $x \geq 0$,

$$\frac{2}{\sum_{k=0}^{n/2} b(B, k) x^{n-2k}} \leq \frac{2}{(h + 1)^2} < 1 .$$

Consequently, the integrand in Eq. (18) is negative for all values of x and therefore ef(B, Z) is positive.

# 5 Bounds for Total π-Electron Energy

In 1971 Bernard J. McClelland communicated the first upper and lower bounds for total π-electron energy [71]. The work [71] seems to caused a turning point in the development of the theory of total π-electron energy: it shifted the interest of the researchers from identities and approximate expressions to inequalities. Since

1971, numerous estimates of E have been discovered, many of them being applicable exclusively to benzenoid systems. In this section, we are concerned only with those bounds for E which are of relevance for benzenoids. Bounds for total π-electron energy of general conjugated molecules or of general alternant hydrocarbons are mentioned only when necessary; for more details on them see [13–15].

## 5.1 The McClelland Inequality

McClelland [71] showed that for arbitrary conjugated hydrocarbons whose total π-electron energy satisfies Eq. (5), E is less than or equal to $E_{MC}$, where

$$E_{MC} = \sqrt{2mn} \ . \tag{19}$$

Recently the present author [72] complemented McClelland's result and so we now have the following

*Theorem 5.* For a benzenoid system with n vertices and m edges, and for $E_{MC}$ being given by Eq. (19),

$$(16/27)^{1/2} E_{MC} \leq E < E_{MC} . \tag{20}$$

Because of the importance of the bounds (20) we provide a complete proof of Theorem 5.

*Proof.* (a) It is evident that

$$\sum_{i=1}^{n} \sum_{j=1}^{n} (|x_i| - |x_j|)^2 \geq 0 \tag{21}$$

and that equality occurs only if all eigenvalues $x_1, x_2, \ldots, x_n$ have equal absolute values. For all graphs,

$$\sum_{i=1}^{n} x_i^2 = 2m . \tag{22}$$

Then, bearing in mind Eqs. (5) and (22), the left-hand side of (21) is immediately transformed as

$$\sum_{i=1}^{n} \sum_{j=1}^{n} (|x_i| - |x_j|)^2 = \sum_{i=1}^{n} \sum_{j=1}^{n} x_i^2 - 2 \sum_{i=1}^{n} \sum_{j=1}^{n} |x_i| \cdot |x_j| + \sum_{i=1}^{n} \sum_{j=1}^{n} x_j^2$$

$$= 2mn - 2E^2 + 2mn$$

and consequently, $4mn - 2E^2 \geq 0$ i.e. $E \leq E_{MC}$. In the case of benzenoid systems not all eigenvalues have equal absolute values and therefore $E < E_{MC}$.

Ivan Gutman

(b) Let $a_1, a_2, \ldots, a_n$ be non-negative real numbers and let

$$\mu_k = \sum_{i=1}^{n} (a_i)^k .$$

It can be shown [72] that for arbitrary numbers $p, q, r$, such that $p \geq q$ and $r \geq 1$,

$$\mu_{pr}/\mu_{qr} \geq (\mu_p/\mu_q)^r .$$

In particular, if $p = 2$, $q = 1$ and $r = 2$ the above general result reduces to

$$\mu_4/\mu_2 \geq (\mu_2/\mu_1)^2 .$$

Choosing $a_i = |x_i|$ and bearing in mind Eqs. (5), (22) and (23),

$$\sum_{i=1}^{n} x_i^4 = 18m - 12n \tag{23}$$

we immediately arrive at

$$(18m - 12n)/(2m) \geq (2m/E)^2$$

from which it follows

$$E \geq e_C \tag{24}$$

where

$$e_C = [4m^3/(9m - 6n)]^{1/2} . \tag{25}$$

Recall that Eq. (23) is just an expression for the fourth spectral moment and that its validity is restricted to benzenoid systems; for more details on the spectral moments of benzenoids see [1].

Inequality (24) was first deduced by Cioslowski [73], but using a different way of reasoning. Eq. (25) is easily transformed into

$$e_C = [d^2/(9d - 12)]^{1/2} E_{MC}$$

where $d = 2m/n$ is the average vertex degree. Obviously, for benzenoid systems, $2 \leq d < 3$. Now, it is elementary to prove that in the interval $[2, 3)$ the minimum of the function $[d^2/(9d - 12)]^{1/2}$ is at $d = 8/3$ and is equal to $(16/27)^{1/2}$. Therefore $e_C \geq (16/27)^{1/2} E_{MC}$ and consequently, $E \geq (16/27)^{1/2} E_{MC}$.

This completes the proof of Theorem 5.

Having in mind the bounds (20) it is not at all surprising that a very good linear proportionality between $E$ and $E_{MC}$ has been observed [71]. This important issue is discussed in due detail in Sect. 6.

## 5.2 Other Inequalities

As already mentioned, quite a few upper and lower bounds for the total π-electron energy of benzenoid hydrocarbons were deduced after the pioneering work [71] of McClelland. The most important of them are briefly outlined in the present section. As before, n, m and K denote the numbers of vertices, edges and Kekulé structures, respectively.

*Theorem 6* [74]. Denote the term $2m - nK^{4/n}$ by $\delta$ and observe that $\delta$ is positive for all benzenoid systems. Then

$$\sqrt{2mn - (n - 2)\,\delta} \le E \le \sqrt{2mn - 2\delta}\;.$$

Recall that $E_{MC} = \sqrt{2mn}$ and compare Theorem 6 with Theorem 5.

*Theorem 7* [75]. Consider the system of equations

$$x^2 + (n/2 - 1)\,y^2 = m$$

$$xy^{n/2-1} = K\,, \qquad K > 0\,.$$

Let $x_1$, $y_1$ be the solution of this system, such that $x_1 > y_1 > 0$. Let $x_2$, $y_2$ be the solution satisfying $y_2 > x_2 > 0$. These solutions exist for all chemically relevant values of n, m and K, and are unique. Then

$$E_{min} \le E < E_{max}$$

where

$$E_{min} = 2x_1 + (n - 2)\,y_1$$

$$E_{max} = 2x_2 + (n - 2)\,y_2\,.$$

The equality $E = E_{min}$ holds only for benzene.

In [75] the bounds $E_{min}$ and $E_{max}$ were approximated as follows:

$$E_{min} = 2T + 2q(K/T)^{1/q} \tag{26}$$

$$E_{max} = 2(mq)^{1/2} + 2K(q/m)^{q/2} \tag{27}$$

where $q = n/2 - 1$ and

$$T = \sqrt{m - qK[m - q(K^2/m)^{1/q}]^{-1/2}}\;.$$

Ivan Gutman

*Theorem 8* [73]. The smallest positive eigenvalue in the spectrum of a benzenoid system is denoted by $x_{n/2}$. Let b be the number of bay regions [1]. Then

$$x_{n/2}(9mn - 6n^2 - 2m^2)/(9m - 6n)$$
$$+ 2\sqrt{2m/n}\ m^2/(9m - 6n) \le E \le (n\sqrt{P} + 2m)\,(Q + 2\sqrt{P})^{1/2}$$

where

$$P = (72mn - 72n^2 - 4m^2 + 48m + 6nb)/(9mn - 6n^2 - 2m^2)$$

$$\tag{29}$$

$$Q = (91mn - 72n^2 - 18m^2 + 24n + 3nb)/(9mn - 6n^2 - 2m^2).$$

$$\tag{30}$$

*Theorem 9* [76]. Let B a benzenoid system and let $E_T$ be defined as

$$E_T = 2\sqrt{m + [\tfrac{1}{2}\,n(n - 2)\,b(B, 2)]^{1/2}}\ .\tag{31}$$

Then $E \le E_T$.

Recall that $b(B, 2)$ is the second coefficient of the characteristic polynomial of B and that it satisfies

$$b(B, 2) = (m^2 - 9m + 6n)/2\ .$$

The result of Theorem 9 was later slightly improved [77]. Let $E_{GTD}$ be defined as

$$E_{GTD} = 2\,[\tfrac{3}{2}\,mE_T + [\tfrac{3}{4}\,n(n - 2)\,(n - 4)\,b(B, 3)]^{1/2} - (8m^3/n)^{1/2}]^{1/3}$$

$$\tag{32}$$

where $b(B, 3)$ is the third coefficient of the characteristic polynomial of B, conforming to the relation:

$$b(B, 3) = \tfrac{1}{6}\,(m^3 - 27m^2 + 158m + 48) + 3n(m - 8) + b\ .\tag{33}$$

Then $E \le E_{GTD}$. In addition to this, $E_{GTD} \le E_T \le E_{MC}$.

*Theorem 10* [78, 79]. Define the function $E_C(x)$ as

$$E_C(x) = 2x\sqrt{2m/n + RS} + (n - 2x)\sqrt{2m/n - R/S}\tag{34}$$

where

$$R = \frac{1}{n}\sqrt{18mn - 12n^2 - 4m^2}$$

$$S = S(x) = \sqrt{(n - 2x)/(2x)}\ .$$

Then $E \le E_C(1)$. Furthermore, $E_C(1) < E_{MC}$.

# 6 Approximate Formulas for Total π-Electron Energy

A plethora of approximate topological formulas for the total π-electron energy have been proposed in the chemical literature. The early works in this area (e.g. [4, 80–82]) were mainly aimed towards obtaining reliable numerical values for E. More recent investigations put the emphasis on the mathematical properties of E and, in particular, on its dependence on molecular structure.

In the case of benzenoid hydrocarbons it is nowadays firmly established that the two most important structural parameters influencing the value of E are n and m, the numbers of carbon atoms and carbon-carbon bonds, respectively. The third-important invariant seems to be K, the number of Kekulé structures. (More about this issue can be found in Sect. 7.)

Approximate expressions for E, depending only on n and m are referred to as formulas of (n,m)-type. Analogously, formulas of (n,m,K)-type are those in which the "topological" parameters are n, m and K.

Other structural invariants have also been considered as parameters influencing the E-value of benzenoid molecules (e.g. the length of linear polyacene fragments [82], the modified Hosoya index [65, 83–85], the number of bay regions [86, 87] etc.). In what follows we focus our attention only to approximate formulas of (n,m)- and of (n,m,K)-type.

## 6.1 The McClelland Formula

In his seminal paper [71] McClelland proposed approximating E by means of the simple (n,m)-type formula.

$$E \approx aE_{MC} \tag{35}$$

where $E_{MC}$ is given by Eq. (19) and where $a$ is an empirical constant, determined by least-squares fitting. (Recall that $E_{MC}$ is an upper bound for E.)

McClelland's formula (35) was criticised by Milun, Trinajstić and the present author [88], who argued that according to (35) all isomers are predicted to have equal E-values. Whereas this objection is certainly true, the criticism turned out to be largely unjustified. In addition to Eq. (35) over 20 isomer-undistinguishing approximate formulas for E have been put forward in the chemical literature and none has been found to be (significantly) better than McClelland's (see the subsequent Section).

In the case of benzenoid hydrocarbons, Eq. (35) reproduces E with an average relative error of only 0.4%. This, in turn, means that more than 99,5% of E is determined by the simple structural parameters n and m and that variations in E-values of isomers are fairly (but not negligibly) small.

As outlined in Sect. 5.2, after the discovery of McClelland's upper bound many other upper bounds for E have been found, all lying closer to E than $E_{MC}$. Nevertheless, none of them shows a better correlation with E than $E_{MC}$. In addition to this, among the numerous lower bounds for E none was found to correlate with E better than $E_{MC}$. This all implies that the gross part of E not only depends on n and m, but that this dependence has the functional form anticipated in Eq. (35).

These somewhat surprising empirical findings were recently rationalized by discovering a McClelland-type lower bound for E of benzenoid hydrocarbons [72]. Bearing in mind Theorem 5, a first guess for the multiplier $a$ in Eq. (35) could be the arithmetic mean of 1 and $(16/27)^{1/2}$, which equals 0.885. This is quite close to the empirical value $a = 0.908$.

## 6.2 Other Formulas of (n,m)-Type

The numerous (n,m)-type approximate formulas for E were examined in three comparative studies [89–91]. Their results, complemented by a few newly obtained approximations are collected in Tables 1 and 2.

Let E* be a mathematical expression of (n,m)-type, either an approximation or an upper bound or a lower bound for the total $\pi$-electron energy. Then we consider the (n,m)-type approximate formulas:

$$E \approx a_1 E^* \tag{36}$$

**Table 1.** Coefficients in the approximate (n,m)-type formulas (36) and (37)

| Equation for E* | Literature source of E* | $a_1$ | $a_2$ | $b_2$ |
|---|---|---|---|---|
| 38 | 71, 93 | 0.908 | 0.898 | 0.45 |
| 39 | 73 | 1.174 | 1.167 | 0.21 |
| 40 | 73 | 0.087 | 0.083 | 1.96 |
| 41 | 76 | 0.919 | 0.899 | 0.90 |
| 42 | 77 | 0.928 | 0.902 | 1.13 |
| 43 | 78 | 0.948 | 0.926 | 0.92 |
| 44 | 79 | 0.959 | 0.931 | 1.13 |
| 45 | 79 | 0.978 | 0.970 | 0.30 |
| 46 | 79 | 1.077 | 1.067 | 0.40 |
| 47 | – | 1.413 | 1.431 | −0.48 |
| 48 | – | 1.167 | 1.125 | 1.42 |
| 49 | 94 | 1.048 | 1.038 | 0.39 |
| 50 | 91 | 1.015 | 1.008 | 0.27 |
| 51 | 91 | 1.031 | 1.019 | 0.44 |
| 52 | 91 | 1.023 | 1.014 | 0.35 |
| 53 | 91 | 1.010 | 1.005 | 0.21 |
| 54 | 91 | 1.030 | 1.018 | 0.44 |
| 55 | 91 | 1.020 | 1.011 | 0.32 |
| 56 | 91 | 1.069 | 1.044 | 0.95 |
| 57 | 95 | 1.326 | 1.322 | 0.13 |
| 58 | 96 | 0.799 | 0.788 | 0.57 |
| 59 | 96 | 0.739 | 0.729 | 0.54 |
| 60 | 96 | 0.702 | 0.693 | 0.51 |
| 61 | 96 | 0.677 | 0.669 | 0.49 |

**Table 2.** Results of numerical testing of approximate (n,m)-type formulas (36) and (37)

| Equation for E* | Eq. (36) | | Eq. (37) | | correlation coefficient |
|---|---|---|---|---|---|
| | mean error (%) | max. error observ. (%) | mean error (%) | max. error observ. (%) | |
| 38 | 0.37 | 1.2 | 0.30 | 1.0 | 0.9998 |
| 39 | 0.33 | 1.3 | 0.31 | 1.2 | 0.9998 |
| 40 | 2.31 | 9.5 | 2.04 | 8.3 | 0.990 |
| 41 | 0.54 | 2.3 | 0.30 | 1.0 | 0.9998 |
| 42 | 0.63 | 2.7 | 0.31 | 1.1 | 0.9998 |
| 43 | 0.54 | 2.1 | 0.30 | 1.1 | 0.9998 |
| 44 | 0.63 | 2.7 | 0.30 | 1.1 | 0.9998 |
| 45 | 0.35 | 1.1 | 0.32 | 1.2 | 0.9998 |
| 46 | 2.73 | 8.8 | 2.77 | 8.8 | 0.98 |
| 47 | 0.66 | 2.7 | 0.58 | 2.6 | 0.9992 |
| 48 | 0.93 | 4.0 | 0.58 | 2.5 | 0.9992 |
| 49 | 0.36 | 1.2 | 0.30 | 1.0 | 0.9998 |
| 50 | 0.33 | 1.2 | 0.31 | 1.1 | 0.9998 |
| 51 | 0.46 | 1.6 | 0.42 | 1.4 | 0.9996 |
| 52 | 0.38 | 1.4 | 0.34 | 1.2 | 0.9997 |
| 53 | 0.33 | 1.3 | 0.31 | 1.2 | 0.9998 |
| 54 | 0.49 | 1.6 | 0.45 | 1.4 | 0.9996 |
| 55 | 0.40 | 1.4 | 0.37 | 1.3 | 0.9997 |
| 56 | 0.92 | 3.2 | 0.87 | 2.8 | 0.9995 |
| 57 | 0.34 | 1.3 | 0.34 | 1.2 | 0.9997 |
| 58 | 0.42 | 1.6 | 0.31 | 1.0 | 0.9998 |
| 59 | 0.44 | 1.6 | 0.35 | 1.0 | 0.9998 |
| 60 | 0.47 | 1.7 | 0.40 | 1.2 | 0.9997 |
| 61 | 0.52 | 1.7 | 0.45 | 1.4 | 0.9996 |

and

$$E \approx a_2 E^* + b_2 \tag{37}$$

and determine $a_1$, $a_2$ and $b_2$ by least-squares fitting. The data base for these calculations is the set of 104 Kekuléan benzenoid hydrocarbons from the book [92], possessing three or more condensed six-membered rings.

The following expressions for E* have been taken into consideration:

$$E^* = E_{MC}, \quad \text{see Eq. (19)} \tag{38}$$

$$E^* = e_C, \quad \text{see Eq. (25)} \tag{39}$$

$$E^* = (n \sqrt{P_0} + 2m)(Q_0 + 2\sqrt{P_0})^{1/2} \tag{40}$$

where $P_0$ and $Q_0$ are obtained from Eqs. (29) and (30), respectively, by setting b = 0;

$$E^* = E_T, \quad \text{see Eq. (31)} \tag{41}$$

$$E^* = E_{GTD}^0 \tag{42}$$

Ivan Gutman

where $E_{GTD}^0$ is obtained from Eqs. (32) and (33) by setting b = 0;

$$E^* = E_C(1), \qquad \text{see Eq. (34)} \tag{43}$$

$$E^* = E_C(2), \qquad \text{see Eq. (34)} \tag{44}$$

$$E^* = E_C(T/2), \qquad \text{see Eq. (34)} \tag{45}$$

$$E^* = E_C(T), \qquad \text{see Eq. (34)} \tag{46}$$

where T is the integer part of $m^2/(9m - 6n)$;

$$E^* = n \tag{47}$$

$$E^* = m \tag{48}$$

$$E^* = [\tfrac{3}{4} n^2/(n^2 - 1)]^{1/2} E_{MC} \tag{49}$$

$$E^* = \frac{n}{4} [[(15m - 10n)/m]^{1/2}$$
$$+ [(24m^2 - 45mn + 30n^2)/(mn)]^{1/2}] \tag{50}$$

$$E^* = \frac{n}{4} [[(553m - 504n + 168)/(45m - 30n)]^{1/2}$$
$$+ [(1080m^2 - 2379mn + 1512n^2 - 504n)/(45m - 30n)]^{1/2}] \tag{51}$$

$$E^* = \frac{n}{4} [(553m - 504n + 168)/(3m)]^{1/4}$$
$$+ [\tfrac{3}{2} mn - (3n^2/16) [(553m - 504n + 168)/(3m)]^{1/2}]^{1/2} \tag{52}$$

$$E^* = 3m[m/(15m - 10n)]^{1/2} \tag{53}$$

$$E^* = 3m[(45m - 30n)/(553m - 504n + 168)]^{1/2} \tag{54}$$

$$E^* = 3m[3m/(553m - 504n + 168)]^{1/4} \tag{55}$$

$$E^* = \frac{n}{2} [[(553m - 504n + 168)/(45m - 30n)]^{1/2}$$
$$+ \frac{1}{n} [(45m - 30n)/(553m - 504n + 168)]^{1/2} (182250m^3$$
$$- 670309m^2 n + 800424mn^2 - 308016n^3 + 169344n^2$$
$$- 185808mn - 28224n)/(24885m^2 - 39270mn + 15120n^2$$
$$+ 7560m - 5040n)] \tag{56}$$

$$E^* = m^{1/3}n^{2/3} \tag{57}$$

$$E^* = [18\,(m - 2n/3)\,n^3]^{1/4} \tag{58}$$

$$E^* = [158\,(m - 72n/79 + 24/79)\,n^5]^{1/6} \tag{59}$$

$$E^* = [1330\,(m - 682n/665 + 352/665)\,n^7]^{1/8} \tag{60}$$

$$E^* = [10762\,(m - 5855n/5381 + 3690/5381)\,n^9]^{1/10} . \tag{61}$$

Table 1 contains the coefficients in Eqs. (36) and (37) as well as the literature source of E\*. Table 2 contains data revealing the quality of the approximations (36) and (37).

The basic conclusions which follow from Tables 1 and 2 have already been summarized in Sect. 6.1. Thus, in spite of its great algebraic simplicity, the McClelland formula is not inferior to any of the much more sophisticated (n,m)-type expressions propposed to approximate the total π-electron energy of benzenoid hydrocarbons. Few of them, namely when E\* is given by Eqs. (39), (45), (49), (50), (53) and (57), provide slightly better results than the McClelland formula, but the gain in accuracy and/or reliability is meager and insignificant.

Formulas worse than $E = a_1n$ or $E = a_2n + b_2$ must be considered as absolutely useless. These are the expressions based on Eqs. (40), (46), (48) and (56).

## 6.3 Formulas of (n,m,K)-Type

The problem of the dependence of E on the number of Kekulé structures is examined in some detail in the subsequent section. Nowadays it is fairly well established that this dependence is linear. In former times, when this simple fact was not conceived, an astonishing variety of mathematical expressions was proposed for the description of the K-dependence of E. A comparative study of these early (in most cases unsuccessful) attempts is reported in [89]. The data collected in Tables 3 and 4 are analogous to those given in Tables 1 and 2. They are obtained using the same sample of 104 benzenoids from the book [92]. The following (n,m,K)-type expressions for E\* have been examined:

$$E^* = E_{min}, \quad \text{see Eq. (26)} \tag{62}$$

$$E^* = E_{max}, \quad \text{see Eq. (27)} \tag{63}$$

$$E^* = \frac{n}{2}\,(4m/n + 2K^{4/n})^{1/2} \tag{64}$$

$$E^* = n + [2\,(m - n/2)\ln K]^{1/2} \tag{65}$$

$$E^* = E_{MC} - (\alpha^3\beta)^{1/4} \tag{66}$$

Ivan Gutman

where

$$\alpha = (n - 2)\, m^2/(2n) - (m^2 - 9m + 6n)/2$$

$$\beta = \frac{n}{2} \ln (2m/n) - 2 \ln K .$$

In addition to Eqs. (62)–(66), in Tables 3 and 4 are included the results of numerical testing of some further approximate (n,m,K)-formulas which do not comply with the form of Eqs. (36) and (37):

$$E = an + b \ln K + c \tag{67}$$

$$E = an + bm + cn^{1/6}(\ln K)^{5/6} + d \tag{68}$$

$$E = an + bm + cKd^{m-n} \tag{69}$$

$$E = (an + bm + 2 \ln K)/(3cm/n + d)$$
$$+ cn^{-2}(an + bm + 2 \ln K)^3/(3cm/n + d)^4 \tag{70}$$

$$E = (a + bx)\, E_{MC} \tag{71}$$

$$E = (a + bx + cx^2)\, E_{MC} \tag{72}$$

$$E = (a + bx + cx^2 + dx^3)\, E_{MC} \tag{73}$$

**Table 3.** Coefficients in the approximate (n,m,K)-type formulas (36), (37) and (67)–(76)

| Equation for E* | Literature source of E* | $a_1$ | $a_2$ | $b_2$ | |
|---|---|---|---|---|---|
| 62 | 75 | 1.054 | 1.091 | −1.38 | |
| 63 | 75 | 0.940 | 0.912 | 1.19 | |
| 64 | 89, 97, 98 | 1.009 | 1.015 | −0.32 | |
| 65 | 99 | 1.021 | 1.037 | −0.63 | |
| 66 | 99 | 1.165 | 1.193 | −0.96 | |
| Equation for E | Literature source of E | $a$ | $b$ | $c$ | $d$ |
| 67 | 4 | 1.379 | 0.585 | 0.747 | |
| 68 | 100, 101 | 0.760 | 0.494 | 0.342 | 0.115 |
| 69 | 102, 103 | 0.442 | 0.788 | 0.34 | 0.632 |
| 70 | 104 | 0.965 | 2.326 | 0.422 | 1.810 |
| 71 | 105 | 0.758 | 0.190 | | |
| 72 | 106 | 1.291 | −1.168 | 0.864 | |
| 73 | 106 | 1.306 | −1.175 | 0.811 | 0.049 |
| 74 | 106 | 0.725 | 0.330 | −0.082 | |
| 75 | 106 | 1.574 | −1.087 | −0.592 | |
| 76 | 106 | 0.893 | 0.186 | 11 | |

$$E = (a + bx + cx^{-1}) E_{MC} \tag{74}$$

$$E = (a + bx + c \ln x) E_{MC} \tag{75}$$

$$E = (a + bx^c) E_{MC} \tag{76}$$

where $x = (2m/n)^{-1/2} K^{2/n}$.

The only remark about the data in Tables 3 and 4 which we will make here is that the expected error of certain multiparameter approximate formulas of (n,m,K)-type is well below 0.1%. The theory lying behind these accurate formulas is outlined in Sect. 7.2.

**Table 4.** Results of numerical testing of approximate (n,m,K)-type formulas (36), (37) and (67)–(76)

| Equation for E* | Eq. (36) | | Eq. (37) | | correlation coefficient |
|---|---|---|---|---|---|
| | mean error (%) | max. error observ. (%) | mean error (%) | max. error observ. (%) | |
| 62 | 1.32 | 4.3 | 1.04 | 3.9 | 0.998 |
| 63 | 0.78 | 2.9 | 0.37 | 2.4 | 0.9998 |
| 64 | 0.43 | 1.7 | 0.41 | 1.5 | 0.9996 |
| 65 | 1.07 | 4.6 | 0.99 | 4.4 | 0.998 |
| 66 | 0.85 | 4.3 | 0.55 | 2.0 | 0.9995 |

| Equation for E | mean error (%) | max. error observ. (%) | correlation coefficient |
|---|---|---|---|
| 67 | 0.47 | 1.8 | 0.9995 |
| 68 | 0.10 | 0.6 | 0.99998 |
| 69 | 0.14 | 0.6 | 0.99997 |
| 70 | 0.59 | 2.1 | 0.9997 |
| 71 | 0.092 | 0.45 | 0.999983 |
| 72 | 0.075 | 0.42 | 0.999986 |
| 73 | 0.074 | 0.42 | 0.999986 |
| 74 | 0.078 | 0.43 | 0.999986 |
| 75 | 0.075 | 0.42 | 0.999986 |
| 76 | 0.072 | 0.42 | 0.999986 |

# 7 Dependence of Total π-Electron Energy on Molecular Structure

## 7.1 The (n,m)-Dependence

The problem of the dependence of the total π-electron energy of benzenoid hydrocarbons on the size of the molecule i.e. on the parameters n and m seems to be essentially settled by means of Theorem 5 and by recognizing the high precision of the McClelland approximation. The conclusions drawn in Sect. 6.1 need no further comments.

## 7.2 The K-Dependence

As early as in 1949, Carter [4] expressed the view that the number of Kekulé structures (K) influences the E-value of a benzenoid hydrocarbon. There is little doubt that the exploration of the $E - K$ dependence was much stimulated by the existence of two closely analogous relations [1]:

$$E = \sum_{i=1}^{n} |x_i|; \qquad K^2 = \prod_{i=1}^{n} |x_i|.$$

In the course of these researches three main standpoints have been advocated:
(a) E is a logarithmic function of K.
(b) E is a function of a structural invariant x, $x = (2m/n)^{-1/2} K^{2/n}$.
(c) E is a linear function of K.

Point (a) is a necessary consequence of the requirement that the total $\pi$-electron energy of a system $M_1 \cup M_2$ composed of two disjoint molecules $M_1$ and $M_2$ is equal to the sum of the total $\pi$-electron energies of $M_1$ and $M_2$:

$$E(M_1 \cup M_2) = E(M_1) + E(M_2). \tag{77}$$

Because $K\{M_1 \cup M_2\} = K\{M_1\} K\{M_2\}$, no relation between K and E other than logarithmic will satisfy Eq. (77).

In spite of this nice argument, empirical testings reveal that the total $\pi$-electron energy of benzenoid hydrocarbons shows no sign of having the desired logarithmic dependence on K (see, for instance, Fig. 3). Therefore, this direction of research [4, 68, 104, 107, 108] seems nowadays to be completely abandoned.

Some time ago Cioslowski [105] put forward a remarkable idea, namely that the eigenvalues of all benzenoid systems obey the same distribution law. This was named the "universal distribution approach" (UDA) and its elaboration resulted in the following (approximate) formula for the total $\pi$-electron energy [105]:

$$E \approx F(x) E_{MC} \tag{78}$$

where $E_{MC}$ is the McClelland expression, Eq. (19), $x = (2m/n)^{-1/2} K^{2/n}$ whereas $F(x)$ is a universal function (i.e. same for all benzenoids) whose actual form within UDA remains unspecified.

The UDA was eventually extended to various other $\pi$-electron properties of benzenoid hydrocarbons [109, 110].

We call Eq. (78) the Cioslowski formula. (Cioslowski himself [105] named it "the generalized McClelland formula".) Evidently, the choice $F(x) = $ const. reduces Eq. (78) to McClelland's approximation, Eq. (35).

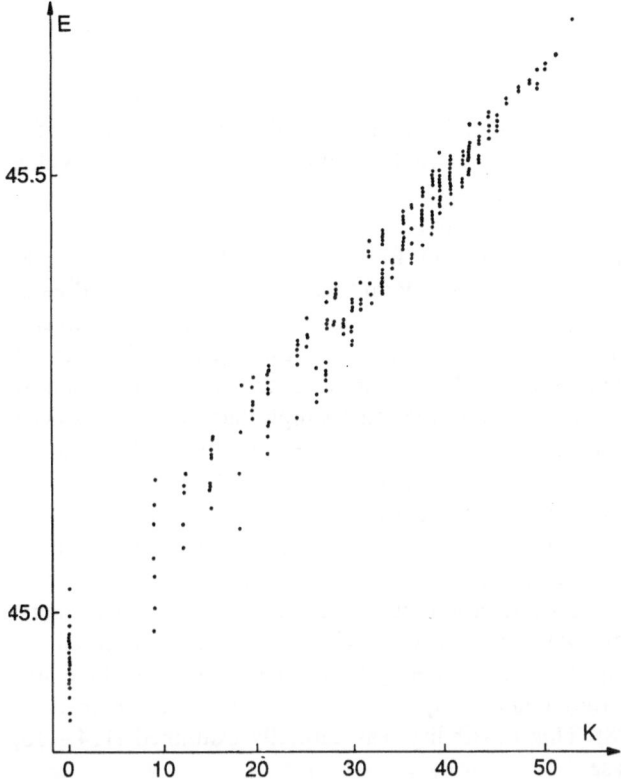

**Fig. 3.** Total π-electron energy vs number of Kekulé structures for a family of 328 benzenoid isomers $C_{32}H_{18}$; reproduced from Ref. [114]

For any application of the Cioslowski formula we must know at least something about the function F(x). In [110] it was established that F(1) = 1. In [106] trial-and-error attempts were made to guess the (approximate) form of F(x), but the results obtained were not conclusive. Quite recently Cioslowski proposed [111] what he calls *"a final solution of the problem"* by adjusting F(x) to the eigenvalue distribution of annulenes ($F_1$) and linear polyenes ($F_2$). He deduced [111]

$$F_1(x) = \frac{u(x)}{\sqrt{2}\,\sin\,[\pi u(x)/4]}$$

where

$$u(x) = \log_2\,(2x^2)$$

and

$$F_2(x) = \frac{2x^2 - 1}{x}\,[1/\sin\,[\pi v(x)/2] - 1]$$

55

where

$$v(x) = (2x^2 - 1)/(4x^2 - 1).$$

In spite of different algebraic forms of $F_1$ and $F_2$, their numerical values for chemically relevant ranges of x are quite close. By means of $F_1$ and $F_2$ it is possible to reproduce E with an average error of 0.13%.

Cioslowski's theory was recently reviewed in [112].

Irrespective of the form of the function $F(x)$, the Cioslowski formula (78) anticipates a curvilinear dependence between E and K. On the other hand, already in 1973 George Hall [113] claimed that the relation between E and K is linear. Hall's arguments were both elementary and convincing. For a group of benzenoid isomers he plotted the E-values versus K and found that the points lie almost perfectly on a straight line. It is curious that this simple fact was not noticed decades earlier. The linear dependency of E on K was eventually studied in more detail, resulting in the approximate formula (69) [87, 102, 103].

When studying the dependence of E on K it is essential to consider families of isomeric benzenoid systems. Then, namely, the values of n and m are the same for all members of the family and, consequently, the large effects caused by n and m remain constant. In Fig. 3 the E-values are plotted versus K for a family of isomers. Although the points are somewhat spread, no curvilinearity can be observed.

Because the points in Fig. 3 do not perfectly lie on a straight line there still seems to be a chance that they could comply with some curvilinear correlation of Cioslowski-type, Eq. (78). This possibility was carefully examined [114–116] and found not to be the case.

Therefore, irrespective of the beautiful algebraic form of Cioslowski's UDA one must conclude that its predictions are not in complete agreement with empirical findings [115, 116]. Furthermore, it was shown [117] that a hidden assumption behind Eq. (78) is that the coefficients $b(B, k)$ of the characteristic polynomial are not mutually correlated. Empirical testing [118] revealed that, on the contrary, a high degree of correlation exists between these coefficients.

It was also demonstrated [119] that the Cioslowski formula (78) is not the only expression for E which follows from UDA considerstions. By maintaining the basic assumptions of UDA, but using different initial conditions, instead of Eq. (78) a formula of (n,m)-type was deduced, namely [119]:

$$E \approx f(y) E_{MC}.$$

Here f is another unspecified function and $y = (9mn - 6n^2)/(2m^2)$. For all benzenoid systems possessing three or more hexagons y was found [120] to lie in the narrow interval [105/64, 108/64]. Consequently, $f(y)$ is practically a constant, reducing the above formula to the McClelland approximation, Eq. (35).

It seems to be a firmly established empirical fact that the relation between E and K is linear. If so, then it would be beneficial to have some theoretical argument supporting this conclusion. The theoretical demonstration of the linearity of the $E - K$ relation turns out to be a hard problem and is nowadays far from being completely resolved. Nevertheless, some progress in this direction has been achieved recently [121]. It has been established that linear $E - K$ dependence is obtained as a good approximation, provided the two smallest positive eigenvalues $x_{n/2}$ and $x_{n/2-1}$ are sufficiently well separated. Recall that in the Hückel theory $x_{n/2}$ and $x_{n/2-1}$ correspond to the highest occupied and the second highest occupied MO energy levels.

## 7.3 Beyond the (n,m,K)-Dependence

An inspection of Fig. 3 clearly shows that the total π-electron energies of isomeric benzenoid hydrocarbons possessing equal numbers of Kekulé structures still vary to some limited extent. The structural invariants causing this variations are not satisfactorily understood, but „suspect no. 1" is the number of bay regions. (For the definition of bay regions see [1] or [12].)

This expectation is based on the fact that the sixth spectral moment $M_6$ (see Ref. [1]) as well as the third coefficient $b(B, 3)$ of the characteristic polynomial (cf Eq. (33)) are functions of the parameters n, m and b. (Recall that $M_2$ and $M_4$ as well as $b(B, 1)$ and $b(B, 2)$ are determined solely by n and m.) A rough theoretical rationalization of the b-dependency of E is obtained when E is expanded in terms of spectral moments [122–124].

In Fig. 4 the total π-electron energies of a family of benzenoid isomers with $K = 9$ are plotted versus the respective b values. Observe that the line obtained in this way has a slight curvature.

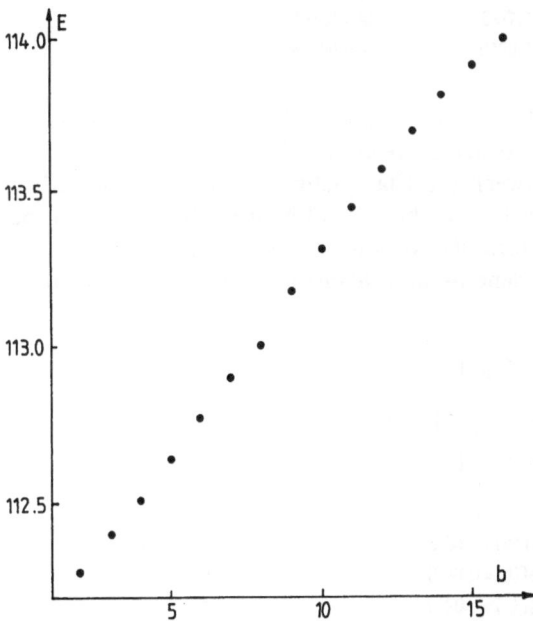

Fig. 4. Total π-electron energy vs number of bay regions for a family of benzenoid isomers possessing 9 Kekulé structures; these h = 20 systems are obtained by attaching phenalene fragments to both ends of benzenoid chains containing 14 hexagons

The data from Fig. 4 indicate that E might be an increasing (almost linear) function of b. Such a conclusion was supported by further empirical studies [87, 125].

Various types of bays (simple bays, coves, fjords [1]) have somewhat different effects on E. Work on the elucidation of these details is in progress.

# 8 Cyclic Conjugation in Benzenoid Hydrocarbons

In Sect. 3 the theory of cyclic conjugation was briefly outlined. The quantity ef(Z), defined via Eq. (10), measures the effect of the conjugation along the cycle Z. At the same time, ef(Z) is the effect of the cycle Z on the total $\pi$-electron energy of the respective conjugated molecule.

Extensive calculations of the ef-values of benzenoid hydrocarbons were recently reported [39, 40, 64]. In what follows we consider only the case when Z is a hexagon. As customary, we then call Z a ring.

In most cases the ef-values closely follow the conjugation pattern anticipated by the Clar aromatic sextet theory [12, 126]. Three typical examples of this kind are perylene (1), dibenzo[g,p]chrisene (2) and coronene (3):

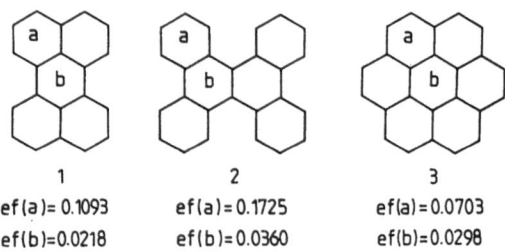

| 1 | 2 | 3 |
|---|---|---|
| ef(a) = 0.1093 | ef(a) = 0.1725 | ef(a) = 0.0703 |
| ef(b) = 0.0218 | ef(b) = 0.0360 | ef(b) = 0.0298 |

Recall that in all Clar formulas of 1, 2 and 3 the rings labeled a may possess an aromatic sextet whereas the rings labeled b are empty.

Especially good agreements between the Clar picture and ef are observed in the case of fully benzenoid systems [39]. (A benzenoid hydrocarbon is said to be fully benzenoid if it has a Clar formula without double bonds.) Two typical examples are provided by triphenylene (4) and dibenzo[fg,op]naphthacene (5).

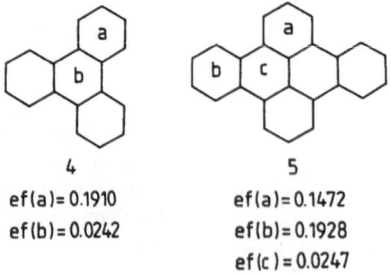

| 4 | 5 |
|---|---|
| ef(a) = 0.1910 | ef(a) = 0.1472 |
| ef(b) = 0.0242 | ef(b) = 0.1928 |
| | ef(c) = 0.0247 |

For fully benzenoid systems the following two rules were formulated [39].

*Rule 1.* The effect of empty rings on the total π-electron energy of fully benzenoid hydrocarbons is nearly the same. The corresponding ef-values vary in the narrow interval (0.02, 0.03).

*Rule 2.* (a) The main factor influencing the effect of a full ring on total π-electron energy is the number of adjacent empty rings. The smaller the number of adjacent empty rings, the larger is the ef-value. (b) A pair of nonadjacent empty rings decreases the ef-value of a full ring more than a pair of adjacent empty rings.

Depending on the number and arrangement of the neighboring empty rings, the ef-values of full rings vary quite a lot, between 0.06 and 0.20.

Rules 1 and 2 are special cases of some more general regularities which seem to hold for all benzenoid hydrocarbons.

*Rule 3.* If the ef-value of a ring is large then the ef-values of the neighboring rings are small, and vice versa. If a certain structural factor increases the ef-value of a ring, then it decreases the ef-values of the neighboring rings, and vice versa.

*Rule 4.* Each biphenyl fragment additionally increases the ef-values of its two hexagons.

For example, the above "biphenyl rule" (together with Rule 3) explains the anomalously small ef-value of the ring *c* in peropyrene (6). The biphenyl fragments, increasing ef of the *b*-rings and consequently decreasing ef in the ring *c*, are indicated on the diagrams 7 and 8. According to the (unique) Clar formula of peropyrene (9), the b-rings are empty whereas the ring *c* possesses a fixed aromatic sextet and is therefore expected to have a major contribution to cyclic conjugation.

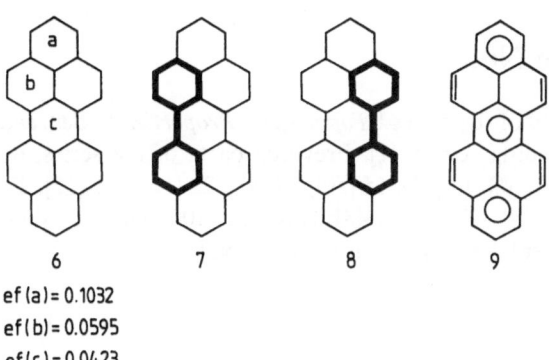

6       7       8       9

ef (a) = 0.1032
ef (b) = 0.0595
ef (c) = 0.0423

The effect of the "biphenyl rule" is small and is usually screened by much stronger conjugation modes (e.g. those taken into account by resonance, conjugated-circuit and/or Clar-aromatic-sextet theories [64]). In some exceptional cases, however, the "biphenyl rule" can completely invert the conjugation pattern anticipated by the classical theories. This particularly occurs in benzenoid hydrocarbons with many fixed single and double bonds [64] for which the classical view predicts no conjugation at all.

Two convincing examples of these "anomalies" are provided by the benzenoid systems 10 and 11.

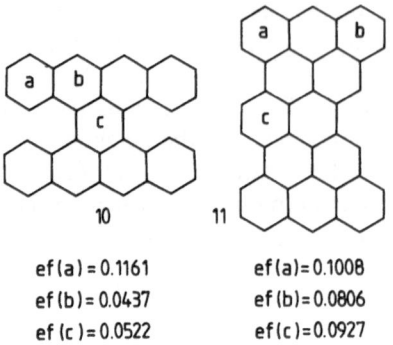

| ef(a) = 0.1161 | ef(a) = 0.1008 |
| ef(b) = 0.0437 | ef(b) = 0.0806 |
| ef(c) = 0.0522 | ef(c) = 0.0927 |

In both 10 and 11 the rings labeled by *a* and *b* may possess an aromatic sextet whereas the rings labeled by *c* are empty. Nevertheless, ef(*c*) is greater than ef(*b*). One should observe that in both cases the ring *c* belongs to four biphenyl fragments.

The analysis of the ef-values of benzenoid systems gives useful information about the conjugation modes in these molecules and reveals the range of the applicability of the resonance/conjugated-circuit/Clar-aromatic-sextet theoretical approaches. Deliberations along these lines [40, 64] go, however, beyond the ambits of the present work.

# 9 Bibliographic Note

This article is Part 79 of the series *"Topological Properties of Benzenoid Systems"*. Part 78 is the preceding review [1]. Previous parts of the series, quoted in this article are 7 [101], 12 [52], 14 [51], 15 [50], 18 [53], 22 [103], 26 [75], 27 [89], 31 [85], 43 [90], 45 [79], 48 [114], 48a [115], 48b [116], 55 [106], 64 [117], 66 [91], 67 [120], 68 [64], 69 [39], 72 [96], 73 [72], 74 [40] and 77 [95].

# 10 References

1. Gutman I (1992) Topics Curr Chem 162 (this issue)
2. Coulson CA (1940) Proc Cambridge Phil Soc 36: 201
3. Coulson CA, Longuet-Higgins HC (1947) Proc Roy Soc London A191: 39
4. Carter PC (1949) Trans Faraday Soc 45: 597
5. Yvan P (1952) J Chim Phys 49: 457
6. Coulson CA (1954) J Chem Soc: 3111
7. Hall GG (1955) Proc Roy Soc London A229: 251

8. Ruedenberg K (1961) J Chem Phys 34: 1884
9. Marcus RA (1963) J Chem Phys 43: 2643
10. England W, Ruedenberg K (1973) J Amer Chem Soc 95: 8769
11. Jiang YS, Tang AC, Hoffmann R (1984) Theor Chim Acta 66: 183
12. Gutman I, Cyvin SJ (1989) Introduction to the theory of benzenoid hydrocarbons, Springer, Berlin Heidelberg New York
13. Gutman I, Polansky OE (1986) Mathematical concepts in organic chemistry, Springer, Berlin Heidelberg New York
14. Gutman I (1978) Bull Soc Chim Beograd 43: 761
15. Gutman I (1978) Berichte Math-Statist Sekt Forschungsz Graz 103: 1
16. Hückel E (1940) Grundzüge der Theorie ungesättigter und aromatischer Verbindungen, Verlag Chemie, Berlin
17. Coulson CA, O'Leary B, Mallion RB (1978) Hückel theory for organic chemists, Academic, London
18. Graovac A, Gutman I, Trinajstić N (1977) Topological approach to the chemistry of conjugated molecules (Lecture Notes in Chemistry 4), Springer, Berlin Heidelberg New York
19. Cvetković D, Doob M, Sachs H (1980) Spectra of graphs − theory and application, Academic, New York
20. Marsh KN et al (1987) TRC thermodynamic tables − hydrocarbons, Thermodynamic Research Center, College Station, Texas
21. Hites RA, Simonsick WJ (1987) Calculated molecular properties of polycyclic aromatic hydrocarbons, Elsevier, Amsterdam
22. Herndon WC, Connor DA, Lin PP (1990) Pure Appl Chem 62: 435
23. Ruedenberg K (1977) J Chem Phys 66: 375
24. Schaad LJ, Robinson BH, Hess BA (1977) J Chem Phys 67: 4616
25. March NH (1977) J Chem Phys 67: 4418
26. Ichikawa H, Ebisawa Y (1985) J Amer Chem Soc 107: 1161
27. Gutman I, Milun M, Trinajstić N (1973) Chem Phys Letters 23: 284
28. Schaad LJ, Hess BA (1972) J Amer Chem Soc 94: 3068
29. Hiberty PC (1990) Topics Curr Chem 153: 27
30. Breslow R, Moháchi E (1963) J Amer Chem Soc 85: 431
31. Dewar MJS (1965) Chem Eng News 43: 86
32. Dewar MJS (1966) Tetrahedron Suppl: 75
33. Dewar MJS, de Llano C (1969) J Amer Chem Soc 91: 789
34. Aihara J (1976) J Amer Chem Soc 98: 2750
35. Gutman I, Milun M, Trinajstić N (1977) J Amer Chem Soc 99: 1692
36. Gutman I, Bosanac S (1977) Tetrahedron 33: 1809
37. Gutman I, Polansky OE (1981) Theor Chim Acta 60: 203
38. Gutman I (1984) Theor Chim Acta 66: 43
39. Gutman I (1990) Rep Mol Theory 1: 115
40. Gutman I, Agranat I (1991) Polyc Arom Comp 2: 63
41. Gutman I (1979) Math Chem 6: 75
42. Godsil CD, Gutman I (1981) J Graph Theory 5: 137
43. Cvetković D, Doob M, Gutman I, Torgašev A (1988) Recent results in the theory of graph spectra, North-Holland, Amsterdam, p 103−122
44. Gutman I, Trinajstić N (1976) Acta Chim Hung 91: 203
45. Gutman I, Milun M, Trinajstić N (1976) Math Chem 1: 171
46. Gutman I (1979) Chem Phys Letters 66: 595
47. Gutman I, Mohar B (1980) Chem Phys Letters 69: 375
48. Gutman I (1980) Theor Chim Acta 56: 89
49. Gutman I, Mohar B (1981) Chem Phys Letters 77: 567
50. Gutman I, Mohar B (1982) Croat Chem Acta 55: 375
51. Gutman I (1983) J Chem Soc Faraday Trans II 79: 337
52. Gutman I (1982) Z Naturforsch 37a: 248
53. Gutman I, Teodorović AV (1982) Bull Soc Chim Beograd 47: 579

54. Godsil CD, Gutman I (1979) Z Naturforsch 34a: 776
55. Bosanac S, Gutman I (1977) Z Naturforsch 32a: 10
56. Gutman I (1979) J Chem Soc Faraday Trans II 75: 799
57. Gutman I (1980) Croat Chem Acta 53: 581
58. Aihara J (1977) J Amer Chem Soc 99: 2048
59. Herndon WC (1982) J Amer Chem Soc 104: 3541
60. Gutman I, Herndon WC (1984) Chem Phys Letters 105: 281
61. Gutman I, Mizoguchi N (1990) J Math Chem 5: 81
62. Mizoguchi N (1990) Bull Chem Soc Japan 63: 765
63. Lepović M, Gutman I, Petrović M, Mizoguchi N (1990) J Serb Chem Soc 55: 193
64. Gutman I (1990) Pure Appl Chem 62: 429
65. Hosoya H, Hosoi K, Gutman I (1975) Theor Chim Acta 38: 37
66. Gutman I, Trinajstić N (1976) J Chem Phys 64: 4921
67. Gutman I (1977) Theor Chim Acta 45: 79
68. Gutman I, Graovac A (1977) Croat Chem Acta 49: 453
69. Gutman I (1979) Bull Soc Chim Beograd 44: 173
70. Mizoguchi N (1989) Chem Phys Letters 158: 383
71. McClelland BJ (1971) J Chem Phys 54: 640
72. Gutman I (1990) J Chem Soc Faraday Trans 86: 3373
73. Cioslowski J (1988) Intern J Quantum Chem 34: 217
74. Gutman I (1974) Chem Phys Letters 24: 283
75. Gutman I, Teodorović AV, Nedeljković L (1984) Theor Chim Acta 65: 23
76. Türker L (1984) Math Chem 16: 83
77. Gutman I, Türker L, Dias JR (1986) Math Chem 19: 147
78. Cioslowski J (1985) Z Naturforsch 40a: 1167
79. Cioslowski J, Gutman I (1986) Z Naturforsch 41a: 861
80. Brown RD (1950) Trans Faraday Soc 46: 1013
81. Green AL (1956) J Chem Soc: 1886
82. Hakala RW (1967) Intern J Quantum Chem Suppl 1: 640
83. Aihara J (1976) J Org Chem 41: 2488
84. Gutman I (1976) Bull Soc Chim Beograd 41: 159
85. Gutman I, Teodorović AV, Bugarčić Ž (1984) Bull Soc Chim Beograd 49: 521
86. Balaban AT (1970) Rev Roum Chim 15: 1243
87. Hall GG (1991) Intern J Quantum Chem 39: 605
88. Gutman I, Milun M, Trinajstić N (1973) J Chem Phys 59: 2772
89. Gutman I, Nedeljković L, Teodorović AV (1983) Bull Soc Chim Beograd 48: 495
90. Gutman I, Marković S, Teodorović AV, Bugarčić Ž (1986) J Serb Chem Soc 51: 145
91. Gutman I, Graovac A, Vuković S, Marković S (1989) J Serb Chem Soc 54: 189
92. Zahradnik R, Pancir J (1970) HMO energy characteristics, Plenum, New York
93. Bochvar DA, Stankevich IV (1980) Zh Strukt Khim 21: 61
94. Gutman I (1983) Math Chem 14: 71
95. Gutman I, Utvić D, Mukherjee AK (1991) J Serb Chem Soc 56: 59
96. Gutman I (1991) Math Chem 26: 123
97. Graovac A, Babić D, Kovačević K (1987) In: King RB, Rouvray DH (eds) Graph theory and topology in chemistry, Elsevier, Amsterdam, p 448
98. Graovac A, Cioslowski J (1988) Croat Chem Acta 61: 797
99. Gutman I (1977) J Chem Phys 66: 1652
100. Gutman I, Trinajstić N (1976) Croat Chem Acta 48: 297
101. Gutman I, Petrović S (1981) Bull Soc Chim Beograd 46: 459
102. Hall GG (1981) Bull Inst Math Appl 17: 70
103. Gutman I, Petrović S (1983) Chem Phys Letters 97: 292
104. Gutman I, Trinajstić N, Wilcox CF (1975) Tetrahedron 31: 143
105. Cioslowski J (1986) Math Chem 20: 95
106. Gutman I, Marković S, Marinković M (1987) Math Chem 22: 277
107. Gutman I (1974) Theor Chim Acta 35: 355
108. Wilcox CF (1975) Croat Chem Acta 47: 87

109. Cioslowski J (1987) Intern J Quantum Chem 31: 581
110. Cioslowski J, Polansky OE (1988) Theor Chim Acta 74: 55
111. Cioslowski J (1990) Math Chem 25: 83
112. Cioslowski J (1990) Topics Curr Chem 153: 85
113. Hall GG (1973) Intern J Math Educ Sci Technol 4: 233
114. Gutman I (1986) Math Chem 21: 317
115. Gutman I, Marković S (1990) Math Chem 25: 141
116. Marković S, Gutman I, Živković S, Stanković M (1989) Coll Sci Papers Fac Sci Kragujevac 10: 61
117. Gutman I (1989) Chem Phys Letters 156: 119
118. Gutman I, Manić D (unpublished results)
119. Gutman I (1987) Math Chem 22: 269
120. Gutman I (1989) J Serb Chem Soc 54: 197
121. Gutman I, Hall GG (1992) Intern J Quantum Chem (in press)
122. Gutman I, Trinajstić N (1972) Chem Phys Letters 17: 535
123. Gutman I, Trinajstić N (1973) Chem Phys Letters 20: 257
124. Jiang Y, Zhu H, Zhang H, Gutman I (1989) Chem Phys Letters 159: 159
125. Gutman I, Hall GG, Marković S, Stanković Z, Radivojević V (1991) Polyc Arom Comp 2: (in press)
126. Clar E (1972) The aromatic sextet, Wiley, London

# Enumeration of Benzenoid Systems
# and Other Polyhexes

Björg N. Cyvin, Jon Brunvoll, and Sven J. Cyvin

Division of Physical Chemistry, The University of Trondheim, N-7034 Trondheim-NTH, Norway

## Table of Contents

1 Introduction . . . . . . . . . . . . . . . . . . . . . . . . 67

2 Definitions and Disposition . . . . . . . . . . . . . . . . . . 68
  2.1 Definition of the Main Systems . . . . . . . . . . . . . . 68
  2.2 Addition Modes . . . . . . . . . . . . . . . . . . . . . 68
  2.3 Circulenes . . . . . . . . . . . . . . . . . . . . . . . 69
  2.4 Helicenic Systems . . . . . . . . . . . . . . . . . . . . 70
  2.5 Survey of the Classes . . . . . . . . . . . . . . . . . . 72
  2.6 Catacondensed and Pericondensed Systems . . . . . . . . . 72
  2.7 Disposition . . . . . . . . . . . . . . . . . . . . . . . 73
  2.8 Contemporary Computer-Enumeration Activity . . . . . . . 74

3 Survey of Enumeration Results for All Polyhexes . . . . . . . . 74
  3.1 Benzenoids and Coronoids . . . . . . . . . . . . . . . . 74
  3.2 Planar Polyhexes (Benzenoids and Planar Circulenes) . . . . . 77
  3.3 Simply Connected Polyhexes (Benzenoids and Helicenes) . . . . 83
  3.4 All Polyhexes (Benzenoids, Helicenes and Circulenes) . . . . . 85
  3.5 Concluding Remarks . . . . . . . . . . . . . . . . . . . 88

4 Additional Definitions for Benzenoids . . . . . . . . . . . . . 88

5 Algebraic Solutions . . . . . . . . . . . . . . . . . . . . . 89
  5.1 Catacondensed Simply Connected Polyhexes (Catafusenes) . . . 89
  5.2 Unbranched Catacondensed Simply Connected Polyhexes
    (Unbranched Catafusenes) . . . . . . . . . . . . . . . . . 90
  5.3 Fibonacenes and Related Systems . . . . . . . . . . . . . 94
  5.4 Gutman Trees (LA-Sequences) . . . . . . . . . . . . . . . 97
  5.5 Generalized Hexagon-Shaped Benzenoids . . . . . . . . . . 98
  5.6 Approximate Formulas . . . . . . . . . . . . . . . . . . 103

Topics in Current Chemistry, Vol. 162
© Springer-Verlag Berlin Heidelberg 1992

**6 Catafusenes** . . . . . . . . . . . . . . . . . . . . . . . 105
  6.1 Introductory Remarks . . . . . . . . . . . . . . . . 105
  6.2 Unbranched Catafusenes . . . . . . . . . . . . . . . 105
  6.3 Branched Catafusenes . . . . . . . . . . . . . . . . 110
  6.4 Special Catacondensed Systems (SCS's) . . . . . . . . . 112
  6.5 Catacondensed Benzenoids with Trigonal Symmetry . . . . . 114
  6.6 Catacondensed Benzenoids with Dihedral Symmetry and
    Centrosymmetry . . . . . . . . . . . . . . . . . . . 115
  6.7 Unbranched Catacondensed Benzenoids with Equidistant
    Segments . . . . . . . . . . . . . . . . . . . . . . 117

**7 Coarse Classifications of Benzenoids** . . . . . . . . . . . . 121
  7.1 Specification . . . . . . . . . . . . . . . . . . . . . 121
  7.2 Catacondensed and Pericondensed Benzenoids . . . . . . . 122
  7.3 Kekuléan and non-Kekuléan Benzenoids: the "neo" Classification 125
  7.4 Color Excess . . . . . . . . . . . . . . . . . . . . . 126
  7.5 Symmetry . . . . . . . . . . . . . . . . . . . . . . 127

**8 Normal Benzenoids** . . . . . . . . . . . . . . . . . . . . 128

**9 Essentially Disconnected Benzenoids** . . . . . . . . . . . . 130

**10 Obvious Non-Kekuléan Benzenoids** . . . . . . . . . . . . . 135
  10.1 Numbers and Forms . . . . . . . . . . . . . . . . . 135
  10.2 Non-Kekuléans with Extremal Properties . . . . . . . . . 138

**11 Concealed Non-Kekuléan Benzenoids** . . . . . . . . . . . . 140

**12 Benzenoids with Specific Symmetries** . . . . . . . . . . . . 143
  12.1 Hexagonal Symmetry: Snowflakes . . . . . . . . . . . 143
  12.2 Trigonal Symmetry . . . . . . . . . . . . . . . . . . 148
  12.3 Dihedral Symmetry and Centrosymmetry . . . . . . . . . 157
  12.4 Mirror Symmetry . . . . . . . . . . . . . . . . . . . 161

**13 All-Benzenoids** . . . . . . . . . . . . . . . . . . . . . 163
  13.1 Some Topological Properties . . . . . . . . . . . . . . 163
  13.2 Catacondensed and Pericondensed All-Benzenoids . . . . . 163
  13.3 Hexagonal Symmetry: All-Flakes . . . . . . . . . . . . 167
  13.4 Trigonal Symmetry . . . . . . . . . . . . . . . . . . 170

**14 Conclusion** . . . . . . . . . . . . . . . . . . . . . . . 172

**15 References** . . . . . . . . . . . . . . . . . . . . . . . 176

The results of enumerations and classifications of polyhexes are reviewed and supplemented with new data. The numbers are collected in comprehensive tables and supplied with a thorough documentation from an extensive literature search. Numerous forms of the polyhexes are displayed, either as dualists or black silhouettes on the background of a hexagonal lattice. In the latter case, the Kekulé structure counts for Kekuléan systems are indicated. Emphasis is laid on the benzenoid systems (planar simply connected polyhexes).

# 1 Introduction

The problem of "cell-growth" [1–5] is a classical problem of enumeration of graphs in mathematics. The biological analogy has been stretched so far as to refer to the benzenoid systems as "hexagonal animals" [2, 5, 6]. The enumeration of hexagonal animals/benzenoid systems may be traced back to Klarner [7], who followed up a suggestion by Golomb [8]. In these works, among others [2, 9, 10], the enumeration problems were viewed in a purely mathematical way. From this point of view, the general problem remains unsolved. Here it is relevant to quote Gutman [11]:

"How many benzenoid systems (with a given number of hexagons) exist?

Harary offers US $ 100 for the solution of this difficult enumeration problem [12]."

As we shall see in the subsequent sections, some partial solutions to this mathematical problem exist.

The first enumeration of benzenoids in a chemical context is probably contained in the paper from 1968 of Balaban and Harary [13], a chemist and a mathematician. After a period of some years with seemingly little activity in this area, the problems were taken up again in pace with the access to modern computers [14, 15]. This is also to be understood in the way that the emphasis gradually shifted from algebraic to numerical solutions. The beginning of this new era can be dated to a paper from 1980 by Balasubramanian et al. [14], from which we cite the passage: "Unfortunately, the enumeration of all benzenoid hydrocarbons containing $n$ [read $h$] rings is the well-known unsolved 'cell-growth problem' with hexagonal animals [4]". A part of this passage was cited elsewhere by Knop and Trinajstić with collaborators, the Düsseldorf-Zagreb group [15]. In 1983 they declared about the paper [14] from which the passage is taken: "This work stimulated our interest in the problem, . . ".

A major event was the appearance of a book in 1985 on generation and enumeration of certain classes of molecules, including benzenoids [16]. It was launched by the Düsseldorf-Zagreb group. We wish also to mention a mini-review with supplements which came from the same school prior to the book [17].

Soon this field of computer-generation and enumeration of benzenoids flared up. A major review with supplements by fourteen authors [18], also referred to as a consolidated report, appeared in 1987 and refers to twenty-one relevant publications from 1984 or later. A great number of works in this area have been published since, as is going to be documented in the subsequent sections. Here we only mention a communication on supplements to the consolidated report [19].

Balaban published an early review [20] and recently a survey of enumeration results for benzenoids and coronoids [21]; these two publications together clearly illustrate the development in the area. The most recent review so far with the most complete quotations of enumeration data for benzenoids and coronoids is found in a monograph by Gutman and Cyvin [22].

Now since the class of coronoids has been mentioned beside the benzenoids we wish to stress the importance of defining the terms so that it should never be doubt about what actually is enumerated. This brings us naturally to the next section. In conclusion of the present section, we only give some general references of relevance to the topic [22–26].

# 2 Definitions and Disposition

## 2.1 Definition of the Main Systems

The opening lines of the chapter on enumeration from the monograph of Gutman and Cyvin [22] read: "By enumeration of benzenoid systems, the counting of all possible non-isomorphic members within a class of benzenoids is understood. Usually, but not always, the number of hexagons $(h)$ is the leading parameter. Thus the enumeration for $h = 1, 2, 3, 4$, etc. is to be executed."

Perhaps the most precise definitions (for hexagonal systems in mathematical chemistry) are attached to the *benzenoid* systems (or simply benzenoids) and *coronoid* systems (or coronoids), although no universally accepted terminology has been established on this point. On the contrary, a "plethora of names" [26] or "surprisingly large number of names" [22] have been used more or less synonymous with what we call a benzenoid (cf. also Gutman [27]) or a coronoid [22]. Here we follow the terminology of some of the cited references [22, 26], both for the two names mentioned above and many other concepts.

A benzenoid is defined in a lucid way by means of a cycle (the perimeter) on a hexagonal lattice [22, 24, 26, 27]. It is a geometrical object consisting of a set of congruent regular hexagons in a plane, which are simply connected. Any two hexagons either share one and only one edge or they are disjoint. In the definition of a coronoid the restriction about connectivity is released. A *single* coronoid is defined in terms of two cycles, one (the outer perimeter) completely embracing the other (the inner perimeter) [21]. Thus a coronoid exhibits a hole, the corona hole. By definition, a corona hole should have a size of at least two hexagons. A *multiple* coronoid is a hexagonal system with more than one hole and defined as a straightforward generalization of single coronoids. One speaks about double coronoids, triple coronoids, and so on; in general g-tuple coronoids (with $g$ holes each).

The term *polyhex* has been used about benzenoids and coronoids together [18]. Benzenoids are polyhexes without holes, while a single or multiple coronoid is a polyhex with one or more holes, respectively. There is also a tendency to use the term polyhex in a more general sense [16]. Here we shall use it as a universal term for all hexagonal systems which are to be considered.

## 2.2 Addition Modes

A specification of the five addition modes is useful for some of the subsequent discussions and definitions. The modes can be vizualized as shown in Fig. 1; (i)–(v). An example of corona-condensation is included (vi); here the addition of a hexagon creates a corona hole.

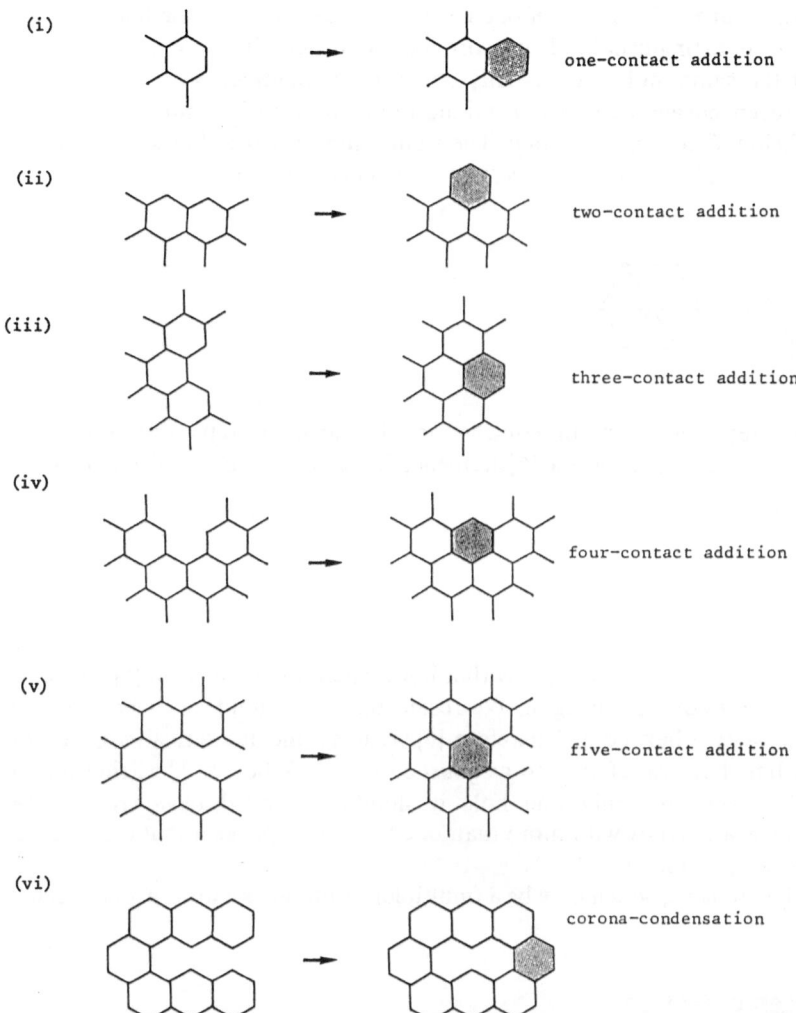

**Fig. 1.** Illustration of the five addition modes, (i)–(v), and an example of corona-condensation (vi). The *added hexagons* are grey. The *pendent lines* symbolize hexagons which may, but need not be present

All polyhexes with $h + 1$ hexagons can be generated by adding one hexagon to a perimeter (outer or inner) each time to all polyhexes with $h$ hexagons according to the five addition modes or a corona-condensation. For simply connected polyhexes the five addition modes applied to the (only) perimeter are sufficient.

## 2.3 Circulenes

Here the term *circulene* is used so that the coronoids form a subclass under circulenes. The smallest circulene which is not a coronoid, is [6]circulene, a system of six hexagons around a cavity of one hexagon. The corresponding hydrocarbon is chemically indistinguishable from coronene. It is actually only a new way of looking at the same molecule. Usually coronene is identified with a benzenoid system of seven hexagons. It has a unique representation in terms of hexagons as shown below (left-hand drawing). The right-hand drawing shows coronene as a dualist. In a dualist each point (vertex) represents a hexagon.

The dualist representation, in contrast to the hexagon representation, does distinguish between coronene and [6]circulene; the latter system is depicted below.

In general, a circulene is a polyhex with a hole, where the hole, as in [6]circulene, may have the size of one hexagon. We recall that a coronoid has a hole with a size of at least two hexagons. Therefore [6]circulene and its derivatives are not coronoids, but they are referred to as *quasi-coronoids*. When emphasizing that a circulene has one and only one hole it should be called *monocirculene*. A *polycirculene* is a polyhex with more than one hole; one speaks about dicirculene, tricirculene, and so on.

Any (poly)circulene, which may be a (multiple) coronoid, is a multiply connected polyhex.

## 2.4  Helicenic Systems

A *helicenic* system is a polyhex with overlapping edges if drawn in a plane. The five addition modes, occasionally supplemented with corona-condensation, should be followed strictly, but irrespective of possible collisions with hexagons already present. Any number of overlapping edges is possible; some examples are shown below. Also for helicenic systems (as for circulenes) it is illuminating to use dualists. Here we show the dualists for the same systems. The left-hand system of the bottom row is interpreted as a substituted [6]circulene.

It should be noticed that any three connected points of the dualist of a polyhex form either (a) a straight line, (b) an angle of 120° or (c) an equilateral triangle.

The collisions of edges are avoided when the helicenic systems are thought of as nonplanar. Then it is imagined that the system during its generation can pass from one layer to another on a multilayer lattice. Therefore it is usual to refer to non-helicenic polyhexes as (geometrically) planar and helicenic as nonplanar.

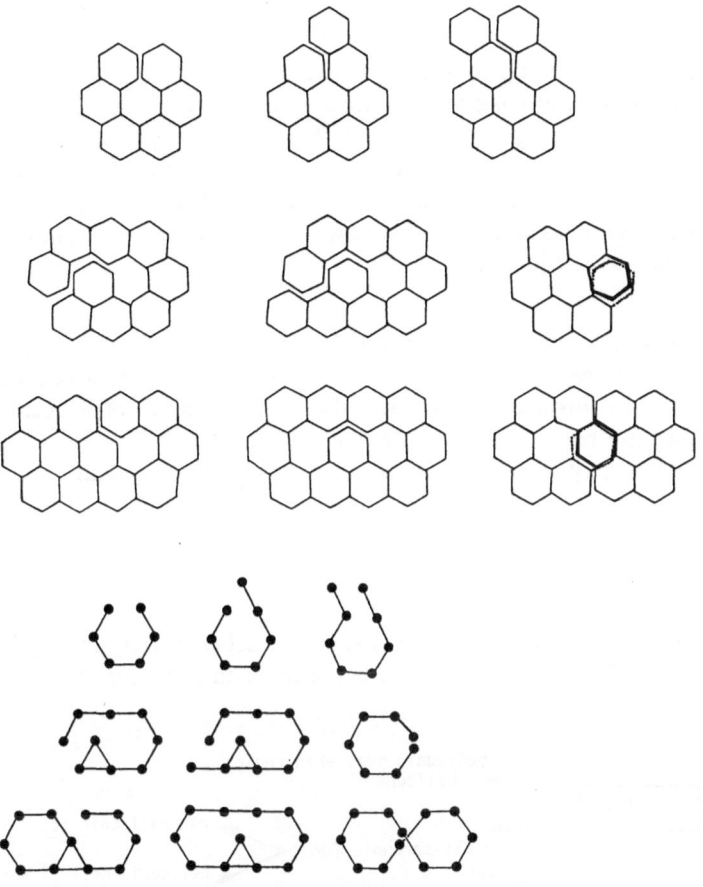

Below we give some examples of geometrical objects which are not (helicenic) polyhexes and therefore not to be considered in the following.

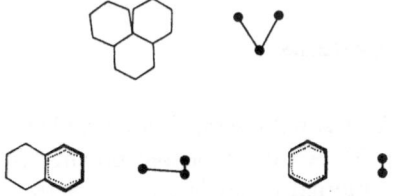

In the first example (top row) the polyhex-like object has a vertex of degree four, which is forbidden. Also the corresponding dualist-like formation is forbidden. In the bottom row the degrees of the vertices are not violated, but yet the systems are obviously not polyhexes. It is not allowed, for instance, to flip naphthalene around the middle edge. In none of the examples above the rules of addition (Fig. 1) have been followed strictly.

*Comment*

It may seem unnecessary to bring in such artifacts as the above examples of non-polyhexes. Yet we find the discussion to be warranted. Ege and Vogler [28, 29], for instance, treated pyrene in a group together with coronene ([6]circulene) and six coronoids. Implicitly, the authors considered pyrene as [4]circulene and ascribed an inner perimeter of two edges to it. That leads to vertices of degree four as in one of the "forbidden" examples above. Therefore we claim that the interpretation of pyrene as [4]circulene is not justified.

A helicenic system which is a simply connected polyhex (without holes) may be called a helicenic *quasi-benzenoid*. For the sake of brevity it is sometimes called just simply a *helicene*. In this section we have also seen examples of helicenic quasi-coronoids. They are referred to as *helicirculenes*.

Any helicenic polyhex may be referred to as (geometrically) non-planar.

## 2.5 Survey of the Classes

The classification of polyhexes treated above is surveyed schematically in Fig. 2. In the following we shall avoid the terms quasi-benzenoid and quasi-coronoid.

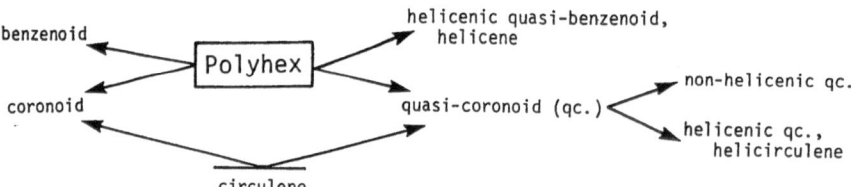

**Fig. 2.** Survey of the classification of polyhexes

## 2.6 Catacondensed and Pericondensed Systems

The distinction between *catacondensed* and *pericondensed* systems is applicable to all the classes of polyhexes treated above (Fig. 2). A catacondensed polyhex is defined by the absence of internal vertices. An internal vertex is a vertex shared by three hexagons. A pericondensed polyhex possesses at least one internal vertex. In terms of dualists a pericondensed polyhex reveals itself by the presence of at least one three-membered cycle (triangle). A dualist of a circulene has a cycle larger than a triangle.

When the title concepts are applied to the simply connected polyhexes we shall sometimes use the brief designations *catafusenes* and *perifusenes* [13, 30]. A dualist of a catafusene is acyclic (a tree).

## 2.7 Disposition

In Sect. 3 some classical results of enumeration of polyhexes are reported. They include "Klarner numbers", "Harary-Read numbers" and "Lunnon numbers", which were so beautifully (but not infallibly) sorted out for the first time by the Düsseldorf-Zagreb group [17]. Reference to the "Düsseldorf-Zagreb numbers" is also contained in Sect. 3. Furthermore, original supplements to the different tables are found therein.

These numbers, when taken together, account for all the classes of polyhexes defined above, viz. benzenoids, helicenes, and circulenes including helicirculenes (Fig. 2). In Sect. 3 the distinction between catacondensed and pericondensed systems is the only one considered for the different classes in question (Fig. 2). Section 4 contains some additional definitions.

In Section 5 the known analytical solutions for different classes of polyhexes are reported.

Section 6 deals with the enumeration of catacondensed simply connected polyhexes, i.e. catacondensed benzenoids and helicenes.

In all the following sections only benzenoids (planar, simply connected polyhexes) are treated. Several subdivisions according to different principles are considered, as explained in the appropriate sections.

In general, this chapter does primarily not deal with methods and principles of enumerations, such as coding and nomenclature. Information on these issues are found in the different individual papers, which are cited, and especially in a recent review [31]. In the present treatise, emphasis is laid on the presentation of results in surveyable tables. The review of relevant literature results, which tends to be as complete as possible, is supplemented by a substantial amount of own results not published before.

The numerical results are to a large extent followed up by illustrations of the forms of the polyhexes of different classes. In Sect. 3, 4 and 6 the dualists are employed. In all the subsequent sections the benzenoids are presented as black silhouettes on the background of a hexagonal lattice. This presentation, which already is encountered in Sect. 5 and 6, emphasizes the fact that a benzenoid can be superimposed on a hexagonal lattice. When appropriate, the silhouettes are supplied with numbers indicating the Kekulé structure counts ($K$ numbers). In these cases the systems are actually ordered according to increasing $K$ numbers. Isoarithmic systems are indicated by a horizontal bracket-like line.

In the following we give a survey of the different centres which at present are engaged in the enumeration of polyhexes by computers.

*Errata*

Corrections to numerical errors in literature data are found throughout this chapter. As a principle, a wrong number from polyhex enumerations is never quoted.

## 2.8 Contemporary Computer-Enumeration Activity

There are five researchers or groups which have been involved in computer-aided enumerations of polyhexes after 1985 [18];

(i)   *Cioslowski*, Jerzy; he has operated from different places in USA — Georgetown University, Washington DC; LANL, Los Alamos, New Mexico; FSU/SCRI, Tallahassee, Florida.

(ii)  *Düsseldorf-Zagreb group*; Jan V. Knop and Klaus Szymanski with others from The University of Düsseldorf (FR Germany), Nenad Trinajstić with others from The Rugjer Bošković Institute, Zagreb (Yugoslavia).

(iii) *He & He*; He Wenchen and He Wenjie, Shijiazhuang (PR China).

(iv)  *Novi Sad*; Ratko Tošić with others from the University of Novi Sad (Yugoslavia).

(v)   *Trondheim*; Jon Brunvoll, Bjørg N. Cyvin and Sven J. Cyvin with others, The University of Trondheim (Norway).

Apart from the preparation of the consolidated report [18] active collaboration employing computers has been going on between Trondheim on one side and (a) Düsseldorf-Zagreb, (b) He & He and (c) Novi Sad on the other. Exchanging of computer programs between different research groups often meets with difficulties in general. To our knowledge the only successful attempt in this direction, as far as the enumeration of polyhexes is concerned, was the transfer of some programs from Novi Sad to Trondheim.

## 3 Survey of Enumeration Results for All Polyhexes

### 3.1 Benzenoids and Coronoids

The Düsseldorf-Zagreb group, from the beginning of their enumeration of polyhexes, concentrated on the planar (non-helicenic) systems. This choice seems to have had an immense impact on the direction of these studies inasmuch as all the contemporary researchers in the field (cf. Sect. 2.8) follow their practice. The planar simply connected polyhexes, viz. benzenoids, are the systems which have been and currently are investigated most extensively with respect to their enumeration and classification. This is also the subject of the main bulk of the present chapter. Coronoids (which also are planar, but not simply connected) are sometimes enumerated together with the benzenoids. Furthermore, a large amount of work has been done in specific studies of coronoids, including their enumeration and classification. Apart from a gross survey in this paragraph, the coronoid systems are not treated in the present chapter.

The grand totals of He and He [32, 33] on one hand and Cioslowski [34] on the other pertain in both cases to benzenoids plus coronoids; cf. Table 1. The totals in this table for $h > 10$ are collected from additional sources [18, 35–37].

In this documentation we have disregarded the possibility of finding the numbers for $h = 13$ and $h = 14$ by combining published numbers of benzenoids and coronoids from different papers. The separate numbers of coronoids are treated in the following and those of benzenoids in subsequent sections.

In Table 2 the numbers of coronoids are listed. Those for $h \leq 13$, which pertain to single coronoids, are documented by different sources [17, 18, 38–40]. The totals for $h > 12$ to be entered in this table are available for $h = 13$ [36, 41], $h = 14$ [36, 41], $h = 15$ [37, 42] and $h = 16$ [37], or more comprehensively from collected data for $h = 8$ to 16 [37, 43, 44]. The numbers for $h = 13$ and 14 were split into those for the catacondensed and pericondensed systems by means of the knowledge about single coronoids [18, 19, 45] and double coronoids [41]. Table 3 shows the pertinent numbers for double coronoids; they are known up to $h = 18$.

So far we have only been speaking about the numbers of single and double coronoids. Triple coronoids do not occur before $h = 18$, in which case there are 2 catacondensed and 2 pericondensed forms of these systems [43, 44].

Numbers for benzenoids separately can be determined as the differences between appropriate numbers in Tables 1 and 2, but this is not the way they were deduced originally. The Düsseldorf-Zagreb numbers (or DZG numbers) is the designation which has been used [16, 17] about the data for catacondensed and pericondensed benzenoids, and benzenoids in total with $h \leq 10$. These data have later been supplemented by other investigators [19, 46, 47], as well as the Düsseldorf-Zagreb

**Table 1.** Numbers of benzenoids with coronoids*

| $h$ | Catacondensed | Pericondensed | Total |
|---|---|---|---|
| 1 | 1[a] | | 1[a] |
| 2 | 1[a] | | 1[a] |
| 3 | 2[a] | 1[a] | 3[a] |
| 4 | 5[a] | 2[a] | 7[a] |
| 5 | 12[a] | 10[a] | 22[a] |
| 6 | 36[a] | 45[a] | 81[a] |
| 7 | 118[a] | 213[a] | 331[a] |
| 8 | 412[a] | 1024[a] | 1436[a] |
| 9 | 1490[b] | 5020[b] | 6510[b] |
| 10 | 5587 | 24542 | 30129[c] |
| 11 | 21177 | 120335 | 141512[d,e] |
| 12 | 81433 | 590105 | 671538[d] |
| 13 | 315511 | 2895109 | 3210620[f,g] |
| 14 | 1231318 | 14212553 | 15443871[f,g] |
| 15 | † | † | 74662005[g] |
| 16 | † | † | 362506902[g] |

* Single coronoids occur at $h \geq 8$; double coronoids at $h \geq 13$.
[a] He and He (1985) [32]; [b] He and He (1986) [33]; [c] Cioslowski (1987) [34]; [d] He and He (1987) [35]; [e] Balaban, Brunvoll, Cioslowski, Cyvin, Cyvin, Gutman, He, He, Knop, Kovačević, Müller, Szymanski, Tošic and Trinajstić (1987) [18]; [f] Müller, Szymanski, Knop, Nikolić and Trinajstić (1990) [36]; [g] Knop, Müller, Szymanski and Trinajstić (1990) [37]; † Unknown

**Table 2.** Numbers of coronoids *

| $h$ | Catacondensed | Pericondensed | Total |
|---|---|---|---|
| 8 | 1[a] | | 1[b] |
| 9 | 3[a] | 2[c] | 5[b] |
| 10 | 15[c] | 28[c] | 43[d] |
| 11 | 62[c] | 221[c] | 283[c] |
| 12 | 312[e] | 1642[e] | 1954[c] |
| 13 | 1436 | 10928[f] | 12364[g] |
| 14 | 6790 | 69504 | 76294[g] |
| 15 | † | † | 454095[h, i] |
| 16 | † | † | 2643124[h] |

* Double coronoids occur at $h \geq 13$.
[a] Dias (1982) [38]; [b] Knop, Szymanski, Jeričević and Trinajstić (1984) [17]; [c] Brunvoll, Cyvin and Cyvin (1987) [40]; [d] Knop, Müller, Szymanski and Trinajstić (1986) [39]; [e] Balaban, Brunvoll, Cioslowski, Cyvin, Cyvin, Gutman, He, He, Knop, Kovačević, Müller, Szymanski, Tošić and Trinajstić (1987) [18]; [f] He, He, Wang, Brunvoll and Cyvin (1988) [19]; [g] Knop, Müller, Szymanski and Trinajstić (1990) [41]; [h] Knop, Müller, Szymanski and Trinajstić (1990) [37]; [i] Brunvoll, Cyvin, Cyvin, Knop, Müller, Szymanski and Trinajstić (1990) [42]; † Unknown

**Table 3.** Numbers of double coronoids

| $h$ | Catacondensed | Pericondensed | Total |
|---|---|---|---|
| 13 | 1[a] | | 1[a] |
| 14 | 5[a] | 6[a] | 11[a] |
| 15 | 33[b] | 116[b] | 149[b] |
| 16 | 211 | 1407 | 1618[c, d] |
| 17 | 1271 | 13852 | 15123[e] |
| 18 | 7243 | 118517 | 125760[e] |

[a] Knop, Müller, Szymanski and Trinajstić (1990) [41]; [b] Brunvoll, Cyvin, Cyvin, Knop, Müller, Szymanski and Trinajstić (1990) [42]; [c] Knop, Müller, Szymanski and Trinajstić (1990) [37]; [d] Cyvin, Brunvoll and Cyvin (1990) [43]; [e] Brunvoll, Cyvin and Cyvin (1990) [44]

group itself [36, 37, 48–50]. A listing of these data with detailed documentation is postponed until Sect. 7, where it marks the start of the treatment of benzenoids exclusively.

*Errata*

The total number of coronoids with $h = 10$ was given erroneously by Knop et al. [16, 17] due to a misprint. The correct number, viz. 43 (see Table 2), obtained from independent computations was reported [34, 40] while Knop et al. [39], in the meantime, had published a correction with explanations.

Dias [38] gave correctly the numbers of $C_{32}H_{16}$ and $C_{36}H_{18}$ coronoid isomers, which correspond to the catacondensed coronoids for $h = 8$ and $h = 9$, respectively (cf. Table 2). However, his numbers for $C_{40}H_{20}$, $C_{44}H_{22}$ and $C_{48}H_{24}$, corresponding to $h = 10$, 11 and 12, respectively, are in error [43, 51].

Knop et al. [17] quoted a wrong number of Dias [38] concerning the closed-shell (Kekuléan) $C_{28}H_{16}$ isomers. Later Dias [52] published the correct number of 62 Kekuléan isomers of $C_{28}H_{16}$. Misled by the first wrong number Knop et al. [17] concluded with a wrong number of biradicalic (non-Kekuléan) $C_{28}H_{16}$ isomers in the very last sentence of the text of their paper [17]. Later, when they correctly had identified the 6 non-Kekuléan $C_{28}H_{16}$ isomers [16], they described the "disturbing" situation and located the error of Dias.

## 3.2 Planar Polyhexes (Benzenoids and Planar Circulenes)

In one of the classical papers on the enumeration of polyhexes, Lunnon [10] found the numbers of all geometrically planar polyhexes with $h$ (number of hexagons) up to 12; see Table 4. In these numbers the helicenic systems are excluded, but all non-helicenic circulenes are included. Hence the numbers pertain to benzenoids plus planar circulenes. The contribution of Balaban and Harary [13] to the Lunnon numbers (cf. footnotes to Table 4) is commented in the next section. The Lunnon numbers have been extended to $h = 13$ and $h = 14$ by means of the data from Müller et al. [36].

The planar (non-helicenic) circulenes have also been considered separately, first by the Düsseldorf-Zagreb group [16, 17, 35]; see Table 5. He and He [33] depicted the forms of these systems for $h \leq 9$; from these figures we have extracted separate numbers for the catacondensed and pericondensed systems up to this $h$ value. For the catacondensed planar monocirculenes with $h \leq 11$ the numbers may be deduced from a work of Brunvoll et al. [54], who enumerated the benzenoid systems composed of coronene with catacondensed appendages up to $h = 12$. On

**Table 4.** Numbers of planar polyhexes*

| $h$ | Catacondensed | Pericondensed | Total |
|---|---|---|---|
| 1 | 1 | | 1[a,b] |
| 2 | 1 | | 1[a,b] |
| 3 | 2 | 1 | 3[a,b] |
| 4 | 5 | 2 | 7[a,b] |
| 5 | 12 | 10 | 22[a,b] |
| 6 | 37 | 45 | 82[a,b] |
| 7 | 119 | 214 | 333[a,b] |
| 8 | 417 | 1031 | 1448[a,b] |
| 9 | 1509 | 5063 | 6572[b] |
| 10 | 5665 | 24825 | 30490[b] |
| 11 | 21507 | 122045 | 143552[b] |
| 12 | 82929 | 600172 | 683101[b] |
| 13 | † | † | 3274826[c] |
| 14 | † | † | 15796897[c] |

* Monocirculenes occur at $h \geq 6$; dicirculenes at $h \geq 10$.
[a] Balaban and Harary (1968) [13]; [b] Lunnon (1972) [10]; [c] Müller, Szymanski, Knop, Nikolić and Trinajstić (1990) [36]; † Unknown

Björg N. Cyvin, Jon Brunvoll, and Sven J. Cyvin

**Table 5.** Numbers of planar circulenes*

| h | Catacondensed | Pericondensed | Total |
|---|---|---|---|
| 6 | 1[a] | | 1[b,c] |
| 7 | 1[a] | 1[a] | 2[b,c] |
| 8 | 6[a] | 7[a] | 13[b,c] |
| 9 | 20[a] | 47[a] | 67[b,c] |
| 10 | 93 | 311 | 404[b,c] |
| 11 | 392 | 1931 | 2323[d] |
| 12 | 1808 | 11709 | 13517[d] |
| 13 | † | † | 76570[d] |
| 14 | † | † | 429320[d] |

* Dicirculenes occur at $h \geq 10$.
[a] He and He (1986) [33]; [b] Knop, Szymanski, Klasinc and Trinajstić (1984) [53]; [c] Knop, Szymanski, Jeričević and Trinajstić (1984) [17]; [d] Müller, Szymanski, Knop, Nikolić and Trinajstić (1990) [36]; † Unknown

removing the central hexagon of coronene one obtains exactly the substituted [6]circulenes up to $h = 11$, which are sought for. Then it is only necessary to add the known numbers of catacondensed coronoids (cf. Table 2) in order to get the numbers to be entered in Table 5. Müller et al. [36] (the Düsseldorf-Zagreb group) have supplied the numbers for the totals in this table up to $h = 14$. In this paper they listed the coronoids and non-coronoids (derivatives of [6]circulene) separately. Their numbers for $h = 11$ and $h = 12$ were reproduced by a new computer run, in which we generated specifically the derivatives of [6]circulene, both catacondensed and pericondensed. These data were again combined with the numbers of coronoids (Table 2) and with the planar polycirculenes. The numbers of planar dicirculenes are entered separately in Table 6.

The smallest tricirculene is a unique system with $h = 13$.

Once the numbers in Table 5 are known all the Lunnon numbers for $h \leq 12$ can be split into those for catacondensed and pericondensed systems. One only has to add the known numbers of benzenoids (see Sect. 3.1).

Figure 3 shows the forms of all circulenes with $h \leq 9$ as dualists. All of them are planar (non-helicenic) monocirculenes. They include the unique coronoid at $h = 8$ and the five coronoids at $h = 9$ (cf Table 2). The dicirculenes with $h \leq 12$ are depicted in Fig. 4. Also these systems are planar, but contain no double coronoid.

**Table 6.** Numbers of planar dicirculenes

| h | Catacondensed | Pericondensed | Total |
|---|---|---|---|
| 10 | 1 | | 1 |
| 11 | 2 | 4 | 6 |
| 12 | 18 | 37 | 55 |
| 13 | 101 | 326 | 427 |

78

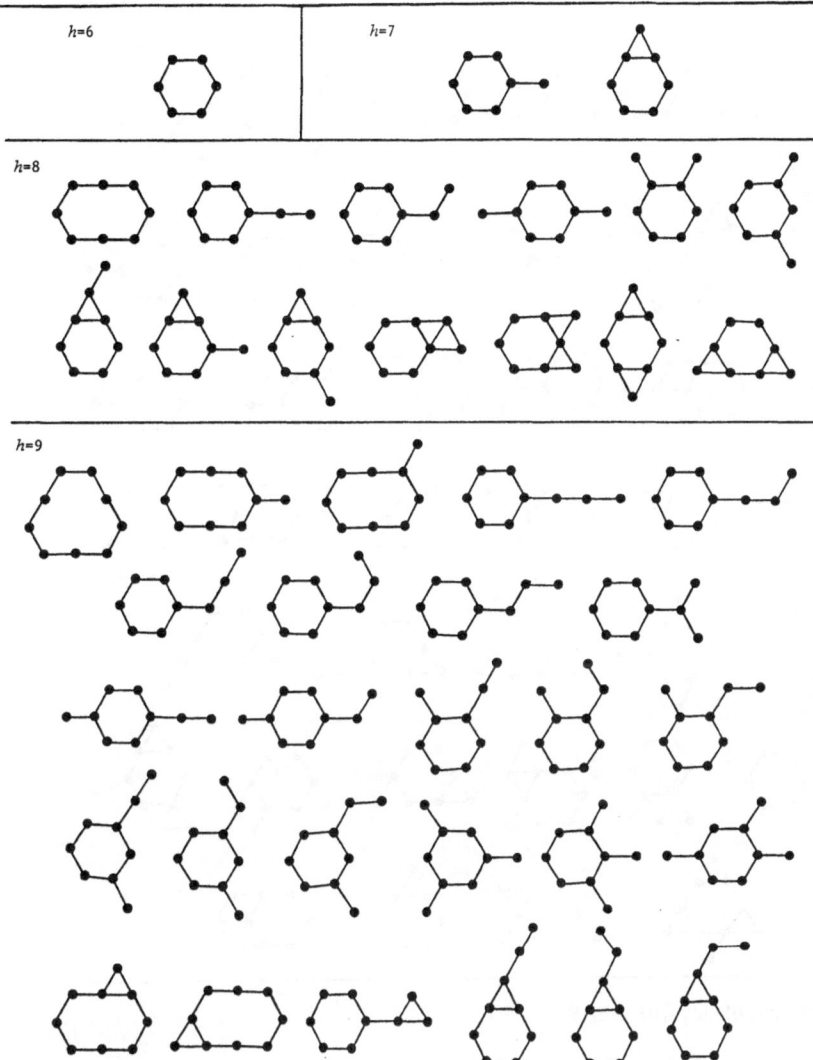

**Fig. 3.** (see next page)

Finally the smallest tricirculenes (for $h = 13$ and 14) were depicted in the form of dualists; see Fig. 5.

*Comments and Errata*

Knop et el. [16, 17] stated that the differences between the Lunnon numbers [10] (which they quoted up to $h = 10$) and Düsseldorf-Zagreb numbers are attributed to the numbers of planar monocirculenes. Thereby they missed the (unique) dicirculene at $h = 10$. The authors [16, 17] also stated erroneously that dicirculenes

(h=9)

**Fig. 3.** All circulenes with $h \leq 9$

do not appear "with less than 15 hexagons", while they actually appear at $h \geq 10$. Nevertheless, in their number 404 (cf. last column of Table 5; $h = 10$), as we have checked, the dicirculene is included. Consequently, the word "mono-circulenes" should be replaced by circulenes in the headings of the appropriate tables (corresponding to the last column for $h \leq 10$ of our Table 5) of the mentioned references [16, 17]. This is also the case for a table in a later publication [39]. Furthermore He and He [33], probably misled by Knop et al. [16], have stated that the Lunnon numbers "included planar mono-circulenes". We take this as a misinterpretation (although the statement is formally true), but it had no consequences for their numbers because they only treated the polyhexes up to $h = 9$.

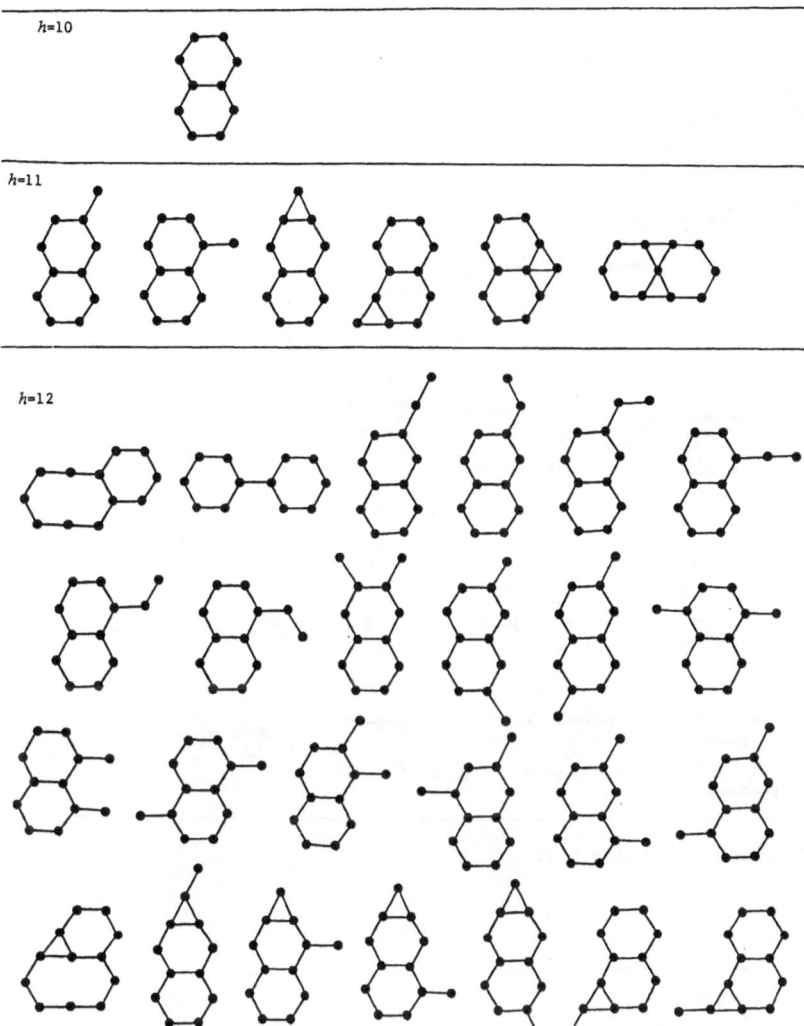

**Fig. 4.** (see next page)

In their illustrations of circulenes as dualists for $h \leq 9$, corresponding to our Fig. 3, He and He [33] represented [6]circulene as coronene with seven vertices. This may just be a way of presentation, but we are inclined to consider it as an error.

We have here given a precise interpretation of the Lunnon numbers, perhaps for the first time since they were published in 1972 [10]; but see also Müller et al. [36]. These numbers pertain to all planar (non-helicenic) polyhexes (benzenoids + circulenes, including dicirculenes). Using this interpretation we have actually reproduced these numbers, also for $h = 11$ and 12. In other words, we have provided an independent check of the Lunnon numbers for the first time.

(h=12)

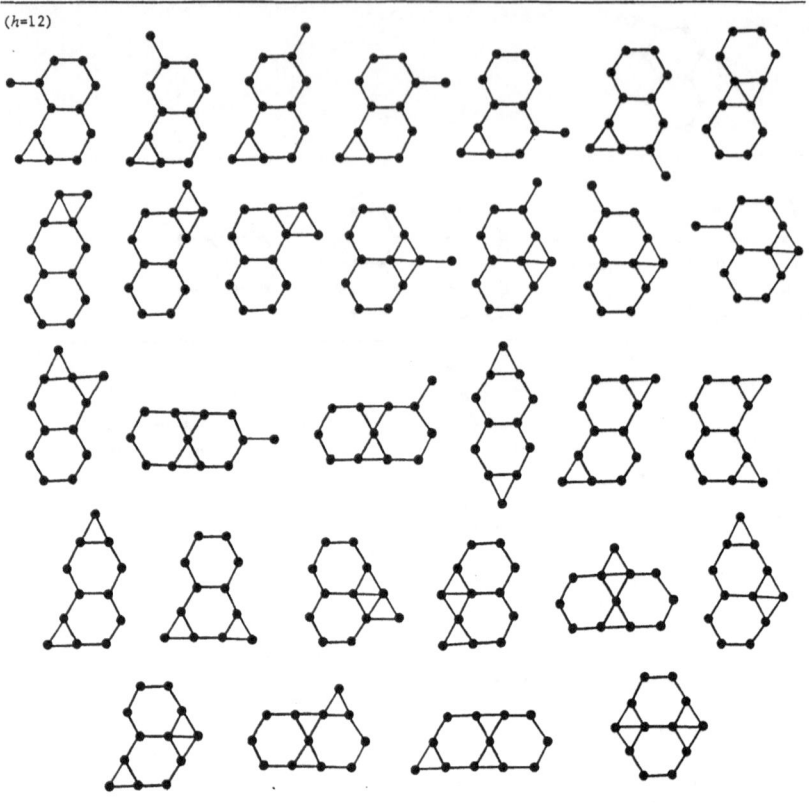

**Fig. 4.** All dicirculenes with $h \leq 12$

h=13

h=14

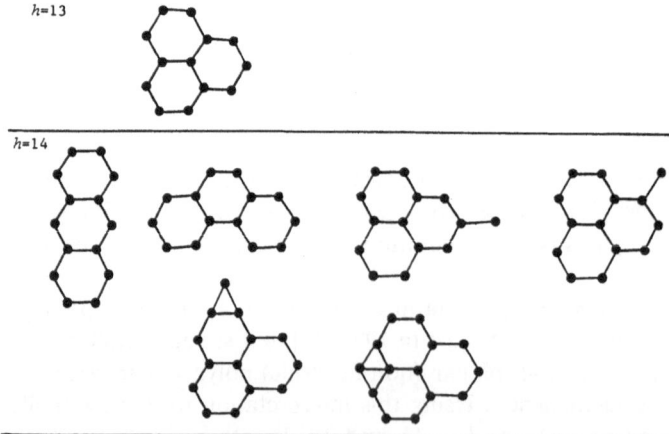

**Fig. 5.** All tricirculenes with $h \leq 14$

## 3.3 Simply Connected Polyhexes (Benzenoids and Helicenes)

The classical Harary-Read numbers [9] are given in the first column of Table 7. Some of them (cf. footnotes to Table 7) are contained in the work of Balaban and Harary [13] (for $h \leq 6$) or given by Balaban [30] (for $h \leq 8$) shortly before the pioneering paper of Harary and Read [9] appeared. These numbers pertain to the catacondensed simply connected polyhexes, viz. catacondensed benzenoids and catacondensed helicenes, while the circulenes are excluded. For short, this class is referred to as catafusenes (Sect. 2.6). The "Dias numbers" [16, 17] were published considerably later [38] and are identical with the Harary-Read numbers, taken up to $h = 8$. Harary and Read [9] in fact derived the numbers of catafusenes up to $h = 40$ by means of a generating function, but gave only a list of the consecutive numbers up to $h = 12$. The last number (for $h = 40$), viz. 256 364 771 375 268 976 315 575, the authors used to check their asymptotic results. Using the same method, we have reproduced numerically all the published Harary-Read numbers, including the twenty-four-digit number above. Here we report, in continuation of Table 7, the numbers of catafusenes for $13 \leq h \leq 20$: 467 262, 1 981 353, 8 487 400, 36 695 369, 159 918 120, 701 957 539, 3 101 072 051 and 13 779 935 438.

The helicenic simply connected polyhexes (helicenes) have also been considered separately. Their small numbers for $h = 6$ and 7 (cf. Table 8) are obtainable from a study of the illustrations in Balaban and Harary [13] and Balaban [30]. But all the entries of Table 8 for $h \leq 10$ have been given by the Düsseldorf-Zagreb group [16, 17].

Now it is an easy matter, with the knowledge of the numbers of benzenoids (cf. Sect. 3.1) to fill out the two last columns in Table 7 up to $h = 10$. We have also determined the five supplementary numbers in Table 8, which have not been given explicitly before.

**Table 7.** Numbers of simply connected polyhexes*

| $h$ | Catacondensed [+] | Pericondensed | Total |
|---|---|---|---|
| 1 | 1 [a, b] | | 1 |
| 2 | 1 [a, b] | | 1 |
| 3 | 2 [a, b] | 1 [a] | 3 |
| 4 | 5 [a, b] | 2 [a] | 7 |
| 5 | 12 [a, b] | 10 [a] | 22 |
| 6 | 37 [a, b] | 45 [a] | 82 |
| 7 | 123 [b, c] | 216 | 339 |
| 8 | 446 [b, c] | (1060) | (1506) |
| 9 | 1689 [b] | (5358) | (7047) |
| 10 | 6693 [b] | (27250) | (33943) |
| 11 | 27034 [b] | † | † |
| 12 | 111630 [b] | † | † |

\* Catafusenes and perifusenes. Helicenes occur at $h \geq 6$.
[+] See the text for $13 \leq h \leq 20$ and for $h = 40$.
[a] Balaban and Harary (1968) [13]; [b] Harary and Read (1970) [9]; [c] Balaban (1969) [30].
† Unknown (Parenthesized numbers are probably wrong)

**Table 8.** Numbers of helicenes

| h | Catacondensed | Pericondensed | Total |
|---|---|---|---|
| 6 | 1[a, b] | | 1[b] |
| 7 | 5[(a, c), b] | 3[b] | 8[b] |
| 8 | 35[b] | (36[b]) | (71[b]) |
| 9 | 200[b] | (342[b]) | (542[b]) |
| 10 | 1121[b] | (2736[b]) | (3857[b]) |
| 11 | 5919 | † | † |
| 12 | 30509 | † | † |
| 13 | 153187 | † | † |
| 14 | 756825 | † | † |
| 15 | 3688195 | † | † |

[a] Balaban and Harary (1968) [13]; [b] Knop, Szymanski, Jeričević and Trinajstić (1984) [17]; [c] Balaban (1969) [30].
† Unknown (Parenthesized numbers are probably wrong)

Figure 6 shows the forms of helicenes for $h \leq 8$ as dualists. Those for the catacondensed systems have been given previously [13, 30, 33], for $h = 8$ by He and He [33] probably for the first time.

Figure 6 includes 35 pericondensed helicenes with $h = 8$, obtained by a systematic search. This number disagrees with Table 8, which prescribes 36 such systems. The controversy represents an open problem; an error in the computer-enumeration [17] càn not be ruled out.

*Comments and Errata*

Balaban and Harary [13] anticipated the Lunnon numbers for $h \leq 8$ as shown in Table 4. As original sources they referred to personal communications from L. S. Kassel for $h = 6$ and 7, and Martin Gardner for $h = 7$ and 8. Balaban [30] repeated these numbers with reference to Kassel and to Klarner [7], where the latter reference as far as we can see is not appropriate. What is worse, however, is that Balaban [30] subtracted Harary-Read numbers (for catacondensed benzenoids + catacondensed helicenes) from Lunnon numbers (for total benzenoids + total planar circulenes), thus producing two "meaningless" numbers (in the characterization of Cyvin et al. [55]) for $h = 7$ and 8 (under the heading "Perifusenes"). Unfortunately Balaban [20] reproduced the two meaningless numbers in his early review on enumeration of polyhexes. They were also quoted by Rouvray [56].

In the consolidated report [18] it is stated erroneously that Lunnon [10] included helicenes in his numbers.

In works of Dias [38, 52] it is often not made clear whether helicenes are included or not in the numbers. Even in his book [25], where a specification is supposed to be given, the classes (with and without helicenes) are mixed up and there are obvious mistakes, also in numerical values.

It seems, strangely enough, that a list of all simply connected polyhexes (catacondensed and pericondensed together), as in the last column of Table 7, has not been given before explicitly.

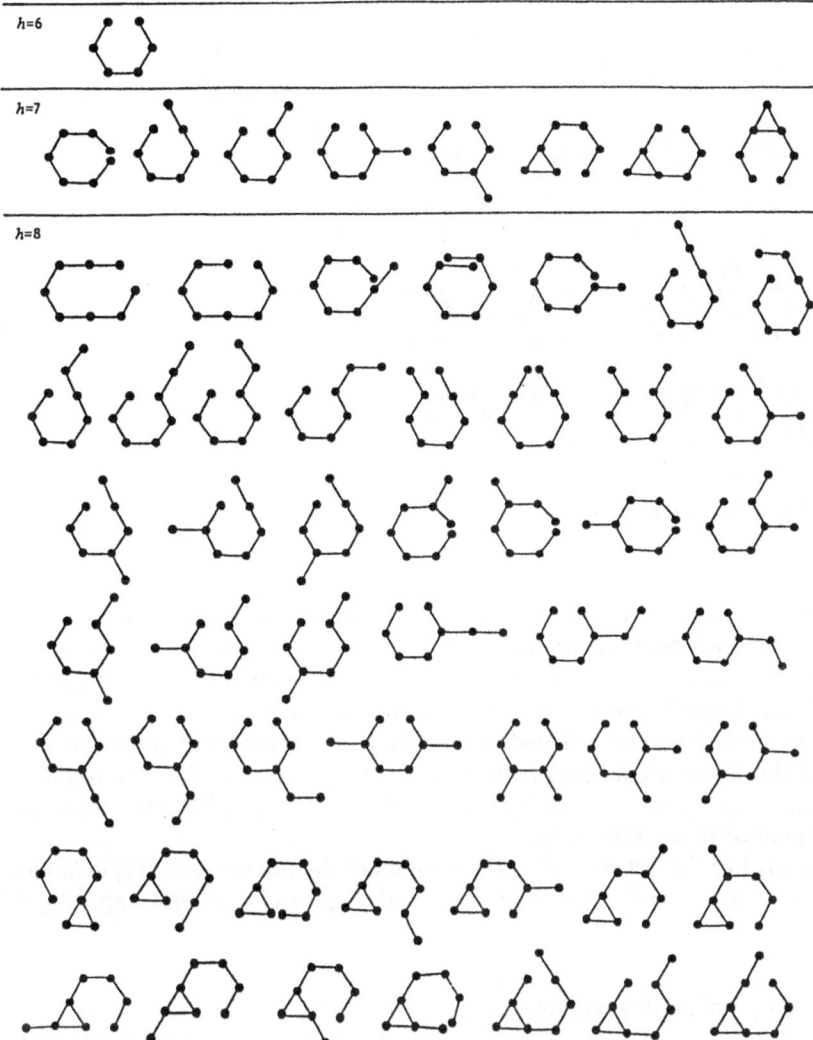

**Fig. 6.** (see next page)

The controversy between Fig. 6 and Table 8 (see above) opens the possibility that the numbers 36 and 71 (for $h = 8$) in Table 8 should be changed to 35 and 70, respectively. This also casts doubt on the correctness of the corresponding numbers for $h > 8$. If these enumeration errors could be proved, also certain numbers in Tables 7 and 9 would have to be corrected.

## 3.4 All Polyhexes (Benzenoids, Helicenes and Circulenes)

"On the Total Number of Polyhexes" is the title of the mini-review [17] which came from the Düsseldorf-Zagreb school and has been to so much inspiration for

Björg N. Cyvin, Jon Brunvoll, and Sven J. Cyvin

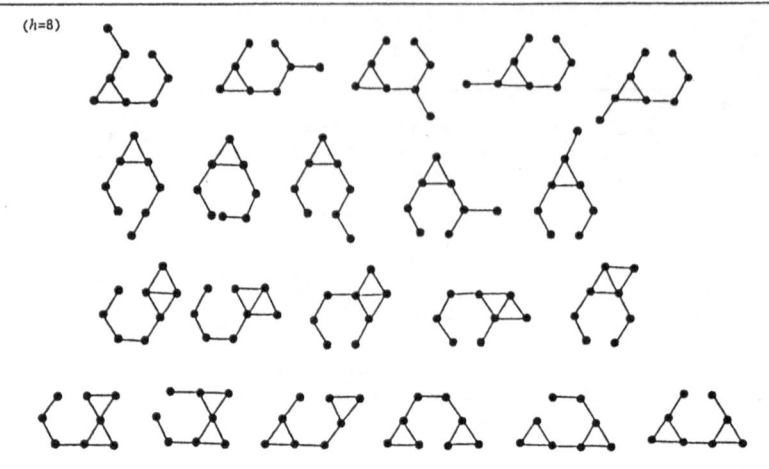

**Fig. 6.** All helicenes with $h \leq 8$

us. The polyhexes, as defined above in Sect. 2.1, encompass benzenoids, helicenes and circulenes, including helicirculenes. Their numbers (up to $h = 10$) are available from the data of the preceding sections if only the numbers of helicirculenes are added; cf. Table 9. The grand totals therein coincide with those of Knop et al. [16, 17] up to $h = 9$. The Düsseldorf-Zagreb group claimed to have extended the list of the classical Klarner numbers [1, 7]. Our grand total for $h = 10$ differs in one unit from the corresponding value of Knop et al. [16, 17]. The reason for this is explained in the following.

With regard to the helicirculenes the authors of the mini-review [17] identified three systems at $h = 10$, which they described as [6]circulene with appendages.

**Table 9.** Numbers of polyhexes in total*

| $h$ | Catacondensed | Pericondensed | Grand total |
|----|---------------|---------------|-------------|
| 1  | 1             |               | 1[a]        |
| 2  | 1             |               | 1[a]        |
| 3  | 2             | 1             | 3[a]        |
| 4  | 5             | 2             | 7[a]        |
| 5  | 12            | 10            | 22[a]       |
| 6  | 38            | 45            | 83[a]       |
| 7  | 124           | 217           | 341[b]      |
| 8  | 452           | (1067)        | (1519[b])   |
| 9  | 1709          | (5405)        | (7114[b])   |
| 10 | 6790          | (27561)       | (34351)     |

* Circulenes occur at $h \geq 6$, helicenes also at $h \geq 6$, and helicirculenes at $h \geq 10$.
[a] Klarner (1965) [7], see also *Comments*; [b] Knop, Szymanski, Jeričević and Trinajstić (1984) [17] (Parenthesized numbers are probably wrong)

The forms are depicted in the book of Knop et al. [16]. They are also reproduced in Fig. 7 (as dualists) together with another system without appendage (the system at the extreme left). Knop et al. [16, 17] either missed or did not consider this system. In our opinion it should be reckoned among helicirculenes because it is undoubtedly a non-planar circulene, not violating the definition of helicenic systems. It should absolutely be accounted for in the grand total for polyhexes.

All the 4 helicirculenes in Fig. 7 are catacondensed. The number of helicirculenes increases rapidly with increasing $h$. For $h = 11$ we have generated 58 systems by hand, 32 catacondensed and 26 pericondensed, but we are not sure that we have not missed any.

## Comments

It is instructive that Lunnon [10] claimed the Klarner number for $h = 6$, viz. 83, to be in error. He says: "[own result] corrects Klarner [1] and Read [A], who both (!) found... $= 83$." (The sign of exclamation is from the original.) Lunnon's reference [A] is a scientific report from 1968, not in our possession. After a fresh scrutiny of the relevant papers of Klarner [1, 7] we are inclined to believe that Lunnon is right. Not in the way that his numbers should coincide with the Lunnon numbers (Table 4), but rather with those of Table 7 (for simply connected polyhexes). The numbers in these two tables are, by the way, coincident up to $h = 6$.

Knop et al. [16, 17], on the other hand, take the Klarner numbers in support of their grand totals (cf. Table 9 with accompanying text). In their own words: "There are, indeed, indications in the literature that these numbers are correct. For example, Klarner [7, 1] reported the total number of polyhexes up to $h = 6$." In our opinion, this statement is somewhat bold since these indications are based on the single number 83. Furthermore, this value seems to be obscure, depending on its interpretation.

Here we have chosen to adhere to the interpretation of the Klarner numbers as grand totals (cf. Table 9) in the spirit of the Düsseldorf-Zagreb group [16, 17], and also in accord with a previous note [55].

In a note with corrections Knop et al. [39] invoked an incorrect summation as the reason for the grand total of $h = 10$ [17] to be in error. However, in the new "grand total" [39] the authors did not include helicirculenes. This "grand total" has also been published later [57].

$h=10$

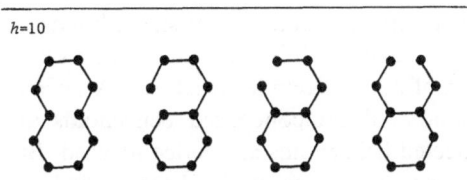

**Fig. 7.** The helicirculenes with $h = 10$

## 3.5 Concluding Remarks

In this section it is supposed that the confusion concerning the interpretation of different classical enumeration results for polyhexes is fully documented. It is also hoped that the matter now is somewhat clarified.

When looking at Tables 1, 4, 7 and 9, for instance, we find that the numbers for $h \leq 5$ are coincident, although their documentations by footnotes vary, since they are viewed in different contexts. They are also identical with the Düsseldorf-Zagreb numbers. This is of course so because no systems other than benzenoids occur at these low $h$ values. For $h = 6$ the pericondensed polyhexes are still only benzenoids, while the numbers of catacondensed polyhexes with $h = 6$ varies from 36 to 38. This is explained by the inclusion of hexahelicene or [6]circulene, occasionally both of them or none. Notice that from the number 37 alone, which is accompanied by the total of 82 ($h = 6$), one can not deduce which interpretation is the actual one.

# 4 Additional Definitions for Benzenoids

In Sect. 2.6 the concepts of catacondensed and pericondensed polyhexes are defined. It is implied that these notions are applicable to benzenoids (cf. Sect. 2.1) in particular. In the following we shall only speak about benzenoids, although all the concepts are applicable to other polyhexes as well.

In preparation to the definitions in the following we need to define the color excess or $\Delta$ value. It is the absolute magnitude of the difference between the numbers of black and white (or starred and unstarred) vertices. Here it is referred to the coloring (or starring) of vertices. It is known that the $\Delta$ value also is the absolute magnitude of the difference between the numbers of valleys and peaks.

Another important quantity for a benzenoid is the Kekulé structure count or $K$ number. A Kekulé structure, being a typical concept from (mathematical) chemistry, corresponds to a 1-factor or perfect matching in mathematics.

Another subdivision of benzenoids (apart from catacondensed/pericondensed) distinguishes between *Kekuléan* and *non-Kekuléan* systems. A Kekuléan benzenoid system possesses Kekulé structures ($K > 0$). A non-Kekuléan benzenoid has no Kekulé structure ($K = 0$). The shorter designations "Kekuléans" and "non-Kekuléans" are often used. All catacondensed benzenoids are Kekuléan; therefore all non-Kekuléans are pericondensed. Any Kekuléan benzenoid has a vanishing color excess; $\Delta = 0$.

A Kekuléan benzenoid may be *normal* or *essentially disconnected*. In an essentially disconnected benzenoid there are fixed double and/or single bonds. A fixed single (resp. double) bond refers to an edge which is associated with a single (resp. double) bond in the same position of all the Kekulé structures. A normal (Kekuléan) benzenoid has no fixed bond. All catacondensed benzenoids are normal; therefore all essentially disconnected benzenoids are pericondensed. But a pericondensed Kekuléan may be either normal or essentially disconnected.

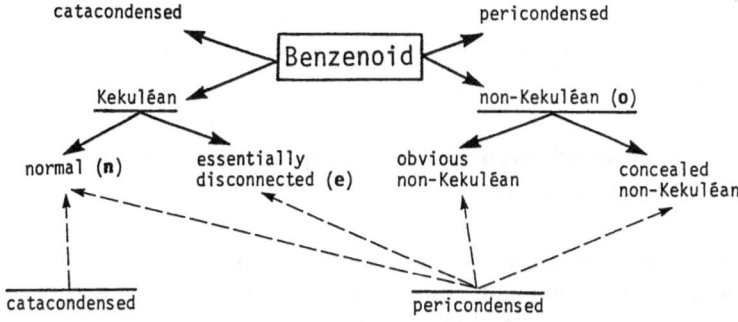

**Fig. 8.** Survey of classes of benzenoids

A non-Kekuléan benzenoid, which necessarily is pericondensed, may be *obvious* non-Kekuléan or *concealed* non-Kekuléan. If $\Delta > 0$ for a benzenoid, then it is obvious non-Kekuléan. If $\Delta = 0$ and $K = 0$, the benzenoid is concealed non-Kekuléan.

The neo classification takes into account all benzenoids; they can be either normal (n), essentially disconnected (e) or non-Kekuléan (o).

A schematic survey of the subclasses of benzenoids encountered in this paragraph is displayed in Fig. 8. Additional classifications, especially with reference to the symmetry groups, are treated in appropriate places of the subsequent sections.

# 5 Algebraic Solutions

## 5.1 Catacondensed Simply Connected Polyhexes (Catafusenes)

In the algebraic enumerations of classes of polyhexes the achievements of Harary and Read [9] exhibit a peak of professional expertise. The work has been quoted several times [14, 30, 58]. Herein the two mathematicians [9] developed a generating function with the explicit form

$$H(x) = (1/24x^2) [12 + 24x - 48x^2 - 24x^3 + (1 - x)^{3/2} (1 - 5x)^{3/2}$$
$$- 3(3 + 5x) (1 - x^2)^{1/2} (1 - 5x^2)^{1/2}$$
$$- 4(1 - x^3)^{1/2} (1 - 5x^3)^{1/2}]. \tag{1}$$

On expanding this function the Harary-Read numbers (see first column of Table 7) emerge as coefficients for the powers of $x$;

$$H(x) = x + x^2 + 2x^3 + 5x^4 + 12x^5 + 37x^6 + 123x^7$$
$$+ 446x^8 + 1689x^9 + 6693x^{10} + \ldots. \tag{2}$$

These numbers, which are defined in Sect. 3.3, pertain to the catacondensed simply connected polyhexes, for short called catafusenes (Sect. 2.6). They include the catacondensed helicenes.

## 5.2 Unbranched Catacondensed Simply Connected Polyhexes (Unbranched Catafusenes)

Explicit formulas for the numbers of the title systems were developed, in a simple combinatorial way, by Balaban and Harary [13]; cf. also Balaban [58]. The derivation [13] involves a treatment of the subclasses of unbranched catafusenes with specific symmetries. It is outlined in the following, basically in the version of Brunvoll et al. [59] (cf. also Balaban et al. [60]), and is supported by illustrations.

Let the numbers of unbranched catafusenes belonging to the different symmetry groups be identified by the below symbols. It is stressed that helicenes are included.

$a$  acenes (linear); $D_{2h}$ for $h > 1$, $D_{6h}$ for $h = 1$ (benzene)
$c$  centrosymmetrical; $C_{2h}$
$m$  mirror-symmetrical; $C_{2v}$
$u$  unsymmetrical; $C_s$

The total number is

$$U = a + c + m + u \tag{3}$$

where the quantities are functions of $h$.

The following recurrence properties are valid.

$$a_{h+1} = a_h \tag{4}$$

$$c_{h+2} = 3c_h + 1 \tag{5}$$

$$m_{h+2} = 3m_h + 1 . \tag{6}$$

Here the number of hexagons ($h = 1, 2, 3, ...$) is indicated by subscripts. The relations (5) and (6) are illustrated in Fig. 9 and Fig. 10, respectively. From each system with $h$ hexagons three systems with $h + 2$ hexagons are generated by adding two hexagons to the ends in three ways. One system, generated from the linear acene, must be added to the set.

From the initial condition $a_1 = 1$ Eq. (4) gives

$$a = 1 ; \quad h = 1, 2, 3, 4, ... \tag{7}$$

i.e. independent of $h$. This reflects the simple fact that there is one and only one linear acene for a given number of hexagons ($h$). From Eq. (5) we obtain the following explicit formulas for $c$ by means of the initial conditions $c_4 = c_5 = 1$

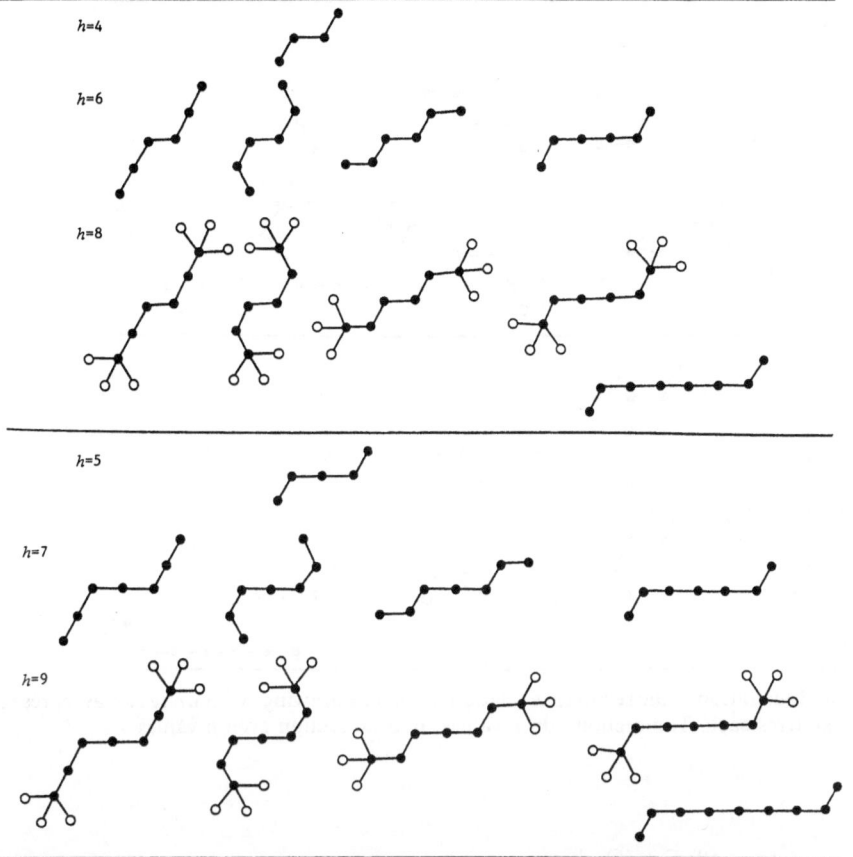

**Fig. 9.** Illustrations of the recurrence relation for $c$. The drawings with *white circles* represent three systems each. Top section even $h$ values; bottom section odd $h$ values

(or $c_2 = c_3 = 0$).

$$c = (1/2) [3^{(h-3)/2} - 1]; \qquad h = 3, 5, 7, \dots \qquad (8a)$$

while $c_1 = 0$, and

$$c = (1/2) [3^{(h-2)/2} - 1]; \qquad h = 2, 4, 6, \dots . \qquad (8b)$$

In compressed form:

$$c = (1/2) \, 3^{\lfloor (h-2)/2 \rfloor} - (1/2); \qquad h = 2, 3, 4, 5, \dots \qquad (8)$$

where the special brackets in the exponent denote the floor function; $\lfloor x \rfloor$ is the largest integer smaller than or equal to $x$. Similarly, by means of $m_3 = m_4 = 1$

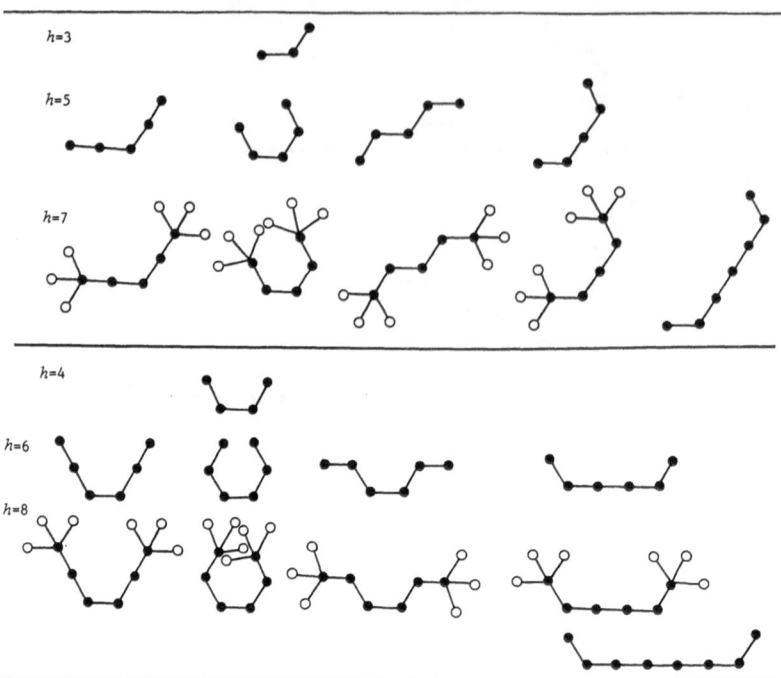

**Fig. 10.** Illustration of the recurrence relation for $m$. The drawings with *white circles* represent three systems each. Top section odd $h$ values; bottom section even $h$ values

(or $m_1 = m_2 = 0$) Eq. (6) gives

$$m = (1/2)\, 3^{\lfloor (h-1)/2 \rfloor} - (1/2); \qquad h = 1, 2, 3, 4, \dots . \qquad (9)$$

Notice that $c_h = m_h$ for $h = 2, 4, 6, \dots$; cf. Figs. 9 and 10, which illustrate this relation for $h = 4$, 6 and 8. Furthermore, $c_h = m_{h-1}$ for $h = 2, 3, 4, 5, \dots$; this feature is illustrated for $h = 4$ to 9 by the figures.

Next we deduce what we shall call the "crude total". It is obtained by starting with two hexagons and generating three systems recursively from each predecessor in the same way as described above (cf. especially Figs. 9 and 10), but now the hexagons are only added to one end, always to the last added hexagon. In this way $3^{h-2}$ systems are generated for $h \geq 2$. This crude total accounts for all catafusenes, but most of them are generated more than once. Specifically, every centrosymmetrical or mirror-symmetrical catafusene is generated twice and every unsymmetrical four times. Only the acenes are generated once each. In consequence, we obtain the equation

$$3^{h-2} = a + 2(c + m) + 4u; \qquad (h > 1). \qquad (10)$$

**Table 10.** Numbers of unbranched catafusenes*

| $h$ | $a$ | $c$ | $m$ | $u$ | Total ($U$) |
|---|---|---|---|---|---|
| 1 | 1 | | | | 1 |
| 2 | 1 | | | | 1 |
| 3 | 1 | | 1 | | 2 |
| 4 | 1 | 1 | 1 | 1 | 4 |
| 5 | 1 | 1 | 4 | 4 | 10 |
| 6 | 1 | 4 | 4 | 16 | 25 |
| 7 | 1 | 4 | 13 | 52 | 70 |
| 8 | 1 | 13 | 13 | 169 | 196 |
| 9 | 1 | 13 | 40 | 520 | 574 |
| 10 | 1 | 40 | 40 | 1600 | 1681 |
| 11 | 1 | 40 | 121 | 4840 | 5002 |
| 12 | 1 | 121 | 121 | 14641 | 14884 |
| 13 | 1 | 121 | 364 | 44044 | 44530 |
| 14 | 1 | 364 | 364 | 132496 | 133225 |
| 15 | 1 | 364 | 1093 | 397852 | 399310 |
| 16 | 1 | 1093 | 1093 | 1194649 | 1196836 |
| 17 | 1 | 1093 | 3280 | 3585040 | 3589414 |
| 18 | 1 | 3280 | 3280 | 10758400 | 10764961 |
| 19 | 1 | 3280 | 9841 | 32278480 | 32291602 |
| 20 | 1 | 9841 | 9841 | 96845281 | 96864964 |

* From algebraic equations (Section 5.2); cf. also (for $h \leq 8$): Balaban and Harary (1968) [13]; Balaban (1969) [30]. Abbreviations: $a$ (linear) acenes; $c$ centrosymmetrical; $m$ mirror-symmetrical; $u$ unsymmetrical

On combining Eqs. (3), (8), (9) and (10), we arrive at the following explicit formulas for $u$ and $U$.

$$u = (1/4) [3^{h-2} + 1] - 3^{(h-3)/2}; \qquad h = 3, 5, 7, \ldots \qquad (11a)$$

while $u_1 = 0$, and

$$u = (1/4) [3^{(h-2)/2} - 1]^2; \qquad h = 2, 4, 6, \ldots. \qquad (11b)$$

Furthermore,

$$U = (1/4) [3^{h-2} + 1] + 3^{(h-3)/2}; \qquad h = 3, 5, 7, \ldots \qquad (12a)$$

while $U_1 = 1$, and

$$U = (1/4) [3^{(h-2)/2} + 1]^2; \qquad h = 2, 4, 6, \ldots. \qquad (12b)$$

Gutman [61] pointed out that the numbers $U$ follow a third-order recurrence relation. The same form is also obeyed by $u$ [59]. One has:

$$u_{h+6} = 13u_{h+4} - 39u_{h+2} + 27u_h; \qquad h = 1, 2, 3, 4 \ldots \qquad (13)$$
$$U_{h+6} = 13U_{h+4} - 39U_{h+2} + 27U_h; \qquad h = 2, 3, 4, 5 \ldots. \qquad (14)$$

93

Björg N. Cyvin, Jon Brunvoll, and Sven J. Cyvin

It was also deduced that

$$s_{h+2} = 3s_h; \qquad h = 2, 3, 4, 5 \ldots \tag{15}$$

where

$$s = a + c + m \tag{16}$$

The explicit formulas for $s$, the number of symmetrical catafusenes, read:

$$s = 2 \times 3^{(h-3)/2}; \qquad h = 3, 5, 7, \ldots \tag{17a}$$

while $s_1 = 1$, and

$$s = 3^{(h-2)/2}; \qquad h = 2, 4, 6, \ldots \tag{17b}$$

Numerical values of $a$, $c$, $m$, $u$ and $U$ for $h \leq 20$ are collected in Table 10.

## 5.3 Fibonacenes and Related Systems

A *fibonacene* is a special catafusene, which consists of 2-segments only (and therefore is "all-kinked"). The name is explained by the feature that it is a nonlinear acene of which the Kekulé structure count is a Fibonacci number. As an alternative definition, a fibonacene is a single chain of hexagons where all of them, apart from the two terminal ones, are angularly annelated. The smallest fibonacene has $h = 3$. For a given number of hexagons we find one zigzag chain among the fibonacenes; where the annelations go left-right-left-right- .... As the other extreme one finds the helix-shaped fibonacene with annelations left-left-left- ... (isomorphic with right-right-right- ...). The above definition implies that all pertinent helicenes are included among fibonacenes.

Balaban [62] derived explicit formulas for the numbers of fibonacenes as functions of $h$, both for the totals and for the subclasses belonging to the symmetry groups $C_{2h}$, $C_{2v}$ and $C_s$. Other symmetries are not possible for fibonacenes. Balaban [62] employed two methods, one of them being close to the derivation for unbranched catafusenes, which is described in detail in the preceding paragraph. Below we give a version which follows the preceding derivation in a perfectly analogous way.

Let the numbers of centrosymmetrical ($C_{2h}$), mirror-symmetrical ($C_{2v}$) and unsymmetrical ($C_s$) fibonacenes be denoted by $c^f$, $m^f$ and $u^f$, respectively. The total is

$$U^f = c^f + m^f + u^f \tag{18}$$

The numbers of symmetrical fibonacenes are governed by very simple recurrence relations, viz. $c^f_{h+2} = 2c^f_h$ and $m^f_{h+2} = 2m^f_h$, while the initial conditions are $c^f_3 = 0$,

94

**Table 11.** Numbers of fibonacenes*

| $h$ | $c^f$ | $m^f$ | $u^f$ | Total $(U^f)$ |
|-----|-------|-------|-------|---------------|
| 3 | | 1 | | 1 |
| 4 | 1 | 1 | | 2 |
| 5 | 0 | 2 | 1 | 3 |
| 6 | 2 | 2 | 2 | 6 |
| 7 | 0 | 4 | 6 | 10 |
| 8 | 4 | 4 | 12 | 20 |
| 9 | 0 | 8 | 28 | 36 |
| 10 | 8 | 8 | 56 | 72 |
| 11 | 0 | 16 | 120 | 136 |
| 12 | 16 | 16 | 240 | 272 |
| 13 | 0 | 32 | 496 | 528 |
| 14 | 32 | 32 | 992 | 1056 |
| 15 | 0 | 64 | 2016 | 2080 |
| 16 | 64 | 64 | 4032 | 4160 |
| 17 | 0 | 128 | 8128 | 8256 |
| 18 | 128 | 128 | 16256 | 16512 |
| 19 | 0 | 256 | 32640 | 32896 |
| 20 | 256 | 256 | 65280 | 65792 |
| 21 | 0 | 512 | 130816 | 131328 |

* From algebraic equations (Section 5.3); cf. also (for $h \leq 10$):
Balaban (1989) [62]. Abbreviations: see Table 10

$c_4^f = 1$ and $m_3^f = m_4^f = 1$. This leads to the explicit equations

$$c^f = 0; \qquad h = 3, 5, 7, \ldots \tag{19a}$$

$$c^f = 2^{(h-4)/2}; \qquad h = 4, 6, 8, \ldots \tag{19b}$$

and

$$m^f = 2^{\lfloor(h-3)/2\rfloor}; \qquad h = 3, 4, 5, 6, \ldots . \tag{20}$$

Notice that $c_h^f = m_h^f = m_{h-1}^f; h = 4, 6, 8, \ldots$. For the crude total in this case one has

$$2^{h-2} = 2(c^f + m^f) + 4u^f \tag{21}$$

which now makes it feasible to deduce the following explicit formulas for $u^f$ and $U^f$.

$$u^f = 2^{h-4} - 2^{\lfloor(h-4)/2\rfloor}; \qquad h = 4, 5, 6, 7, \ldots \tag{22}$$

while $u_3^f = 0$, and

$$U^f = 2^{h-4} + 2^{\lfloor(h-4)/2\rfloor}; \qquad h = 4, 5, 6, 7, \ldots \tag{23}$$

while $U_3^f = 1$. The following regularities have been pointed out [62].

$$u_{h+1}^f = 2u_h^f, \qquad U_{h+1}^f = 2U_h^f; \qquad h = 3, 5, 7, \ldots . \tag{24}$$

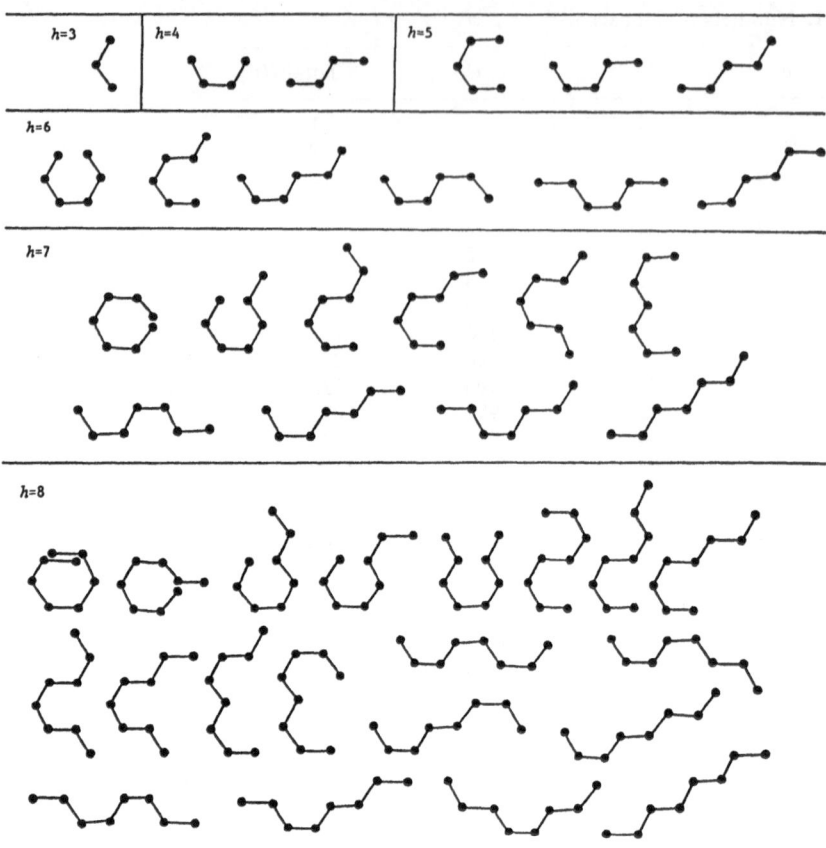

**Fig. 11.** All fibonacenes for $h \leq 8$

Numerical values of $c^f$, $m^f$, $u^f$ and $U^f$ for $h \leq 21$ are collected in Table 11.

Figure 11 shows the dualist representations of all fibonacenes up to $h = 8$. They were also depicted by Balaban [62].

All the formulas in the present paragraph also apply to *generalized fibonacenes*, viz. unbranched catafusenes consisting of equidistant segments, which are not necessarily 2-segments.

Very recently, Balaban and Artemi [63] generalized the treatment of fibonacenes. They considered the numbers $N(h, s)$, which pertain to the unbranched catafusenes with $h$ hexagons and having a longest segment of the length $s$. It is emphasized that the helicenes are included. Fibonacenes are the special cases for $s = 2$. The authors [63] arrived at recurrence formulas for $N(h, s)$. Furthermore, they deduced an explicit formula for $N(h, s)$ in the case of $h \leq 2s - 1$. In that case $N(h, s)$ no longer depends on $h$ and $s$ independently, but on the difference $h - s$.

## 5.4 Gutman Trees (*LA*-Sequences)

To any unbranched catafusene an *LA*-sequence (equivalent with the L-transform of a three-digit code) is associated. If the system has $h$ hexagons, then the *LA*-sequence is a string of $h$ letters $L$ or $A$, where $L$ indicates a linearly annelated (also terminal) hexagon and $A$ an angularly annelated. A given *LA*-sequence is often compatible with several non-isomorphic polyhexes, but in that case they have the same $K$ number and are called isoarithmic. The subject of the present paragraph is the enumeration of different *LA*-sequences with given $h$ values. If an *LA*-sequence is read forwards and backwards with different results, the two strings do not, by definition, represent "different" *LA*-sequences and are in other words counted as one. Notice that we are here concerned with an enumeration of polyhex classes rather than individual (non-isomorphic) polyhexes. The fibonacenes is such a class, which corresponds to the *LA*-sequence usually written as $LA^{h-2}L$. In the preceding paragraph the non-isomorphic systems within this class were counted; in the present paragraph this class counts as one.

The enumeration of *LA*-sequences is equivalent to the enumeration of Gutman trees, which in the mathematical literature are called caterpillar trees (or caterpillars). These objects correspond to special trees in the graph-theoretical sense.

Harary and Schwenk [64] derived an explicit formula for the number of caterpillars/Gutman trees as a function of $h$ by two methods. One of these methods, the simpler one, is basically analogous with the treatments above (Sects. 5.2 and 5.3). Below we give a derivation which closely follows these treatments.

Let a Gutman tree be called symmetrical if its *LA*-sequence is the same read forewards and backwards. Otherwise it is said to be unsymmetrical. The numbers of symmetrical and unsymmetrical Gutman trees are identified by the symbols $s^c$ and $u^c$, respectively. The total is

$$U^c = s^c + u^c \tag{25}$$

For the numbers of symmetrical Gutman trees one has $s^c_{h+2} = 2s^c_h$ and $s^c_1 = s^c_2 = 1$. Consequently,

$$s^c = 2^{\lfloor (h-1)/2 \rfloor}; \qquad h = 1, 2, 3, 4, \ldots \tag{26}$$

The crude total is

$$2^{h-2} = s^c + 2u^c \tag{27}$$

by means of which we arrive at the following explicit equations for $u^c$ and $U^c$.

$$u^c = 2^{h-3} - 2^{\lfloor (h-3)/2 \rfloor}; \qquad h = 3, 4, 5, 6, \ldots \tag{28}$$

while $u^c_1 = u^c_2 = 0$, and

$$U^c = 2^{h-3} + 2^{\lfloor (h-3)/2 \rfloor}; \qquad h = 3, 4, 5, 6, \ldots \tag{29}$$

**Table 12.** Numbers of Gutman trees*

| $h$ | $s^c$ | $u^c$ | Total ($U^c$) |
|---|---|---|---|
| 1 | 1 | | 1 |
| 2 | 1 | | 1 |
| 3 | 2 | | 2 |
| 4 | 2 | 1 | 3 |
| 5 | 4 | 2 | 6 |
| 6 | 4 | 6 | 10 |
| 7 | 8 | 12 | 20 |
| 8 | 8 | 28 | 36 |
| 9 | 16 | 56 | 72 |
| 10 | 16 | 120 | 136 |
| 11 | 32 | 240 | 272 |
| 12 | 32 | 496 | 528 |
| 13 | 64 | 992 | 1056 |
| 14 | 64 | 2016 | 2080 |
| 15 | 128 | 4032 | 4160 |
| 16 | 128 | 8128 | 8256 |
| 17 | 256 | 16256 | 16512 |
| 18 | 256 | 32640 | 32896 |
| 19 | 512 | 65280 | 65792 |
| 20 | 512 | 130816 | 131328 |

* From algebraic equations (Section 5.4). Abbreviations:
$s$ symmetrical; $u$ unsymmetrical

while $U_1^c = U_2^c = 1$. Some regularities are observed, viz.

$$u_{h+1}^c = 2u_h^c, \qquad U_{h+1}^c = 2U_h^c; \qquad h = 2, 4, 6, \ldots \tag{30}$$

Numerical values of $s^c$, $u^c$ and $U^c$ for $h \leq 20$ are collected in Table 12.

Let a Gutman tree or $LA$-sequence be represented by an unbranched catacondensed benzenoid drawn in a standard way: start from left to right with a horizontal row and alternating kinks left-right-left-right- .... Furthermore, the first row should be as long as possible, and in general a lexicographic order should be followed, where $L$ has preference before $A$. Under these conventions the fibonacenes are represented by a zigzag chain. Figure 12 shows the standard dualists representing all Gutman trees up to $h = 8$.

## 5.5 Generalized Hexagon-Shaped Benzenoids

A free edge of a benzenoid is an edge (on the perimeter) between two vertices of degree two. It is known that any benzenoid has at least six free edges. If it has exactly six we shall call it a *generalized hexagon-shaped* benzenoid. A hexagon-shaped benzenoid (or "hexagon", not to be confused with a hexagonal unit here called a hexagon) in its original sense has also exactly six free edges, while its two and two opposite rows of hexagons are parallel. This last condition is released in the definition of a generalized hexagon-shaped benzenoid. It is clear that the hexagon-shaped benzenoids constitute a subclass of generalized hexagon-shaped benzenoids.

An algebraic solution has been found for the numbers of "hollow hexagons", a class of primitive coronoids [40, 45, 65]. The solution was adapted to a class of special [2p]annulenes [66]. It may also be adapted, even more directly, to the

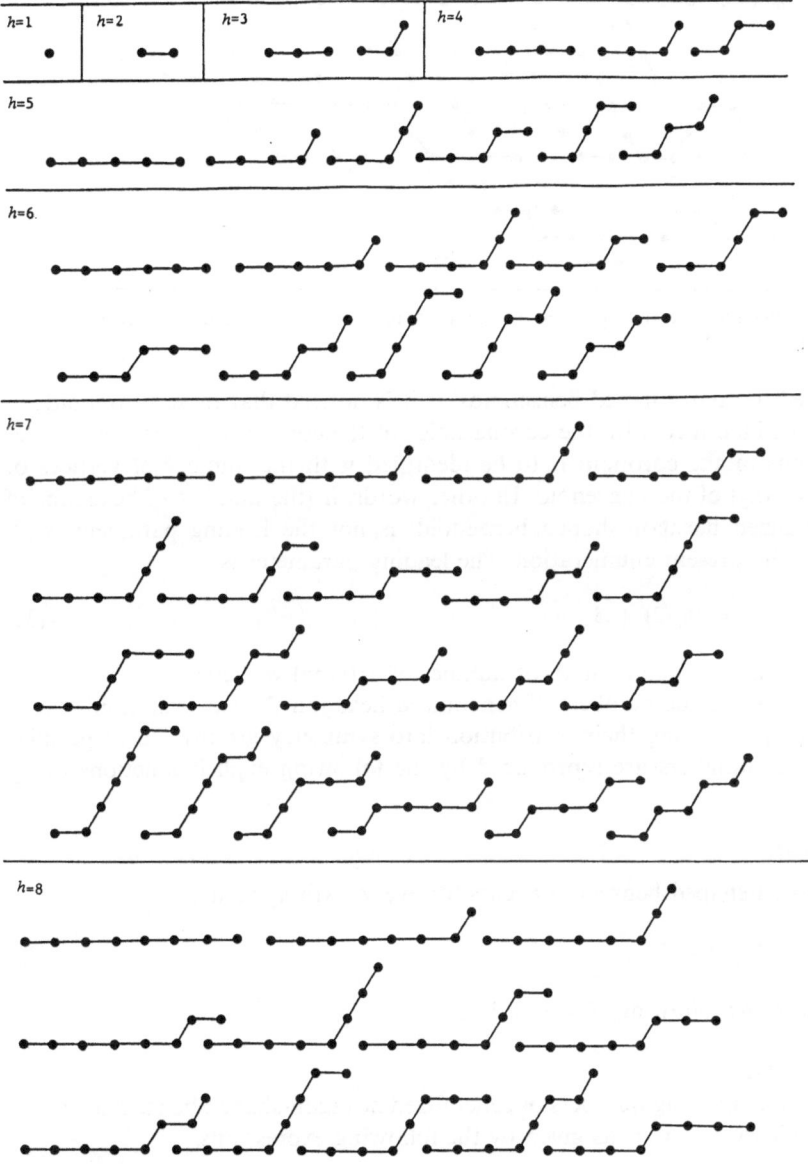

**Fig. 12.** (see next page)

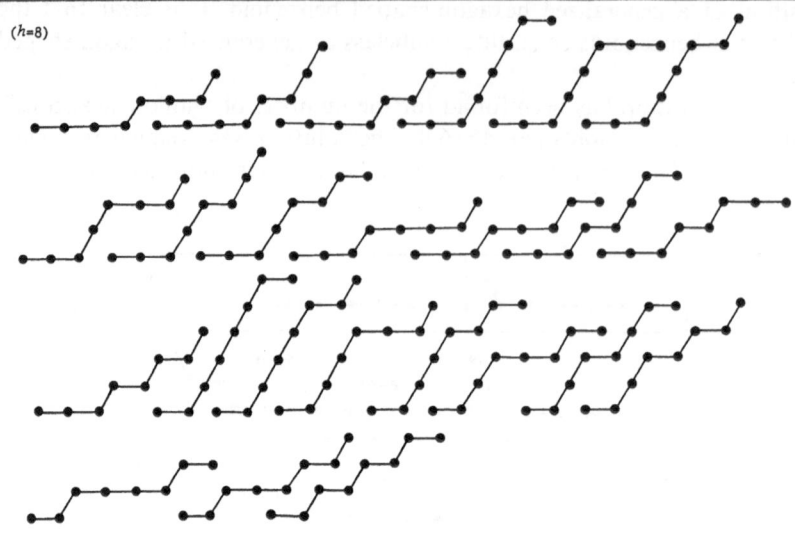

(h=8)

**Fig. 12.** Unbranched catacondensed benzenoids representing all Gutman trees (or *LA*-sequences) for $h \leq 8$

generalized hexagon-shaped benzenoids, if it is noticed that these benzenoids in fact may be identified with the corona holes of the coronoids. Then the number of hexagons of the coronoid is to be identified with the number of vertices of degree two ($n_2$) of the benzenoid. In other words, $h$ (the number of hexagons of the generalized hexagon-shaped benzenoid) is not the leading parameter with respect to the present enumeration. The leading parameter is

$$n_2 = (n_e/2) + 3 \tag{31}$$

where $n_e$ is the perimeter length or number of external vertices.

Table 13 shows the numbers of generalized hexagon-shaped benzenoids up to $n_2 = 30$ [45], including their distribution into symmetry groups. These peculiar sequences of numbers are reproduced by the following explicit functions of $n_2$, say $N(n_2)$.

*Symmetry $D_{6h}$*

One hexagon-shaped benzenoid occurs for every sixth $n_2$ value:

$$N(6j) = 1 \tag{32}$$

Here and in the following, $j = 1, 2, 3, \ldots$

*Symmetry $D_{3h}$*

The only nonvanishing numbers of generalized hexagon-shaped benzenoids occur at every third value of $n_2$ as given by the following expressions.

$$N(6j) = j - 1, \qquad N(6j + 3) = j \tag{33}$$

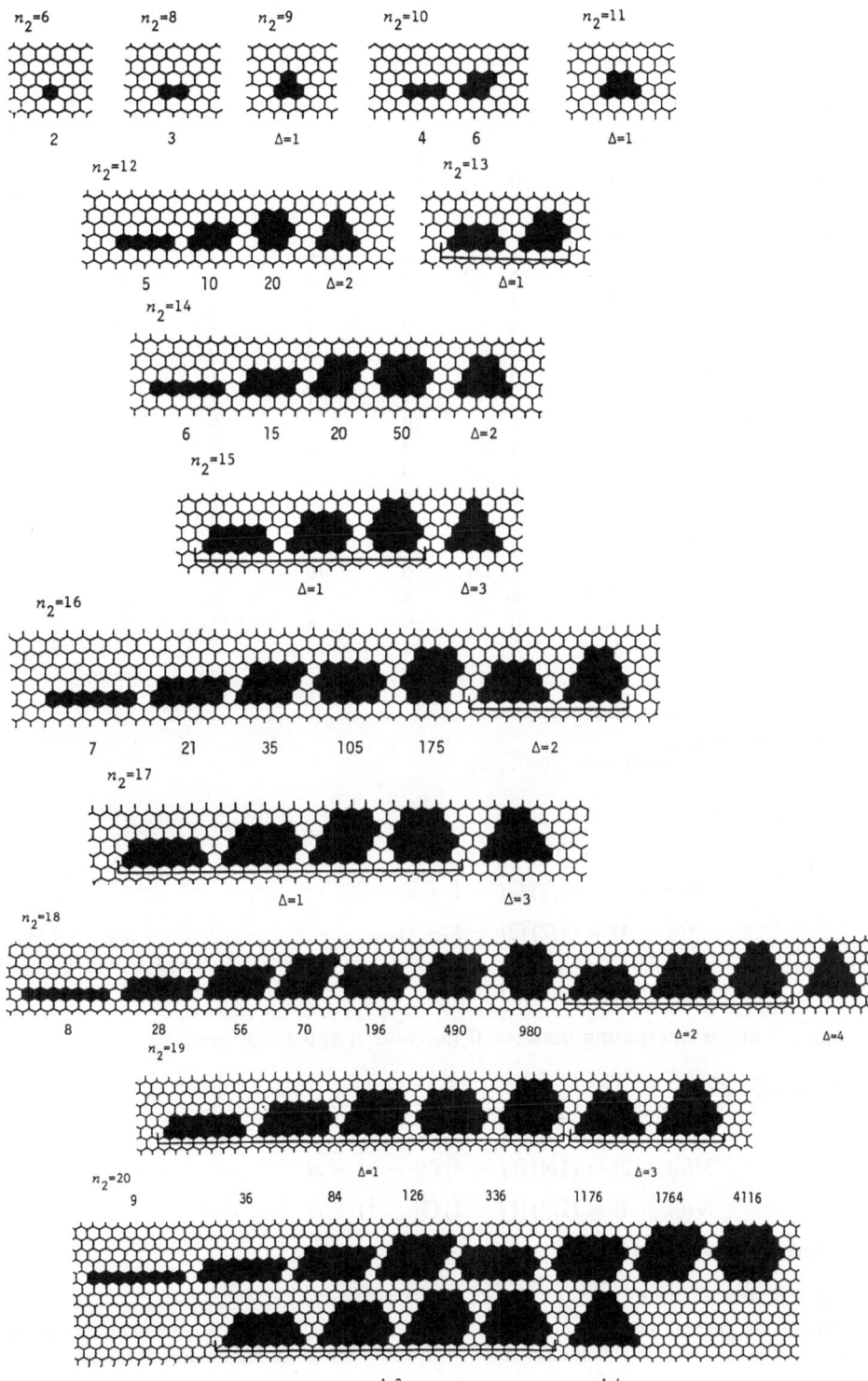

$n_2=6$      $n_2=8$      $n_2=9$      $n_2=10$      $n_2=11$

2      3      $\Delta=1$      4  6      $\Delta=1$

$n_2=12$      $n_2=13$

5  10  20  $\Delta=2$      $\Delta=1$

$n_2=14$

6  15  20  50  $\Delta=2$

$n_2=15$

$\Delta=1$      $\Delta=3$

$n_2=16$

7  21  35  105  175  $\Delta=2$

$n_2=17$

$\Delta=1$      $\Delta=3$

$n_2=18$

8  28  56  70  196  490  980  $\Delta=2$  $\Delta=4$

$n_2=19$

$\Delta=1$      $\Delta=3$

$n_2=20$

9  36  84  126  336  1176  1764  4116

$\Delta=2$      $\Delta=4$

101

Björg N. Cyvin, Jon Brunvoll, and Sven J. Cyvin

**Table 13.** Numbers of generalized hexagon-shaped benzenoids[*]

| $n_2$ | $D_{6h}$ | $D_{3h}$ | $D_{2h}$ | $C_{2h}$ | $C_{2v}$ | $C_s$ | Total ($N^h$) |
|---|---|---|---|---|---|---|---|
| 6 | 1 | | | | | | 1 |
| 7 | 0 | | | | | | 0 |
| 8 | 0 | | 1 | | | | 1[a] |
| 9 | 0 | 1 | 0 | | | | 1[a] |
| 10 | 0 | 0 | 2 | | | | 2[a] |
| 11 | 0 | 0 | 0 | | 1 | | 1[a] |
| 12 | 1 | 1 | 1 | 1 | 0 | | 4[a] |
| 13 | 0 | 0 | 0 | 0 | 2 | | 2[a] |
| 14 | 0 | 0 | 3 | 1 | 1 | | 5[a] |
| 15 | 0 | 2 | 0 | 0 | 1 | 1 | 4[a] |
| 16 | 0 | 0 | 3 | 2 | 2 | 0 | 7[a] |
| 17 | 0 | 0 | 0 | 0 | 4 | 1 | 5[a] |
| 18 | 1 | 2 | 3 | 3 | 1 | 1 | 11[a] |
| 19 | 0 | 0 | 0 | 0 | 5 | 2 | 7[a] |
| 20 | 0 | 0 | 4 | 4 | 4 | 1 | 13[a] |
| 21 | 0 | 3 | 0 | 0 | 4 | 4 | 11 |
| 22 | 0 | 0 | 5 | 5 | 5 | 2 | 17 |
| 23 | 0 | 0 | 0 | 0 | 8 | 5 | 13 |
| 24 | 1 | 3 | 4 | 7 | 4 | 4 | 23 |
| 25 | 0 | 0 | 0 | 0 | 10 | 7 | 17 |
| 26 | 0 | 0 | 6 | 8 | 8 | 5 | 27 |
| 27 | 0 | 4 | 0 | 0 | 8 | 11 | 23 |
| 28 | 0 | 0 | 6 | 10 | 10 | 7 | 33 |
| 29 | 0 | 0 | 0 | 0 | 14 | 13 | 27 |
| 30 | 1 | 4 | 6 | 12 | 8 | 11 | 42 |

[*] Cf. Raragraph 5.4.
[a] Brunvoll, Cyvin and Cyvin (1987) [40]

*Symmetry* $D_{2h}$

$$N(6j) = (1/2) [3(j - 1) - \varepsilon] \tag{34a}$$

$$N(6j + 2) = (1/2) (3j - 1 + \varepsilon) \tag{34b}$$

$$N(6j + 4) = (1/2) (3j + 1 - \varepsilon) \tag{34c}$$

where

$$\varepsilon = (1/2) [1 + (-1)^j] \tag{35}$$

viz. a number alternating between 0 (for odd $j$) and 1 (for even $j$).

*Symmetry* $C_{2h}$

$$N(6j) = (1/4) [3(j - 1)^2 + \varepsilon] \tag{36a}$$

$$N(6j + 2) = (1/4) [(j - 1) (3j - 1) - \varepsilon] \tag{36b}$$

$$N(6j + 4) = (1/4) [(j - 1) (3j + 1) + \varepsilon] \tag{36c}$$

where $\varepsilon$ is given by (35).

◄

**Fig. 13.** All generalized hexagon-shaped benzenoids for $n_2 \leq 20$. $K$ numbers are given for the Kekuléan systems (which have $\Delta = 0$); $\Delta$ values are given for the non-Kekuléans

*Symmetry $C_{2v}$*

$$N(6j) = (1/4)\,[(j-1)\,(3j-7) + \varepsilon] \qquad (37a)$$

$$N(6j+1) = N(6j+4) = (1/4)\,[(j-1)\,(3j+1) + \varepsilon] \qquad (37b)$$

$$N(6j+2) = N(6j+3) = (1/4)\,[(j-1)\,(3j-1) - \varepsilon] \qquad (37c)$$

$$N(6j+5) = (1/4)\,[(j+1)\,(3j-1) + \varepsilon] \qquad (37d)$$

where $\varepsilon$ again is given by (35).

*Symmetry $C_s$*

$$N(6j) = (1/8)\,[(j-1)\,(2j^2 - 7j + 7) - \varepsilon] \qquad (38a)$$

$$N(6j+1) = N(6j+4) = (1/8)\,[(j-1)\,(2j^2 - 3j - 1) - \varepsilon] \qquad (38b)$$

$$N(6j+2) = (1/8)\,[(j-1)\,(2j^2 - 5j + 1) + \varepsilon] \qquad (38c)$$

$$N(6j+3) = (1/8)\,[(j-1)\,(2j^2 - j + 1) + \varepsilon] \qquad (38d)$$

$$N(6j+5) = (1/8)\,[(j-1)\,(j+1)\,(2j-1) - \varepsilon] \qquad (38e)$$

where $\varepsilon$ is given by (35) as before.

*Total*

When the expressions from (32)–(34) and (36)–(38) are added appropriately the total number of (non-isomorphic) generalized hexagon-shaped benzenoids, say $N^h(n_2)$, is obtained. The result was rendered into the form:

$$N^h(6j) = (1/8)\,\{(j+1)\,(2j^2 + j + 1) - (1/2)\,[1 + (-1)^j]\} \qquad (39a)$$

$$N^h(6j+2) = (1/8)\,\{(j+1)\,(2j^2 + 3j - 1) + (1/2)\,[1 + (-1)^j]\} \qquad (39b)$$

$$N^h(6j+4) = (1/8)\,\{(j+1)\,(2j^2 + 5j + 1) - (1/2)\,[1 + (-1)^j]\} \qquad (39c)$$

$(j = 1, 2, 3, \ldots)$. The expressions (39) account for the even $n_2$ values and are supplemented by

$$N^h(n_2 + 3) = N^h(n_2); \qquad n_2 = 6, 8, 10 \ldots \qquad (40)$$

The last equation (40) expresses the fact that $N^h$ attains the same value on adding three units to $n_2$.

In Fig. 13 all the non-isomorphic generalized hexagon-shaped benzenoids with $n_2$ up to 20 are depicted.

## 5.6 Approximate Formulas

For the numbers of catafusenes as a function of $h$, say $C_h$, an exact asymptotic behavior for large values of $h$ is known [9] and has the form

$$C_h \approx (1/2)\, 5^{1/2}\,(2h-1)!\,[(h-1)!\,(h+2)!]^{-1}\,(5/4)^h \qquad (41)$$

Björg N. Cyvin, Jon Brunvoll, and Sven J. Cyvin

Using the Stirling approximation, Eq. (41) was transformed to [67]

$$C_h \approx (5/16\pi)^{1/2} h^{-5/2} 5^h \tag{42}$$

It is recalled that $C_h$, which pertains to catafusenes, include the catacondensed helicenes; $C_h$ are, in other words, the Harary-Read numbers (cf. Sect. 3.3). Based on the form (42), Gutman [67] assumed

$$C'_h \approx ah^p b^h \tag{43}$$

and attempted to fit the empirical constants therein, viz. $a$, $b$ and $p$, to the number of catacondensed benzenoids (without helicenes), $C'_h$. By means of the exact values up to $C'_{11}$, which were known at the time Gutman made this analysis, he estimated the constants to $a = 0.049$, $b = 4.27$, $p = -5/4$. In the same work Gutman [67] produced the approximate formula

$$B'_h \approx 0.045 h^{-3/2} (5.4)^h \tag{44}$$

for the total number of benzenoids ($B'_h$), again without helicenes. Here the empirical constants were based on the known exact values of $B'_h$ up to $B'_{11}$. In the same way, Cyvin et al. [68] derived

$$N'_h \approx 0.0242 h^{-0.9} (4.5)^h \tag{45}$$

for the numbers of normal benzenoids, based on exact $N'_h$ values up to $N'_{11}$.

After the appearance of a computational result for $B'_{12}$ Aboav and Gutman [69] realized that Eq. (44) does not have the desired precision. They improved the approximation by producing a recursive formula of the form

$$B'_{h+1}/B'_h \approx b(1 - qh^{-2}) \tag{46}$$

and estimated the empirical constants to $b = 4.98$, $q = 5.77$ by means of the exact $B'_h$ values up to $B'_{12}$. The authors [69] also proposed an approximate formula for $C'_{h+1}/C'_h$ of the same form as (46), but here we do not quote the reported numerical parameters because they were fitted to a wrong $C'_{13}$ value [47].

For the number of unbranched catacondensed benzenoids, $U'_h$, Gutman [61] launched the very simple approximate formula

$$U'_h = 0.0400 (2.869)^h \tag{47}$$

where the numerical parameters were fitted to known exact $U'_h$ values up to $U'_{20}$. A more sophisticated analysis by Aboav and Gutman [70] led to the following recursive formula.

$$U'_{h+1} = (1 - \alpha)(U'_h)^2/U'_{h-1} + (-1)^h \beta(U'_h)^{1/2} \tag{48}$$

104

The same material of exact $U'_h$ values as previously ($h \leq 20$) was used to estimate the parameters in (48) to $\alpha = 0.000714$, $\beta = 0.75$.

# 6 Catafusenes

## 6.1 Introductory Remarks

In the present section the catafusenes (catacondensed simply connected polyhexes; cf. Sect. 5.1) are treated. However, in contrast to the Harary-Read numbers (first column of Table 7) we shall be interested in the numbers of unbranched and branched systems separately. The numbers of unbranched catafusenes (Table 10) are known from algebraic formulas (cf. Sect. 5.2), but now we are interested in the unbranched catacondensed benzenoids and helicenes separately. Likewise we shall treat the numbers of branched catacondensed benzenoids and helicenes separately.

After the definition and enumeration of different "special catacondensed systems" (SCS's) the catacondensed benzenoids belonging to the symmetries $D_{3h}$, $C_{3h}$, $D_{2h}$ and $C_{2h}$ are treated in particular. Those of the $D_{3h}$ and $D_{2h}$ symmetries were enumerated by an algorithm invoking SCS's.

Finally some results for unbranched catacondensed benzenoids with equidistant segments are reported. These systems are the benzenoids (without helicenes) belonging to fibonacenes and generalized fibonacenes.

## 6.2 Unbranched Catafusenes

Let the numbers of unbranched catacondensed benzenoids and unbranched catacondensed helicenes be denoted by $U'$ and $U^*$, respectively. Then

$$U = U' + U^* \tag{49}$$

where $U$ has the same meaning as in Sect. 5.2. We write also for the symmetry groups $C_{2h}$, $C_{2v}$ and $C_s$ separately:

$$c = c' + c^*, \qquad m = m' + m^*, \qquad u = u' + u^* \tag{50}$$

respectively. One has, of course

$$U' = a + c' + m' + u', \qquad U^* = a + c^* + m^* + u^* \tag{51}$$

Table 14 shows the numbers for all of the above classes when $h \leq 20$. The documentations (see footnotes of the table) pertaining to the smallest $h$ values are difficult and ambiguous if they are supposed to share the credit among authors properly. Here we have credited Balaban and Harary [13] for the smallest numbers which could be taken from their paper before the helicenes start to interfere. A

105

Björg N. Cyvin, Jon Brunvoll, and Sven J. Cyvin

report by Brunvoll et al. [71] is incorporated among the footnotes. The largest numbers, which first appeared in the consolidated report [18], were produced by Tošić and Kovačević [72], and published almost simultaneously by these authors as a full report with description of their methods. These numbers have also been

**Table 14.** Numbers of unbranched catacondensed benzenoids; unbranched catacondensed helicenes in parentheses[+]

| $h$ | $a$ | $c'(c^*)$ | $m'(m^*)$ | $u'(u^*)$ | Total: $U'(U^*)$ |
|---|---|---|---|---|---|
| 1 | 1 | | | | 1[a] |
| 2 | 1 | | | | 1[a] |
| 3 | 1 | | 1[a] | | 2[a] |
| 4 | 1 | 1[a] | 1[a] | 1[a] | 4[a] |
| 5 | 1 | 1[a] | 4[a] | 4[a] | 10[a] |
| 6 | 1 | 4[a] | 3[b](1) | 16[a] | 24[c](1) |
| 7 | 1 | 4[a] | 12[b](1) | 50[b](2) | 67[c](3) |
| 8 | 1 | 13[a] | 10[b](3) | 158[b](11) | 182[c](14) |
| 9 | 1 | 13[a] | 34[b](6) | 472[b](48) | 520[d](54) |
| 10 | 1 | 39[b](1) | 28[b](12) | 1406[e](194) | 1474[e](207) |
| 11 | 1 | 39[b](1) | 97[b](24) | 4111[e](729) | 4248[e](754) |
| 12 | 1 | 116[b](5) | 81[b](40) | 11998[e](2643) | 12196[e](2688) |
| 13 | 1 | 115[f](6) | 271[f](93) | 34781[e](9263) | 35168[e](9362) |
| 14 | 1 | 339[f](25) | 226[f](138) | 100660[e](31836) | 101226[e](31999) |
| 15 | 1 | 336[f](28) | 764[e](329) | 290464[e](107388) | 291565[e](107745) |
| 16 | 1 | 988[e](105) | 638[e](455) | 837137[e](357512) | 838764[e](358072) |
| 17 | 1 | 977[e](116) | 2141[e](1139) | 2408914[e](1176126) | 2412033[e](1177381) |
| 18 | 1 | 2866[e](414) | 1787[e](1493) | 6925100[e](3833300) | 6929754[e](3835207) |
| 19 | 1 | 2832[e](448) | 6025[e](3816) | 19888057[e](12390423) | 19896915[e](12394687) |
| 20 | 1 | 8298[e](1543) | 5030[e](4811) | 57071610[e](39773671) | 57084939[e](39780025) |

[+] Abbreviations: see footnote to Table 10.
[a] Balaban and Harary (1968) [13]; [b] Brunvoll, Cyvin and Cyvin (1987) [54]; [c] He and He (1985) [32]; [d] He and He (1986) [33]; [e] Balaban, Brunvoll, Cioslowski, Cyvin, Cyvin, Gutman, He, He, Knop, Kovačević, Müller, Szymanski, Tošić and Trinajstić (1987) [18]; [f] Brunvoll, Cyvin and Cyvin (1987) [71]

**Table 15.** Numbers of symmetrical unbranched catacondensed benzenoids; corresponding helicenes in parentheses[+]

| $h$ | $m'(m^*)$ | $c'(c^*)$ |
|---|---|---|
| 21 | 16924(12600) | 8185(1656) |
| 22 | 14116(15408) | 23969(5555) |
| 23 | 47623(40950) | 23623(5901) |
| 24 | 39727(48846) | 69109(19464) |
| 25 | 133934(131786) | 68057(20516) |
| 26 | 111655(154065) | 198995(66725) |
| 27 | 377003(420158) | 195806(69914) |
| 28 | 314297(482864) | 572216(224945) |
| 29 | 1061064(1330420) | 562675(234486) |
| 30 | 884199(1507285) | 1642975(748509) |

[+] In continuation of Table 14. From: Brunvoll, Tošić, Kovačević, Balaban, Gutman and Cyvin (1990) [59]

submitted for publication elsewhere [59]; herein the parenthesized numbers of Table 14 (for $c^*$, $m^*$, $u^*$ and $U^*$) are given explicitly. These numbers, which pertain to the unbranched catacondensed helicenes, are of course obtainable by subtraction from the known numbers for unbranched catafusenes (cf. Sect. 5.2 and especially Table 10).

For the symmetrical ($C_{2h}$ and $C_{2v}$) catafusenes the enumerations were supplemented up to $h = 30$ by Brunvoll et al [59]; cf. Table 15.

The unbranched catacondensed benzenoids (without helicenes) up to $h = 7$ are represented as dualists in Fig. 14. These forms were first given by Balaban and

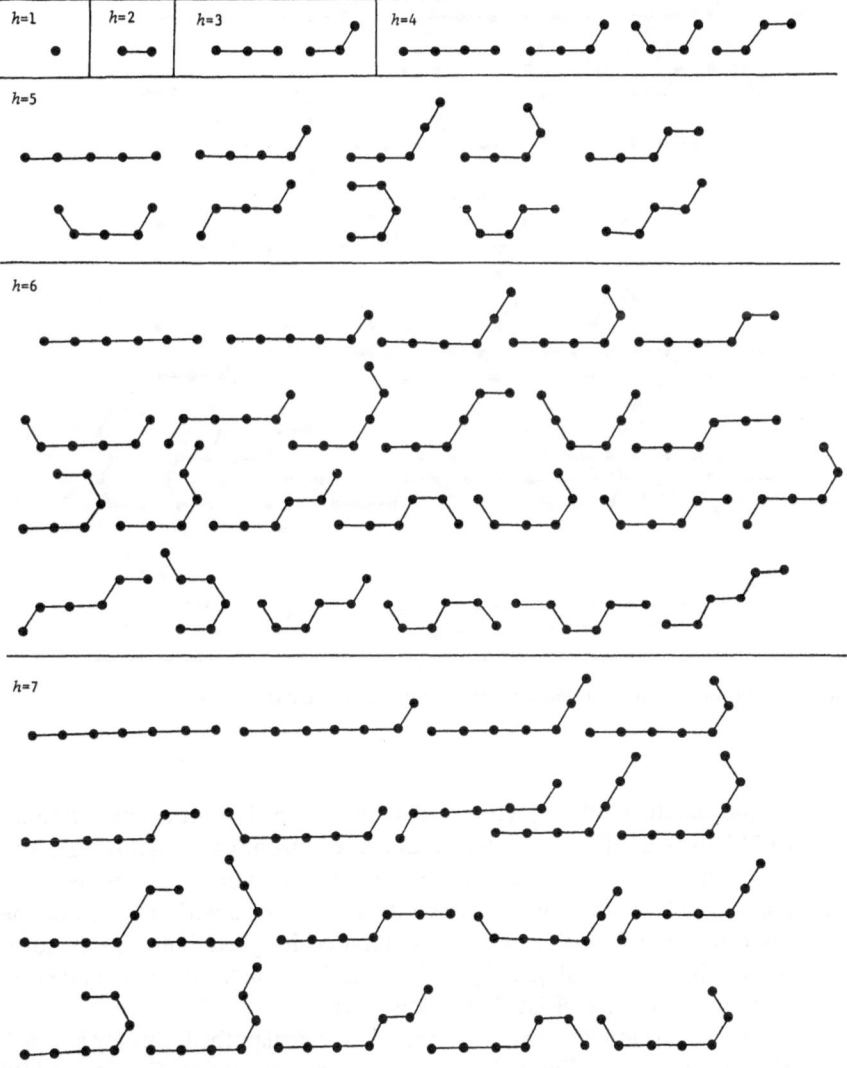

**Fig. 14.** (see next page)

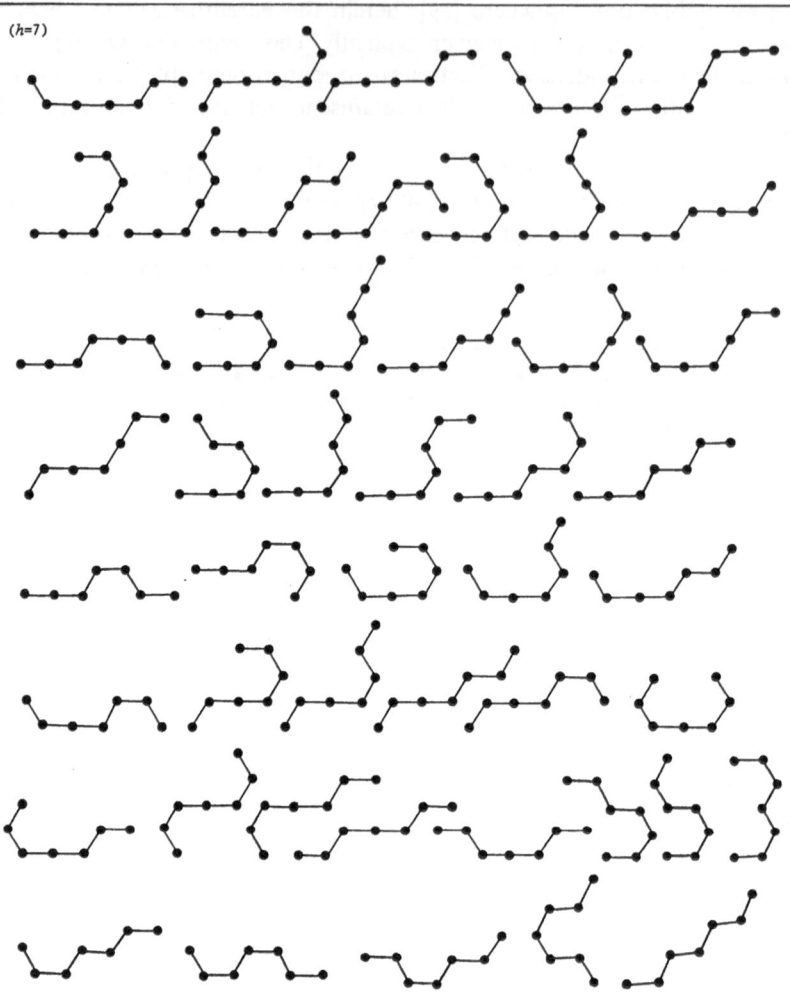

(h=7)

**Fig. 14.** All unbranched catacondensed benzenoids (as dualists) for $h \leq 7$

Harary [13] and Balaban [30, 73], who included helicenes. For an enumeration up to $h = 6$ with figures in different representations, see Džonova-Jerman-Blažič and Trinajstić [74]. In his studies of resonance energies Gutman [75] compiled data for benzenoids which include all the catacondensed systems with $h \leq 5$. For the catacondensed benzenoids with $h = 7$ also computer-designed figures in the form of mini-hexagons have been published [76]. El-Basil [77] depicted the catacondensed $h \leq 5$ systems in his studies of labelling sequences.

For the unbranched catacondensed helicenes in particular, the forms for $h = 6, 7$ and 8 are found as parts of Fig. 6. For $h = 9$ the 6 symmetrical [26, 54] and 48 unsymmetrical [54] such systems have been depicted in the form of dualists, but

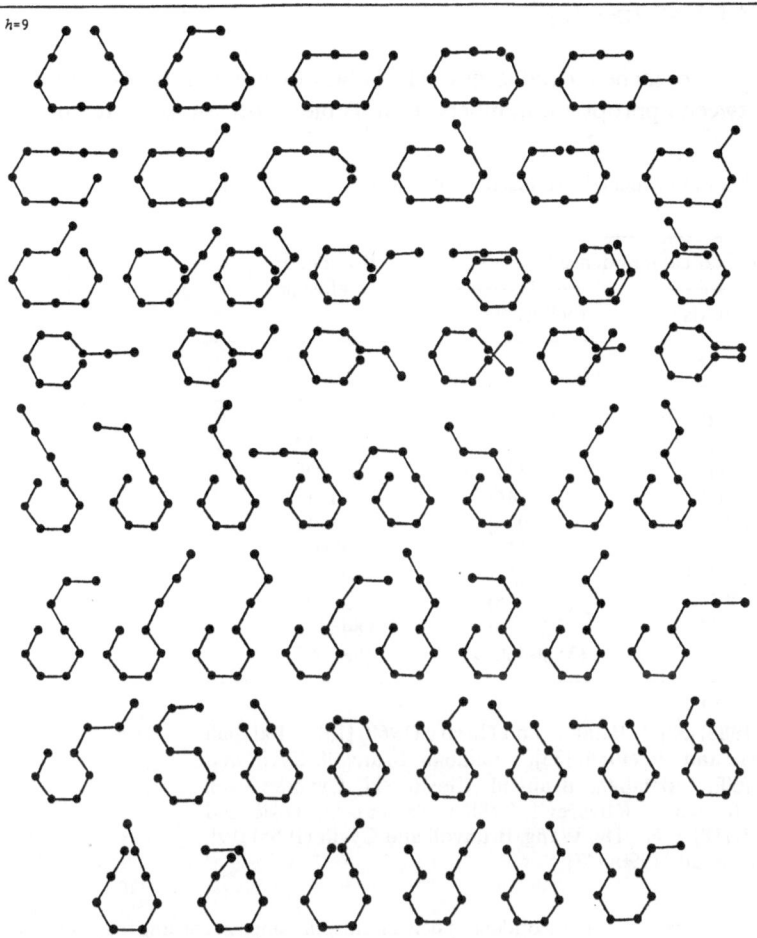

**Fig. 15.** All unbranched catacondensed helicenes with $h = 9$

the latter set with a mistake (see below). The total of 54 unbranched catacondensed helicenes with $h = 9$ are shown in Fig. 15.

*Errata*

Balaban and Harary [9] have a misprint in their number of unsymmetrical ($u$) unbranched catafusenes with $h = 6$, while their figure shows correctly all the 16 systems. Another minor misprint: in the depiction of unbranched catafusenes with $h = 7$ by Balaban and Harary [13] one point (representing a hexagon) is omitted.

In the dualist representation of the unsymmetrical unbranched catacondensed helicenes Brunvoll et al. [54] missed one system, while two of their systems are isomorphic.

## 6.3 Branched Catafusenes

The numbers of branched catafusenes (cf. Table 16) are consistent with the differences between appropriate numbers from Table 7 with supplements in text

**Table 16.** Numbers of branched catafusenes: benzenoids and helicenes (in parentheses)

| $h$ | Branched catacondensed | | Branched catafusenes |
|---|---|---|---|
| | benzenoids* | (helicenes) | |
| 4 | 1[a] | | 1[b] |
| 5 | 2[a] | | 2[b] |
| 6 | 12[a] | | 12[b] |
| 7 | 51[a] | (2) | 53[c] |
| 8 | 229[a] | (21) | 250[c] |
| 9 | 969[d] | (146) | 1115[e] |
| 10 | 4098[f] | (914) | 5012[e] |
| 11 | 16867[f] | (5165) | 22032[e] |
| 12 | 68925[e,g] | (27821) | 96746[e] |
| 13 | 278907[h] | (143825) | 422732 |
| 14 | 1123302[h] | (724826) | 1848128 |
| 15 | 4507640 | (3580450) | 8088090 |

* See also Table 17.
[a] He and He (1985) [32]; [b] Balaban and Harary (1968) [13]; [c] Balaban (1969) [30]; [d] He and He (1986) [33]; [e] Balaban, Brunvoll, Cyvin and Cyvin (1988) [60]; [f] Balaban, Brunvoll, Cioslowski, Cyvin, Cyvin, Gutman, He, He, Knop, Kovačević, Müller, Szymanski, Tošić and Trinajstić (1987) [18]; [g] He, He, Wang, Brunvoll and Cyvin (1988) [19]; [h] Cyvin and Brunvoll (1990) [47]

**Table 17.** Numbers of branched catacondensed benzenoids, classified according to symmetry

| $h$ | $D_{3h}$ | $C_{3h}$ | $D_{2h}$ | $C_{2h}$ | $C_{2v}$ | $C_s$ | Total* |
|---|---|---|---|---|---|---|---|
| 4 | 1[a] | | | | | | 1[b] |
| 5 | 0 | | | | 1[a] | 1[a] | 2[b] |
| 6 | 0 | | 1[a] | | 4[a] | 7[a] | 12[b] |
| 7 | 1[a] | 1[a] | 1[a] | | 4[a] | 44[a] | 51[b] |
| 8 | 0 | 0 | 1[a] | 4[a] | 18[a] | 206[a] | 229[b] |
| 9 | 0 | 0 | 1[a] | 4[a] | 27[a] | 937[a] | 969[c] |
| 10 | 2[a] | 4[a] | 3[a] | 25[a] | 67[a] | 3997[a] | 4098[a] |
| 11 | 0 | 0 | 4[a] | 26[a] | 118[a] | 16719[a] | 16867[a] |
| 12 | 0 | 0 | 4[a] | 132[a] | 269[a] | 68520[d] | 68925[d,e] |
| 13 | 2[a] | 15[a] | 4[a] | 140[a] | 507[f] | 278239[f] | 278907[f] |
| 14 | 0 | 0 | 9[a] | 620 | 1041 | 1121632 | 1123302[f] |
| 15 | 0 | 0 | 11[a] | 658 | 2096 | 4504875 | 4507640 |

* See also first column of Table 16.
[a] Brunvoll, Cyvin and Cyvin (1987) [71]; [b] He and He (1985) [32]; [c] He and He (1986) [33]; [d] Balaban, Brunvoll, Cyvin and Cyvin (1988) [60]; [e] He, He, Wang, Brunvoll and Cyvin (1988) [19]; [f] Cyvin and Brunvoll (1990) [47]

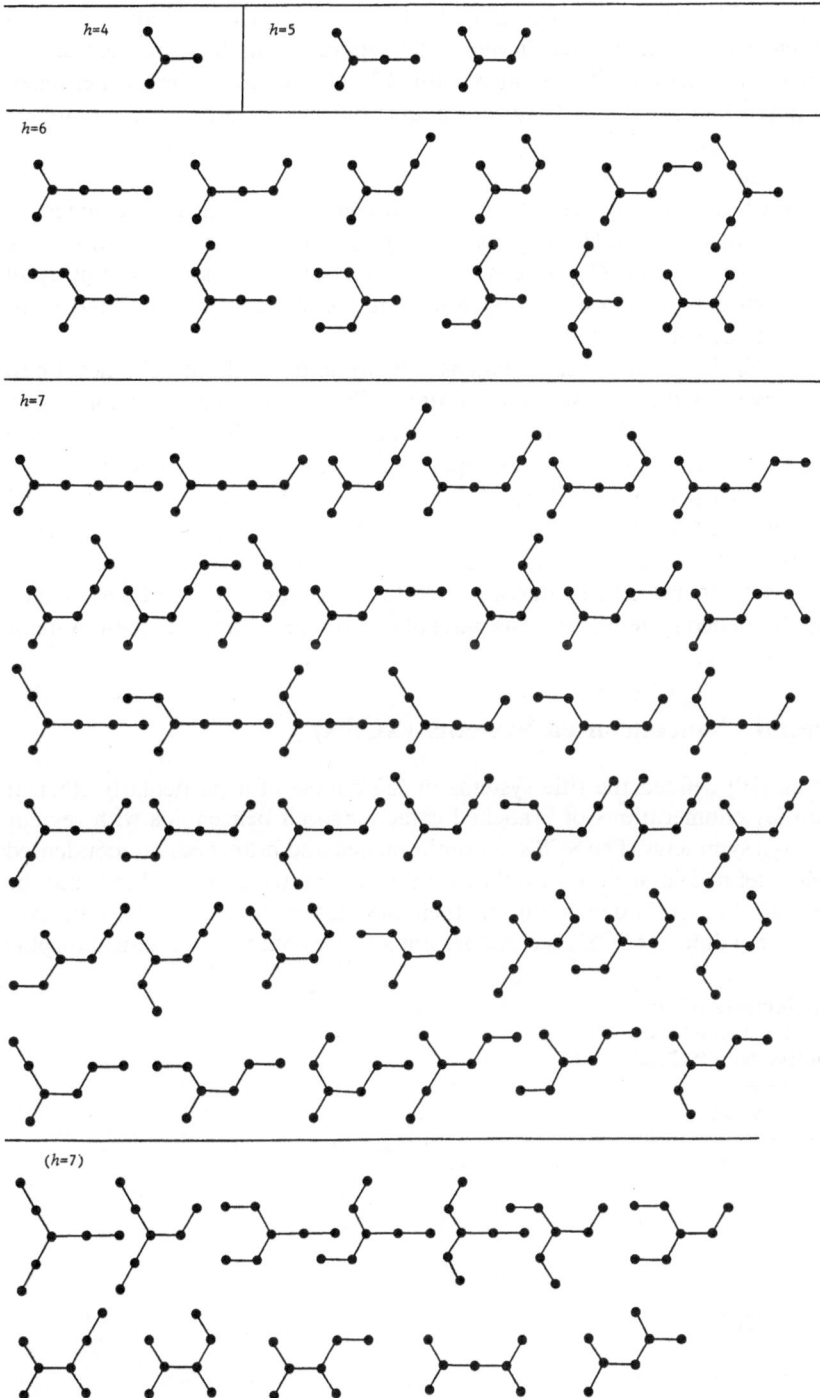

**Fig. 16.** All branched catacondensed benzenoids for $h \leq 7$

and Table 10. This class separates into the branched catacondensed benzenoids and branched catacondensed helicenes. The separate numbers are included in Table 16; for the benzenoids, see also Table 17. The numbers for the helicenes, given in parentheses in Table 16, are of course differences between the numbers of the two other columns therein. These numbers are consistent with the appropriate differences from Tables 8 and 14.

In Table 17, a detailed account on the numbers of branched catacondensed benzenoids is displayed, including the distribution into symmetry groups. Here the numbers for $D_{3h}$ and $C_{3h}$ at $h \leq 13$ are also obtainable from a scrutiny of figures published by Cyvin et al. [78]. Supplements to Table 17 are found in or from some of the subsequent tables.

In Fig. 16 all the branched catacondensed benzenoids (without helicenes) up to $h = 7$ are given in the dualist representation. They have been given for $h \leq 6$, with helicenes included, by Balaban and Harary [13] and by Balaban [73]. The forms for $h = 7$ are found in an other paper by Balaban [30]. The above mentioned works of Džonova-Jerman-Blažič and Trinajstić [74], by Gutman [75], Trinajstić et al. [76] and by El-Basil [77] display the figures of both branched and unbranched benzenoids, with $h \leq 6$, $h \leq 5$, $h = 7$ and $h \leq 5$, respectively.

The forms of the branched catacondensed helicenes for $h = 7$ and 8 are again, like the unbranched systems, found as parts of Fig. 6. The numbers of these systems are 2 and 21, respectively.

## 6.4 Special Catacondensed Systems (SCS's)

Tošić et al. [79] defined the title systems in the course of a particularly efficient algorithm for enumerations of branched catacondensed benzenoids with regular trigonal ($D_{3h}$) symmetry. The SCS's are (unbranched and branched) catacondensed benzenoids defined in such a way that isomorphic systems of this kind may be reckoned as "different", depending on their orientation with respect to an axis. The counting of different SCS's is therefore not a single counting of non-isomorphic

**Table 18.** Numbers of different special catacondensed (benzenoid) systems (SCS's)*

| $h$ | SCS's |
|---|---|
| 1 | 1 |
| 2 | 3 |
| 3 | 9 |
| 4 | 29 |
| 5 | 99 |
| 6 | 348 |
| 7 | 1260 |
| 8 | 4625 |

* From Tošić, Budimac, Brunvoll and Cyvin (1990) [79]

systems. For example, naphthalene ($h = 2$) gives rise to three different SCS's. The somewhat complicated definition is explained in the following. The symbols $L$ and $A$ denote a linearly and an angularly annelated hexagon, respectively.

(i) Start with the phenanthrene system, $LAL'$, which by convention is drawn from left to right.
(ii) Define a "horizon" by extending the two first hexagons ($LA$) infinitely to both sides into a linear chain of hexagons.

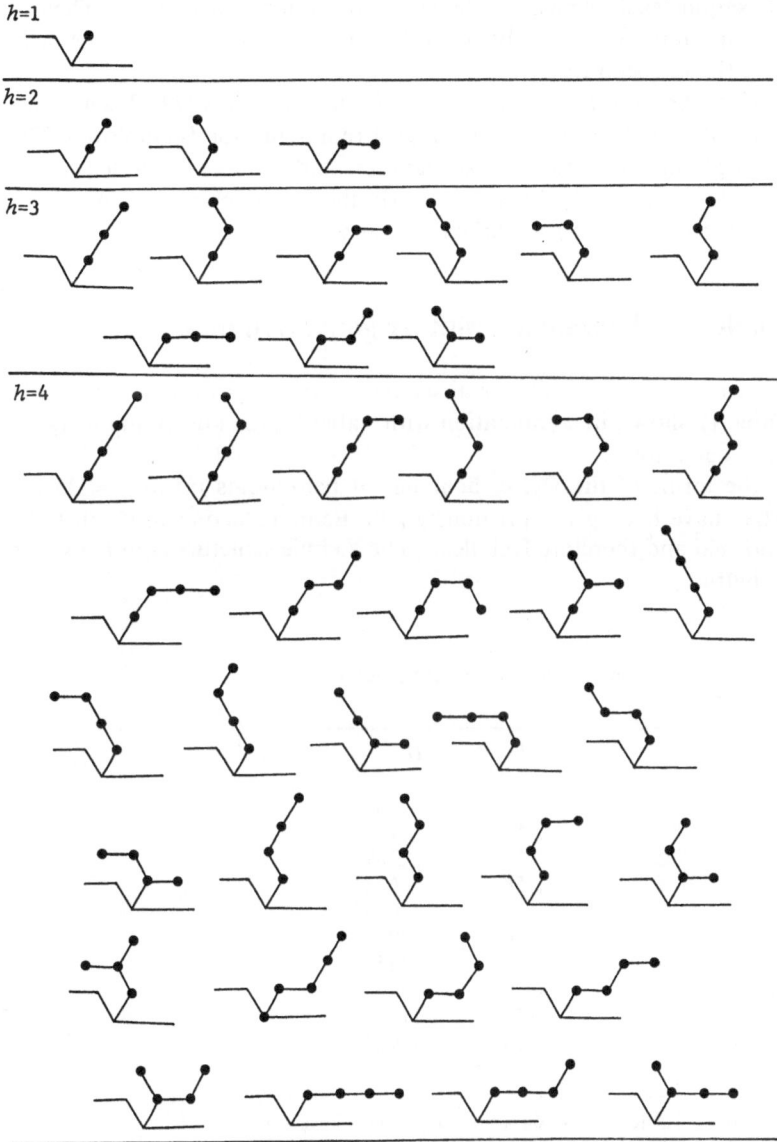

**Fig. 17.** All the different special catacondensed systems (SCS's) for $h \leq 4$

(iii) Add $h - 1$ hexagons to the last hexagon $(L')$ of the phenanthrene system so that a catacondensed benzenoid $C(LAL' \ldots)$ of $h + 2$ hexagon emerges, which should not have any hexagon on the horizon to the right of $A$ and should not come in contact with the horizon to the left of $A$. Otherwise this system is arbitrary $(h = 1, 2, 3, \ldots)$. It may occasionally be branched. The definition includes $h = 1$ as the degenerate case, where C is the original $LAL'$ system.

(iv) Delete the first two hexagons $(LA)$ to obtain an SCS.

Let $N$ be the number of all the non-isomorphic catacondensed benzenoids (C) with $h + 2$ hexagons each, which can be constructed according to (iii). Then $N$ is taken by definition to be the number of "different" SCS's with $h$ hexagons each, as obtained by the last step (iv).

Table 18 shows the results of enumeration of different SCS's [79]. Their forms for $h$ up to 4 are depicted in Fig. 17. The dualist representation is employed. The horizon to the right of $A$ is drawn as a straight line, indicating a forbidden region (on and below this line) for the SCS's. Similarly the forbidden region to the left of $A$ is indicated by a straight line shifted one step up.

## 6.5 Catacondensed Benzenoids with Trigonal Symmetry

A catacondensed benzenoid with trigonal symmetry, viz. $D_{3h}$ or $C_{3h}$, is necessarily branched. Table 19 shows, in combination with Table 17, the known numbers for these classes of benzenoids.

In Fig. 18 the forms of the $D_{3h}$ catacondensed benzenoids up to $h = 25$ are displayed. They have been given previously [79]. Being catacondensed, all these systems are normal and therefore Kekuléan. The Kekulé structure counts $(K)$ are given in the figure.

**Table 19.** Numbers of catacondensed (branched) benzenoids of trigonal symmetry*

| $h$ | $D_{3h}$ | $C_{3h}$ | $h$ | $D_{3h}$ |
|-----|----------|----------|-----|----------|
| 16 | 4 | 55 | 49 | 893 |
| 19 | 5 | 203 | 52 | 1876 |
| 22 | 9 | 755 | 55 | 2899 |
| 25 | 12 | 2855 | 58 | 6140 |
| 28 | 24 | † | 61 | 9630 |
| 31 | 32 | † | 64 | 20563 |
| 34 | 65 | † | 67 | 32565 |
| 37 | 94 | † | 70 | 69741 |
| 40 | 191 | † | 73 | 111626 |
| 43 | 283 | † | 76 | 239831 |
| 46 | 588 | † | | |

* In continuation of Table 17. All data for $D_{3h}$ from: Tošić, Budimac, Brunvoll and Cyvin (1990) [79];
† Unknown

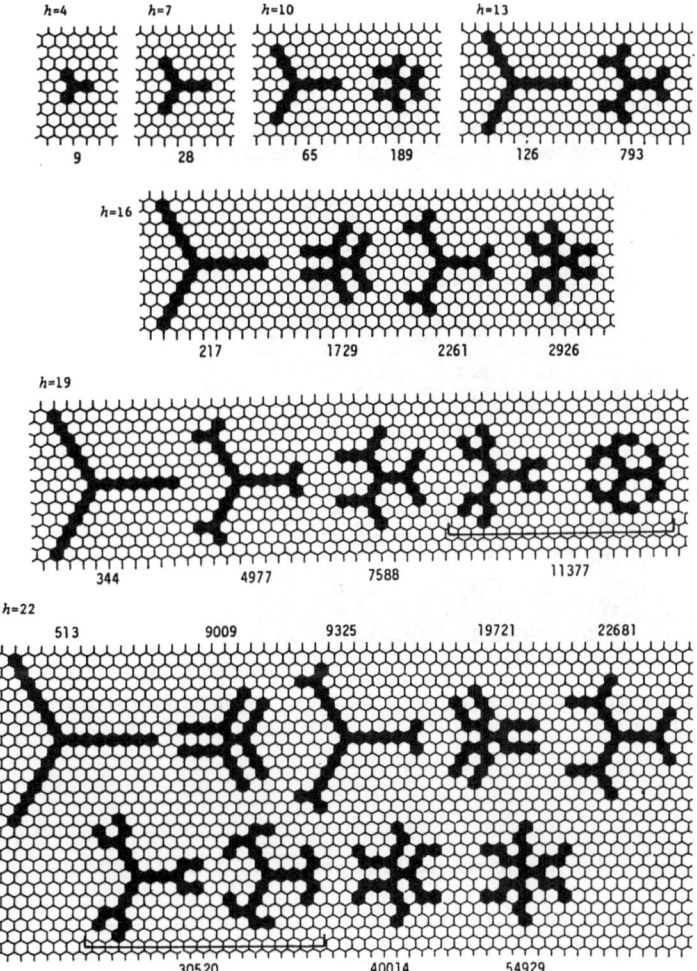

**Fig. 18.** (see next page)

## 6.6 Catacondensed Benzenoids with Dihedral Symmetry and Centrosymmetry

A catacondensed benzenoid with dihedral symmetry, viz. $D_{2h}$, is either a branched system or an (unbranched) linear acene. A centrosymmetrical ($C_{2h}$) catacondensed benzenoid is either branched or unbranched. The $D_{2h}$ systems under consideration have been enumerated by the efficient algorithm invoking SCS's (cf. Sect. 6.4) [80]. Table 20, in combination with Table 17, shows the known numbers for the branched catacondensed $D_{2h}$ and $C_{2h}$ benzenoids. The numbers of unbranched catacondensed benzenoids with $C_{2h}$ symmetry are found under the designation $c'$ in Tables 14 and 15 for $h \le 20$ and $21 \le h \le 30$, respectively.

Figure 19 displays the forms of the $D_{2h}$ catacondensed benzenoids up to $h = 17$.

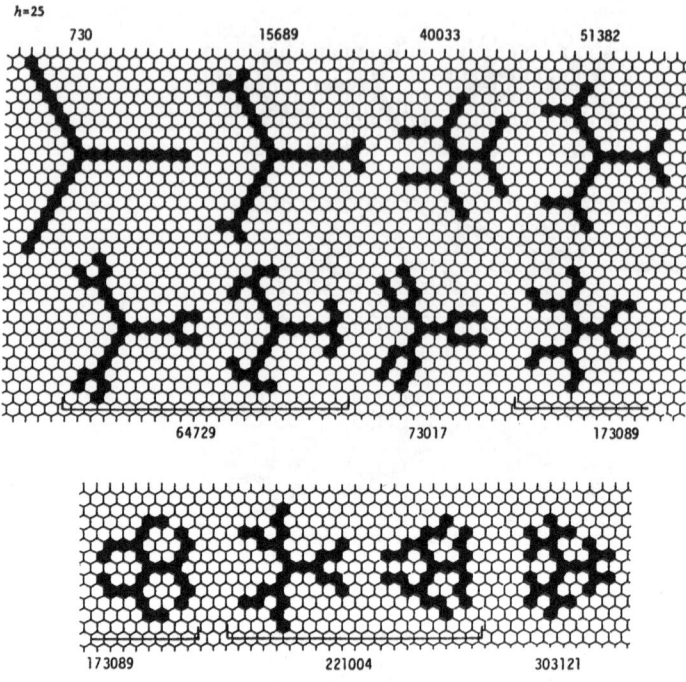

h=25

730    15689    40033    51382

64729    73017    173089

173089    221004    303121

**Fig. 18.** All (branched) catacondensed benzenoids with regular trigonal ($D_{3h}$) symmetry and $h < 28$. $K$ numbers are given

**Table 20.** Numbers of branched catacondensed benzenoids with dihedral symmetry and centrosymmetry*

| h | $D_{2h}$ | $C_{2h}$ | h | $D_{2h}$ | h | $D_{2h}$ |
|---|---|---|---|---|---|---|
| 16 | 12 | 2762 | 29 | 406 | 42 | 33127 |
| 17 | 13 | 2935 | 30 | 820 | 43 | 44702 |
| 18 | 26 | 11890 | 31 | 1074 | 44 | 50339 |
| 19 | 33 | 12640 | 32 | 1205 | 45 | 59247 |
| 20 | 36 | † | 33 | 1376 | 46 | 117023 |
| 21 | 39 | † | 34 | 2763 | 47 | 159249 |
| 22 | 80 | † | 35 | 3653 | 48 | 178938 |
| 23 | 102 | † | 36 | 4118 | 49 | 212451 |
| 24 | 112 | † | 37 | 4745 | 50 | 417164 |
| 25 | 124 | † | 38 | 9487 | 51 | † |
| 26 | 251 | † | 39 | 12686 | 52 | 640877 |
| 27 | 325 | † | 40 | 14298 | | |
| 28 | 364 | † | 41 | 16672 | | |

* In continuation of Table 17. All data for $D_{2h}$ from: Tošić, Budimac, Cyvin and Brunvoll (1990) [80];
† Unknown

## 6.7 Unbranched Catacondensed Benzenoids with Equidistant Segments

The title classes are the non-helicenic fibonacenes and non-helicenic generalized fibonacenes. (cf. Sect. 5.3).

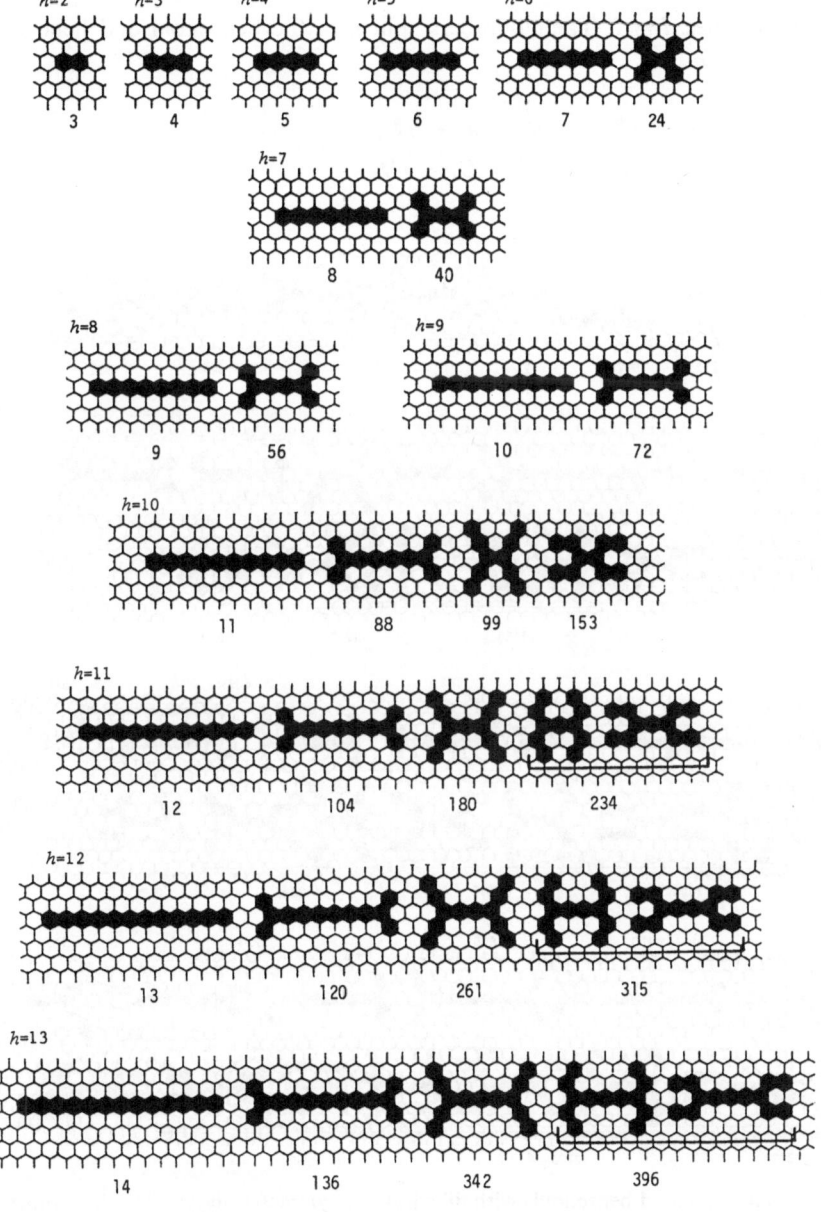

**Fig. 19.** (see next page)

Björg N. Cyvin, Jon Brunvoll, and Sven J. Cyvin

Let a benzenoid of the considered class consist of $S$ segments, each of length $l$. The length is, by definition, the number of hexagons in the linear chain between two angularly annelated ($A$) hexagons or one terminal and one $A$ hexagon, the two end hexagons included. The total number of hexagons is

$$h = 1 + (l - 1) S \qquad (S > 1) \tag{52}$$

For fibonacenes, $l = 2$. Their numbers split into the benzenoid and helicenic systems; we write

$$
\begin{aligned}
c^f &= c_f' + c_f^*, & m^f &= m_f' + m_f^*, \\
u^f &= u_f' + u_f^*, & U^f &= U_f' + U_f^*
\end{aligned}
\tag{53}
$$

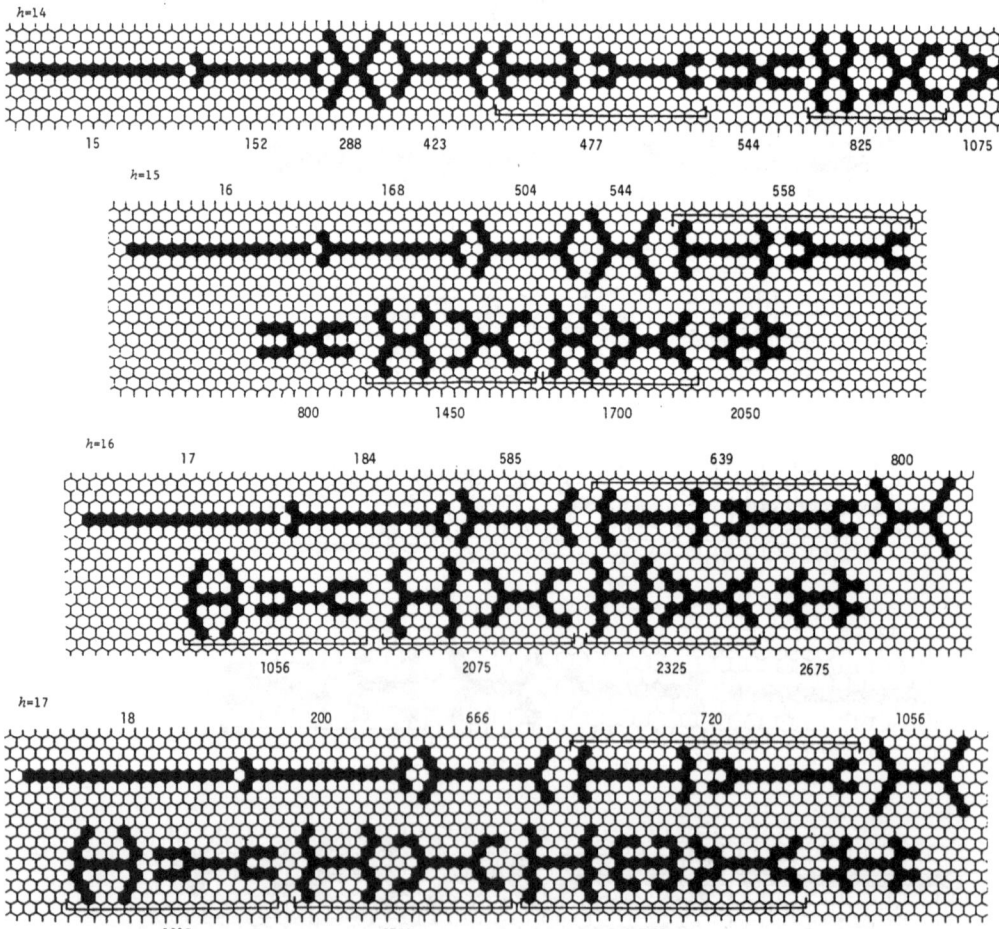

**Fig. 19.** All catacondensed benzenoids with dihedral $D_{2h}$ symmetry and $h \leq 17$. $K$ numbers are given

118

**Table 21.** Numbers of fibonacenes: benzenoids and helicenes (in parentheses)

| $h$ | $c_f'(c_f^*)$ | $m_f'(m_f^*)$ | $u_f'(u_f^*)$ | Total: $U_f'(U_f^*)$ |
|---|---|---|---|---|
| 3 | | 1[a] | | 1[a] |
| 4 | 1[a] | 1[a] | | 2[a] |
| 5 | 0 | 2[a] | 1[a] | 3[a] |
| 6 | 2[a] | 1[a](1) | 2[a] | 5[a](1) |
| 7 | 0 | 3[a](1) | 5[a](1) | 8[a](2) |
| 8 | 4[a] | 2[a](2) | 9[a](3) | 15[a](5) |
| 9 | 0 | 6[a](2) | 19[a](9) | 25[a](11) |
| 10 | 7[a](1) | 4[a](4) | 35[a](21) | 46[a](26) |
| 11 | 0 | 11 (5) | 69 (51) | 80 (56) |
| 12 | 13 (3) | 7 (9) | 125 (115) | 145 (127) |
| 13 | 0 | 18 (14) | 238 (258) | 256 (272) |
| 14 | 24 (8) | 12 (20) | 430 (562) | 466 (590) |
| 15 | 0 | 33 (31) | 800 (1216) | 833 (1247) |
| 16 | 44 (20) | 22 (42) | 1447 (2585) | 1513 (2647) |
| 17 | 0 | 58 (70) | 2662 (5466) | 2720 (5536) |
| 18 | 81 (47) | 36 (92) | 4808 (11448) | 4925 (11587) |
| 19 | 0 | 102 (154) | 8779 (23861) | 8881 (24015) |
| 20 | 147 (109) | 68 (188) | 15848 (49432) | 16063 (49729) |
| 21 | 0 | 183 (329) | 28813 (102003) | 28996 (102332) |

[a] Balaban (1989) [62]

**Table 22.** Numbers of generalized fibonacenes with $l > 2$: benzenoids and helicenes (in parentheses)

| $S^+$ | $c_w'(c_w^*)$ | $m_w'(m_w^*)$ | $u_w'(u_w^*)$ | Total: $U_w'(U_w^*)$ |
|---|---|---|---|---|
| 2 | | 1 | | 1 |
| 3 | 1 | 1 | | 2 |
| 4 | 0 | 2 | 1 | 3 |
| 5 | 2 | 2 | 2 | 6 |
| 6 | 0 | 3(1) | 6 | 9(1) |
| 7 | 4 | 3(1) | 11(1) | 18(2) |
| 8 | 0 | 6(2) | 25(3) | 31(5) |
| 9 | 8 | 6(2) | 47(9) | 61(11) |
| 10 | 0 | 11(5) | 96(24) | 107(29) |
| 11 | 15(1) | 11(5) | 181(59) | 207(65) |
| 12 | 0 | 20(12) | 358(138) | 378(150) |
| 13 | 29(3) | 21(11) | 674(318) | 724(332) |
| 14 | 0 | 36(28) | 1297(719) | 1333(747) |
| 15 | 56(8) | 36(28) | 2445(1587) | 2537(1623) |
| 16 | 0 | 68(60) | 4655(3473) | 4723(3533) |
| 17 | 106(22) | 70(58) | 8762(7494) | 8938(7574) |
| 18 | 0 | 123(133) | 16551(16089) | 16674(16222) |
| 19 | 201(55) | 125(131) | 31129(34151) | 31455(34337) |

[+] For the $h$ value, see Eq. (52)

Björg N. Cyvin, Jon Brunvoll, and Sven J. Cyvin

Then, of course,

$$U'_f = c'_f + m'_f + u'_f, \qquad U^*_f = c^*_f + m^*_f + u^*_f. \tag{54}$$

Table 21 shows the numbers for all of these classes when $h \leq 21$.

For the generalized fibonacenes with $l > 2$ we write, in analogy with the above notation,

$$\begin{aligned} c^f &= c'_w + c^*_w, & m^f &= m'_w + m^*_w, \\ u^f &= u'_w + u^*_w, & U^f &= U'_w + U^*_w \end{aligned} \tag{55}$$

and then:

$$U'_w = c'_w + m'_w + u'_w, \qquad U^*_w = c^*_w + m^*_w + u^*_w. \tag{56}$$

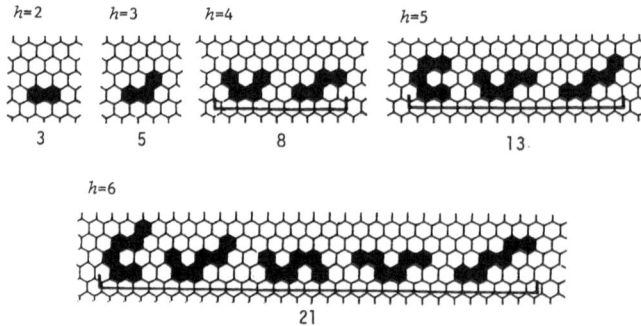

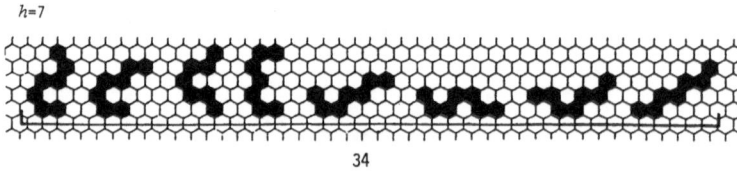

**Fig. 20.** All unbranched catacondensed benzenoids with 2-segments only (non-helicenic fibonacenes) for $h \leq 8$. $K$ numbers are given; they are the Fibonacci numbers $F_{h+1}(F_0 = F_1 = 1)$

120

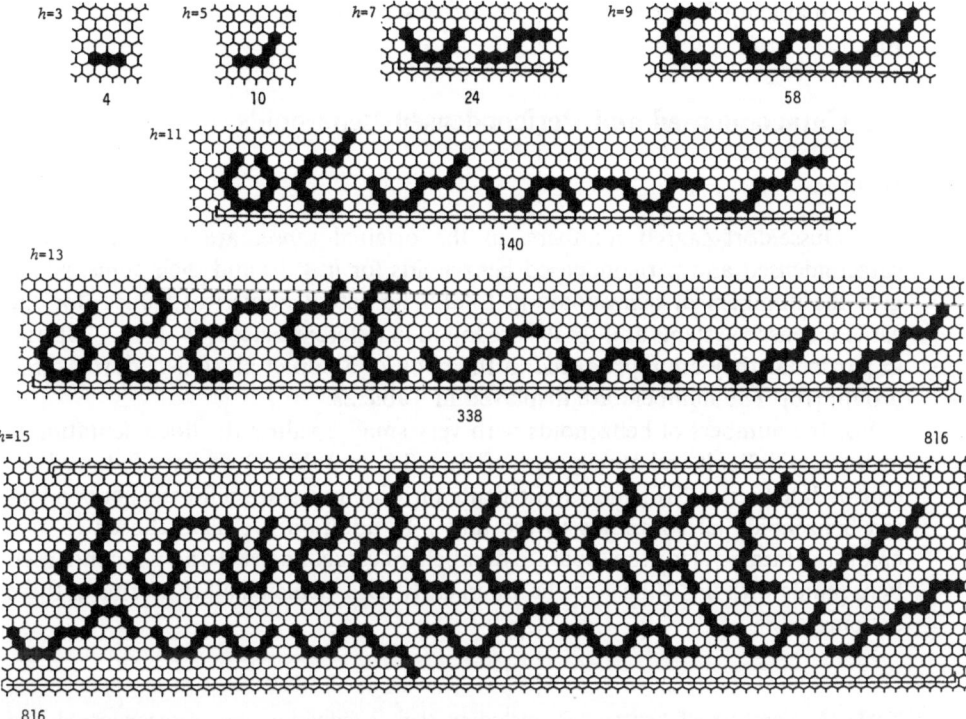

**Fig. 21.** All unbranched catacondensed benzenoids with 3-segments only (non-helicenic generalized fibonacenes with $l = 3$) for $h < 17$. $K$ numbers are given

Numerical values for $S \leq 19$ are shown in Table 22. Notice that these numbers are independent of $l$, if only $l > 2$.

All the numbers in Tables 21 and 22 are consistent with Table 11, which can be reproduced from either of these two tables by additions of appropriate numbers therein.

Figures 20 and 21 display the forms of the unbranched catacondensed benzenoids with only 2-sements or 3-segments, respectively, up to the systems with 7 segments.

# 7 Coarse Classifications of Benzenoids

## 7.1 Specification

In view of the discussions in Sect. 3 (cf. especially the concluding remarks of Sect. 3.5) the reader will certainly excuse us for repeating the specification of the class of polyhexes to be treated in the remainder of this chapter.

Only benzenoid systems are considered. According to the adopted definition they are the planar, simply connected polyhexes. Consequently all circulenes (including coronoids, multiple coronoids and polycirculenes) are excluded, and all helicenic systems (helicenes, helicirculenes) are also excluded.

121

The main subclasses of benzenoids are described in Sect. 4 and summarized in Fig. 8.

## 7.2 Catacondensed and Pericondensed Benzenoids

*Numbers*

The Düsseldorf-Zagreb numbers, in the original sense, are the numbers of catacondensed and pericondensed benzenoids for $h \leq 10$ and their sums, which give the numbers of benzenoids in total. The actual numerical values (for the catacondensed systems and totals) are found in different places [15, 17, 53, 76], and in particular they are reproduced in the book of Knop et al. (Düsseldorf-Zagreb group) [16]. The numbers are displayed in Table 23.

For the numbers of benzenoids with very small $h$ values the documentation is again very difficult and ambiguous; cf. Sect. 3.5. In Table 23 we have followed the consolidated report [18] and supplemented it with a quotation for the pericondensed benzenoids (cf. footnotes to the table). In addition, we have chosen to give credit to Harary [2] for the separate numbers of catacondensed and pericondensed benzenoids with $h \leq 4$ because he, at an early date, depicted all the forms of these systems.

**Table 23.** Numbers of benzenoids, including their subdivision into catacondensed and pericondensed systems

| $h$ | Catacondensed | Pericondensed | Total |
|---|---|---|---|
| 1 | 1[a,b] | | 1[c] |
| 2 | 1[a,b] | | 1[c] |
| 3 | 2[a,b] | 1[a,b] | 3[c] |
| 4 | 5[a,b] | 2[a,b] | 7[c] |
| 5 | 12[b] | 10[b] | 22[c] |
| 6 | 36[d] | 45[b] | 81[d] |
| 7 | 118[d] | 213[d] | 331[d] |
| 8 | 411[d] | 1024[d] | 1435[d] |
| 9 | 1489[d] | 5016[d] | 6505[d] |
| 10 | 5572[d] | 24514[d] | 30086[d] |
| 11 | 21115[e] | 120114[e] | 141229[e] |
| 12 | 81121[f,g] | 588463[g] | 669584[g] |
| 13 | 314075[h] | 2884181 | 3198256[i−m] |
| 14 | 1224528[h] | 14143049 | 15367577[j−m] |
| 15 | 4799205 | 69408705 | 74207910[k,l,m] |
| 16 | † | † | 359863778[m] |

[a] Harary (1967) [2]; [b] Balaban and Harary (1968) [13]; [c] Klarner (1965) [7]; [d] Knop, Szymanski, Jeričević and Trinajstić (1983) [15]; [e] Stojmenović, Tošić and Doroslovački (1986) [46]; [f] Balaban, Brunvoll, Cyvin and Cyvin (1988) [60]; [g] He, He, Wang, Brunvoll and Cyvin (1988) [19]; [h] Cyvin and Brunvoll (1990) [47]; [i] Müller, Szymanski, Knop, Nikolić and Trinajstić (1989) [48]; [j] Müller, Szymanski, Knop, Nikolić and Trinajstić (1990) [36]; [k] Nikolić, Trinajstić, Knop, Müller und Szymanski (1990) [49]; [l] Knop, Müller, Szymanski and Trinajstić (1990) [50]; [m] Knop, Müller, Szymanski and Trinajstić (1990) [37]; † Unknown

The consolidated report [18] summarizes the pertinent data (of Table 23) up to $h = 11$. The information for $h = 12$ is found in the supplements [19] to this report.

With regard to the totals for $h > 12$ they were all produced by the Düsseldorf-Zagreb group; specifically with references: $h = 13$ [36, 37, 48–50], $h = 14$ [36, 37, 49, 50]; $h = 15$ [37, 49, 50], and finally $h = 16$ [50].

*Forms*

The depictions of Harary [2] are mentioned above. The forms of benzenoid systems with given numbers of hexagons have later been depicted many times in different contexts. The altogether 5 systems for $h \leq 3$ are, for instance, found in Harary and Palmer [4]. At the start of a long series of interesting papers on topological properties of benzenoid systems Gutman [81] depicted the 3 benzenoid systems with $h = 3$. Three other parts of this series are cited above [11, 27, 75]. In another part [82] the pericondensed benzenoids are treated, but the listings are complete only for the 3 systems with $h \leq 4$. The altogether 12 benzenoids with $h \leq 4$ have otherwise been depicted in reviews [27, 83], which followed Harary [2], and elsewhere [56]. Some authors [84, 85] omitted pictures of the smallest of these systems. The last reference [85] pertains to a recent work on the ordering of chemical graphs by Kirby, who employed efficiently the depictions by Knop et al. [16] (see below).

The forms of the benzenoids for $h = 5$ or up to $h = 5$ are found in different places [13, 86–88]. Also the forms for $h = 6$ or up to $h = 6$ have been depicted several times [14, 15, 26, 89–91].

Apart from the Düsseldorf–Zagreb group [15, 76] some computer-generated figures in the form of mini-hexagons have been published by Brunvoll et al. [71]. But we must not omit to mention the same kind of depictions which occupy a substantial part of the book by Knop et al. [16]. This material deserves supplementary treatment, which is given in the following.

The mentioned monograph [16] from the Düsseldorf-Zagreb group contains 8386 miniatures, which display the forms of all benzenoids with $h \leq 9$. The pictures are ordered according to the number of internal vertices $(n_i)$. Hence the catacondensed systems $(n_i = 0)$ are sorted out automatically. Furthermore, each picture is supplied with the Kekulé structure count, $K$, occasionally $K = 0$ for the non-Kekuléans. The corresponding mammoth listing of the 30086 benzenoids with $h = 10$ has also been produced, as reported by Knop et al. [17].

We have no illusions that the above survey of benzenoid forms depicted in the literature is complete. To take an example, the really classical work of Pólya [92] from 1936 on algebraic computations of isomers of organic compounds is not mentioned above. Nevertheless it contains pictures of a few benzenoid hydrocarbons.

*Comments and Errata*

By virtue of the impact of the Düsseldorf-Zagreb school the enumeration of polyhexes is often associated just with the numbers of Table 23 (benzenoids, subdivided into catacondensed and pericondensed systems). It is not to be denied that there has been a tendency to competition, trying to overbid the Düsseldorf-Zagreb numbers in a kind of a race.

The record of $h = 10$ achieved by the Düsseldorf-Zagreb group in 1983 stood for three years when it was beaten by Novi Sad; in 1986 Stojmenović et al. [46] published their results for $h = 11$.

For benzenoids with $h = 12$ the first data were published in 1988 as a result of a collaboration between He & He and Trondheim (cf. Sect. 2.8 for a listing of the research centres). The number $C'_{12}$ (for catacondensed benzenoids with $h = 12$) became available after the enumeration of the branched catacondensed $h = 12$ systems in Trondheim, while He & He succeeded in a complete enumeration of benzenoids with this number of hexagons. Consistent $h = 12$ numbers were reported [69] as private communications from He & He and from Cioslowski.

On the other hand, the $C'_{13}$ number of Cioslowski as quoted by Aboav and Gutman [69] in 1988 did not agree with the results of He & He communicated privately to us the same year. The controversy was resolved by an independent computation in Trondheim [47], where also the complete $h = 12$ results were reconfirmed. It was concluded that the He & He result for $C'_{13}$ was correct, while the Cioslowski number looks like a misprint.

The cited work of Cyvin and Brunvoll [47] includes $C'_{14}$. The chemical formula for a catacondensed $h = 14$ benzenoid is given erroneously therein; it should be $C_{58}H_{32}$. The number $C'_{15}$ is a present result.

Eventually the Düsseldorf–Zagreb group took up the challenge, as is documented by the footnotes of Table 23. Trinajstić was able to present the numbers of total benzenoids, not only for $h = 12$, but also for $h = 13$, 14 and 15 at a conference (Galveston, Texas) in 1989 [49], where he was congratulated by Cyvin. A citation from the proceedings of the conference [93]: "The data for $h = 13$ and $h = 14$ ... were reported for the first time (together with the results for $h = 15$) by Trinajstić as an achievement of the Düsseldorf–Zagreb group. Congratulations!" To be precise, the Düsseldorf–Zagreb group communicated the numbers up to $h = 13$ [48] shortly before the mentioned conference.

The Düsseldorf–Zagreb group pursued their success by a computation for $h = 16$ [37], which took 91 days, 7 hours, 24 minutes and 33.69 seconds of computer (CPU) time.

The computations for $h > 12$ became feasible by an exploitation of a code named DAST (dualist angle-restricted directional information) [31, 48]. It is no doubt that the Düsseldorf-Zagreb group has regained the hegemony in benzenoid enumerations.

In the following we point out specifically some errors, apparently all of them misprints. They may seem to be a trifle, but should nevertheless be treated seriously. In the paper of Aboav and Gutman [69] the number of $C'_{13}$ (from Cioslowski) is claimed to be wrong [47]. The same (wrong) number was quoted in a table of the monograph by Gutman and Cyvin [22], but in parentheses as uncertain. The parenthesized number for pericondensed benzenoids with $h = 13$ is also wrong. Furthermore, in the same table, there is a misprint in $C'_8$ and in the total for $h = 15$, the last number therein.

## 7.3 Kekuléan and non-Kekuléan Benzenoids: the "neo" Classification

The neo classification divides all benzenoids into normal (n), essentially disconnected (e) and non-Kekuléans (o), where the n and e systems cover all the Kekuléans. Cyvin and Gutman [26] have advocated for this classification by saying: "From the point of view of the enumeration of Kekulé structures the classification . . . [neo] . . . seems to be a rather appropriate one [94, 87]". However, the distinction between Kekuléan (closed-shell, non-radicalic) and non-Kekuléan (radicalic) benzenoid hydrocarbons was made long before the explicit definition of the neo classification. This practice started with the first (substantial) enumeration of benzenoids in the chemical context by Balaban and Harary [13].

Table 24 shows the known numbers of benzenoids belonging to the different classes of neo and at the same time the numbers of total Kekuléan (n + e) and non-Kekuléan (o) systems. In the documentations of this table (cf. the footnotes) we have taken into account that the Kekuléan and non-Kekuléan benzenoids can be counted separately for $h \leq 9$ from the figures in the book of Knop et al. [16]. That has actually been done, as it was reported [54]. As to the documentations for higher $h$ values the literature needs to be supplemented by two references [95, 96].

**Table 24.** Numbers of benzenoids according to the neo classification*

| | Kekuléan | | | non-Kekuléan |
|---|---|---|---|---|
| $h$ | n | e | Total Kek. | o |
| 1 | 1 [a] | | 1 [a] | |
| 2 | 1 [a] | | 1 [a] | |
| 3 | 2 [a] | | 2 [a] | 1 [a] |
| 4 | 6 [a] | | 6 [a] | 1 [a] |
| 5 | 14 [b-d] | 1 [c,d] | 15 [a] | 7 [a] |
| 6 | 48 [b-d] | 3 [c,d] | 51 [e,f] | 30 [a] |
| 7 | 167 [b,c] | 23 [c] | 190 [e,f] | 141 [e,f] |
| 8 | 643 [c] | 121 [c] | 764 [e,f] | 671 [e,f] |
| 9 | 2531 [c] | 692 [c] | 3223 [f] | 3282 [f] |
| 10 | 10375 [g] | 3732 [h,i] | 14107 [h,i] | 15979 [i] |
| 11 | 42919 [g] | 19960 [h,i] | 62879 [h,i] | 78350 [i] |
| 12 | 180205 | 104713 [k] | 284918 [j] | 384666 [j] |
| 13 | 761599 | 543262 | 1304861 | 1893395 |
| 14 | 3241584 | 2790058 | 6031642 | 9335935 |

* Abbreviations: e essentially disconnected; n normal; o non-Kekuléan.
[a] Balaban and Harary (1968) [13]; [b] Cyvin (1986) [94]; [c] Cyvin, Brunvoll, Cyvin and Gutman (1986) [95]; [d] Cyvin and Gutman (1986) [88]; [e] He and He (1985) [32]; [f] Knop, Müller, Szymanski and Trinajstić (1985) [16]; [g] Cyvin, Brunvoll and Cyvin (1986) [68]; [h] Brunvoll, Cyvin and Cyvin (1987) [71]; [i] Balaban, Brunvoll, Cioslowski, Cyvin, Cyvin, Gutman, He, He, Knop, Kovačević, Müller, Szymanski, Tošić and Trinajstić (1987) [18]; [j] He, He, Wang, Brunvoll and Cyvin (1988) [19]; [k] Brunvoll, Cyvin, Cyvin and Gutman (1989) [96]

A substantial amount of additional enumeration data for normal benzenoids and some data for essentially disconnected benzenoids are available, but shall not be reproduced here. They were produced in the course of the extensive studies of the distribution of $K$, the Kekulé structure count.

These studies of normal benzenoids started with an account by Cyvin [94] on the distribution of $K$ for normal benzenoids up to $h = 7$ in the form of curves. In the same work the enumeration of all normal benzenoids with $K \leq 9$ is reported and illustrated by figures of the 16 systems in question. Here the upper limit for $K$ is equal to the maximum ($K_{max}$) for $h = 4$. The distribution of $K$ for $h = 8$ and $h = 9$ followed [95]. Next the enumerations were extended to $K \leq 24$ ($K_{max}$ for $h = 6$) with illustrations for $K \leq 14$ ($K_{max}$ for $h = 5$) [88]. The distribution of $K$ for $h = 10$ was given graphically [68] and by numerical values [26]; here also the depictions are extended to $K \leq 24$. The studies culminated by a master review of the enumerations of normal benzenoids with supplements up to $K \leq 110$ ($K_{max}$ for $h = 9$) [55]. Figures of all these systems for $K \leq 30$ are found therein. A summary of the distributions of $K$ is under way, with supplements up to $h = 11$ [97]. In this work, and elsewhere [59], computer-generated curves for such distributions are presented for the first time.

Parallel with the studies of normal benzenoids described above the distributions of $K$ for essentially disconnected benzenoids were treated: for $h \leq 9$ [95], $h \leq 10$ [26], $h \leq 11$ [98] and $h = 12$ [96].

Hosoya and Yamaguchi [99] published the sextet polynomials systematically for all Kekuléan benzenoids with $h \leq 5$. This material was supplemented up to $h = 6$ by Ohkami and Hosoya [100].

*Comments and Errata*

As far back as 1968, Balaban and Harary [13] were aware of the unique position of zethrene, which was placed in a class of its own under the pericondensed benzenoids with $h = 6$. It is an essentially disconnected benzenoid. However, these authors did not sort out the other two essentially disconnected benzenoids (annelated perylenes) with the same numbers of hexagons. Neither did they sort out perylene itself, which is the unique essentially disconnected benzenoid with $h = 5$. In the table we are referring to [13], the entry for the classified pericondensed system with $h = 3$ is misplaced.

Knop et al. [91] published a list of the numbers of Kekuléan benzenoids for $h \leq 9$ with a misprint in the number for $h = 9$. Brunvoll et al. [101] reported a wrong number (with reference to private communication from He and He) for the Kekuléan benzenoids with $h = 12$. The curve of the distribution of $K$ for essentially disconnected benzenoids with $h = 9$ [26, 95] in imperfect.

## 7.4 Color Excess

A classification of the benzenoids according to the color excess ($\Delta$) sorts out the obvious non-Kekuléans ($\Delta > 0$) and the systems with vanishing color excess

**Table 25.** Numbers of benzenoids classified according to their Δ values*

| h | Color excess (Δ) | | | | | |
|---|---|---|---|---|---|---|
| | 0 | 1 | 2 | 3 | 4 | 5 |
| 1 | 1[a] | | | | | |
| 2 | 1[a] | | | | | |
| 3 | 2[a] | 1[b] | | | | |
| 4 | 6[a] | 1[b] | | | | |
| 5 | 15[a] | 7[b] | | | | |
| 6 | 51[c,d] | 28[b] | 2[b] | | | |
| 7 | 190[c,d] | 134[b] | 7[b] | | | |
| 8 | 764[c,d] | 619[b] | 52[b] | | | |
| 9 | 3223[d] | 2957[e] | 322[e] | 3[e] | | |
| 10 | 14107[e,f] | 14024[e] | 1916[e] | 39[e] | | |
| 11 | 62887[e] | 67046[e] | 10922[e] | 374[e] | | |
| 12 | 285016[g,h] | 320859[g-i] | 60705[g-i] | 2990[g-i] | 14[e] | |
| 13 | 1305958 | 1540174 | 330238 | 21675[g-i] | 211[g-i] | |
| 14 | 6041446 | 7408410 | 1769625 | 145508 | 2588[g-i] | |
| 15 | † | † | † | † | † | 48[g-i] |

* Concealed non-Kekuléans (Δ = 0, K = 0) occur at h ≥ 11.
[a] Balaban and Harary (1968) [13], [b] Brunvoll, Cyvin and Cyvin (1987) [54]; [c] He and He (1985) [32], [d] Knop, Müller, Szymanski and Trinajstić (1985) [16]; [e] Balaban, Brunvoll, Cioslowski, Cyvin, Cyvin, Gutman, He, He, Knop, Kovačević, Müller, Szymanski, Tošić and Trinajstić (1987) [18]; [f] Brunvoll, Cyvin and Cyvin (1987) [71]; [g] He, He, Wang, Brunvoll and Cyvin (1988) [19]; [h] Brunvoll, Cyvin, Cyvin and Gutman (1988) [101]; [i] Gutman and Cyvin (1988) [102]; † Unknown

(Δ = 0). The latter class consists of the Kekuléans (K > 0) and concealed non-Kekuléans (Δ = 0, K = 0).

A summary of the known data is displayed in Table 25. The key reference to their generation is Brunvoll et al. [101]. Additional computations by Brunvoll, of relevance to these systems, are quoted in Gutman and Cyvin [102].

The benzenoid systems with Δ = 0 coincide with the Kekuléans for h ≤ 10. For h ≥ 11 the concealed non-Kekuléans must be added.

## 7.5 Symmetry

Balaban and Harary [13] distinguished between the centrosymmetrical and mirror-symmetrical unbranched catafusenes. Also in their classical enumerations of polyhexes Harary and Read [9], as well as Lunnon [10], exploited symmetry. The familiar symmetry group designations for benzenoids were employed, perhaps for the first time by Rouvray [56, 86]. Cyvin and Gutman [83] specified explicitly the eight possible symmetries for benzenoids, viz. $D_{6h}$, $C_{6h}$, $D_{3h}$, $C_{3h}$, $D_{2h}$, $C_{2h}$, $C_{2v}$ and $C_s$. Brunvoll et al. [71] were the first who generated specifically benzenoids belonging to different symmetry groups.

Table 26 shows the known numbers of benzenoids with their distributions into symmetry groups for h ≤ 20. Extensions of this table for some of the symmetries are found in a forthcoming section.

Björg N. Cyvin, Jon Brunvoll, and Sven J. Cyvin

*Errata*

A minor misprint occurs as one of the symmetry group specifications by Rouvray [86]; the non-Kekuléan $h = 4$ system should be $C_{1h}$ [$C_s$].

We cite from Cyvin et al. [103]: "In the paper [83] ... it is also stated that the smallest benzenoid with $C_{6h}$ symmetry occurs for $h = 19$, and that only one such system exists. The last part of this statement is wrong. There are exactly two benzenoids with $h = 19$ and $C_{6h}$ symmetry [71]" (cf. Table 26).

**Table 26.** Numbers of benzenoids belonging to the different symmetry groups

| $h$ | $D_{6h}$ | $C_{6h}$ | $D_{3h}$ | $C_{3h}$ | $D_{2h}$ | $C_{2h}$ | $C_{2v}$ | $C_s$ |
|---|---|---|---|---|---|---|---|---|
| 1 | 1[a] | | | | | | | |
| 2 | 0 | | | | 1[a] | | | |
| 3 | 0 | | 1[a] | | 1[a] | | 1[a] | |
| 4 | 0 | | 1[a] | | 2[a] | 1[a] | 1[a] | 2[a] |
| 5 | 0 | | 0 | | 2[b] | 1[b] | 9[b] | 10[b] |
| 6 | 0 | | 1[c,d] | 1[c,d] | 3[c,d] | 7[c,d] | 12[c,d] | 57[c,d] |
| 7 | 1[c,d] | | 1[c,d] | 1[c,d] | 3[c,d] | 7[c,d] | 39[c,d] | 279[c,d] |
| 8 | 0 | | 0 | 0 | 6[c,d] | 35[c,d] | 61[c,d] | 1333[c,d] |
| 9 | 0 | | 1[c,d] | 5[c,d] | 7[c,d] | 36[c,d] | 178[c,d] | 6278[c,d] |
| 10 | 0 | | 4[c,d] | 5[c,d] | 11[c,d] | 169[c,d] | 274[c,d] | 29623[c,d] |
| 11 | 0 | | 0 | 0 | 14[d] | 177[d] | 796[d] | 140242[d] |
| 12 | 0 | | 3[d] | 21[d] | 21[d] | 807[d] | 1251 | 667481 |
| 13 | 2[c,d] | | 4[d] | 26[d] | 23[d] | 859[d] | 3578 | 3193764 |
| 14 | 0 | | 0 | 0 | 41[d] | 3864 | 5692 | 15357980 |
| 15 | 0 | | 3[d] | 95[d] | 50[d] | 4145 | 16290 | 74187327 |
| 16 | 0 | | 12[d] | 118[d] | 80[d] | 18616 | 26069 | 359818883 |
| 17 | 0 | | 0 | 0 | 94[d] | 20098 | † | † |
| 18 | 0 | | 6[d] | 423[d] | 156[d] | † | † | † |
| 19 | 2[c,d] | 2[c,d] | 19[d] | 543[d] | 189[d] | † | † | † |
| 20 | 0 | 0 | 0 | 0 | 310[d] | † | † | † |

[a] Rouvray (1973) [56]; [b] Rouvray (1974) [86]; [c] Brunvoll, Cyvin and Cyvin (1987) [71]; [d] Balaban, Brunvoll, Cioslowski, Cyvin, Cyvin, Gutman, He, He, Knop, Kovačević, Müller, Szymanski, Tošić and Trinajstić (1987) [18]; † Unknown

# 8 Normal Benzenoids

The class of normal benzenoids consists of the catacondensed and the normal pericondensed systems. They are all Kekuléan ($K > 0$) and have $\Delta = 0$.

In Table 27 known numbers of normal benzenoids, the catacondensed (cat) and normal pericondensed (np) systems separately are summarized with their distribution into symmetry groups. The entries under cat are of course obtainable from additions of the appropriate numbers from Tables 14 and 17. With the aid of tables 19 and 20 (containing supplements to Table 17), Table 27 can be extended further as far as the catacondensed systems are concerned. Still more supplementary numbers (for $h > 17$) could be produced with the aid of forthcoming tables for specific symmetries. In the documentation (cf. footnotes to Table 27) we have included a special study of benzenoids with trigonal symmetry [104]; the pertinent (small) values are obtainable from a scrutiny of the figures therein.

Figure 22 shows all the normal benzenoids to $h = 7$ as black silhouettes. Such figures have been given previously for $h \leq 6$ [26] and $h \leq 7$ [97].

**Table 27.** Numbers of normal benzenoids classified according to symmetry[+]

| $h$ | Type* | $D_{6h}$ | $D_{3h}$ | $C_{3h}$ | $D_{2h}$ | $C_{2h}$ | $C_{2v}$ | $C_s$ |
|---|---|---|---|---|---|---|---|---|
| 1 | cat | 1[a] | | | | | | |
| 2 | cat | 0 | | | 1[a] | | | |
| 3 | cat | 0 | | | 1[a] | | 1[b] | |
| 4 | cat | 0 | 1[a] | | 1[a] | 1[b] | 1[b] | 1[b] |
| | np | 0 | 0 | | 1[a] | 0 | 0 | 0 |
| 5 | cat | 0 | 0 | | 1[a] | 1[b] | 5[a] | 5[a] |
| | np | 0 | 0 | | 0 | 0 | 1[a] | 1[a] |
| 6 | cat | 0 | 0 | | 2[a] | 4[b] | 7[a] | 23[a] |
| | np | 0 | 0 | | 1[a] | 2[a] | 3[a] | 6[a] |
| 7 | cat | 0 | 1[a] | 1[a] | 2[a] | 4[b] | 16[a] | 94[a] |
| | np | 1[a] | 0 | 0 | 1[a] | 0 | 6[a] | 41[a] |
| 8 | cat | 0 | 0 | 0 | 2[a] | 17[a] | 28[a] | 364[a] |
| | np | 0 | 0 | 0 | 2[a] | 11[a] | 19[a] | 200[a] |
| 9 | cat | 0 | 0 | 0 | 2[a] | 17[a] | 61[a] | 1409[a] |
| | np | 0 | 0 | 0 | 3[a] | 3[a] | 39[a] | 997[a] |
| 10 | cat | 0 | 2[a] | 4[a] | 4[a] | 64[a] | 95[a] | 5403[a] |
| | np | 0 | 1[a] | 0 | 6[a] | 52[a] | 90[a] | 4654[a] |
| 11 | cat | 0 | 0 | 0 | 5[a] | 65[a] | 215[a] | 20830[a] |
| | np | 0 | 0 | 0 | 6 | 23 | 193 | 21582 |
| 12 | cat | 0 | 0 | 0 | 5[a] | 248[a] | 350[a] | 80518[c] |
| | np | 0 | 0 | 0 | 11 | 248 | 432 | 98393 |
| 13 | cat | 0 | 2[a] | 15[a] | 5[a] | 255[a] | 778 | 313020 |
| | np | 2[a] | 2[d] | 3[d] | 11 | 129 | 896 | 446481 |
| 14 | cat | 0 | 0 | 0 | 10[a] | 959 | 1267 | 1222292 |
| | np | 0 | 0 | 0 | 23 | 1145 | 1934 | 2013954 |
| 15 | cat | 0 | 0 | 0 | 12[a] | 994 | 2860 | 4795339 |
| | np | 0 | 0 | 0 | 26 | 657 | 4048 | † |
| 16 | cat | 0 | 4 | 55 | 13 | 3750 | 4670 | † |
| | np | 0 | 6 | 18 | 45 | 5240 | 8549 | † |
| 17 | cat | 0 | 0 | 0 | 14 | 3912 | † | † |
| | np | 0 | 0 | 0 | 51 | 3216 | † | † |

[+] Normal pericondensed systems with symmetry $C_{6h}$ occur for $h \geq 19$.
* Abbreviations: cat catacondensed; np normal pericondensed.
[a] Brunvoll, Cyvin and Cyvin (1987) [71]; [b] Balaban and Harary (1968) [13]; [c] Balaban, Brunvoll, Cyvin and Cyvin (1988) [60]; [d] Cyvin, Brunvoll and Cyvin (1988) [104]; † Unknown

Björg N. Cyvin, Jon Brunvoll, and Sven J. Cyvin

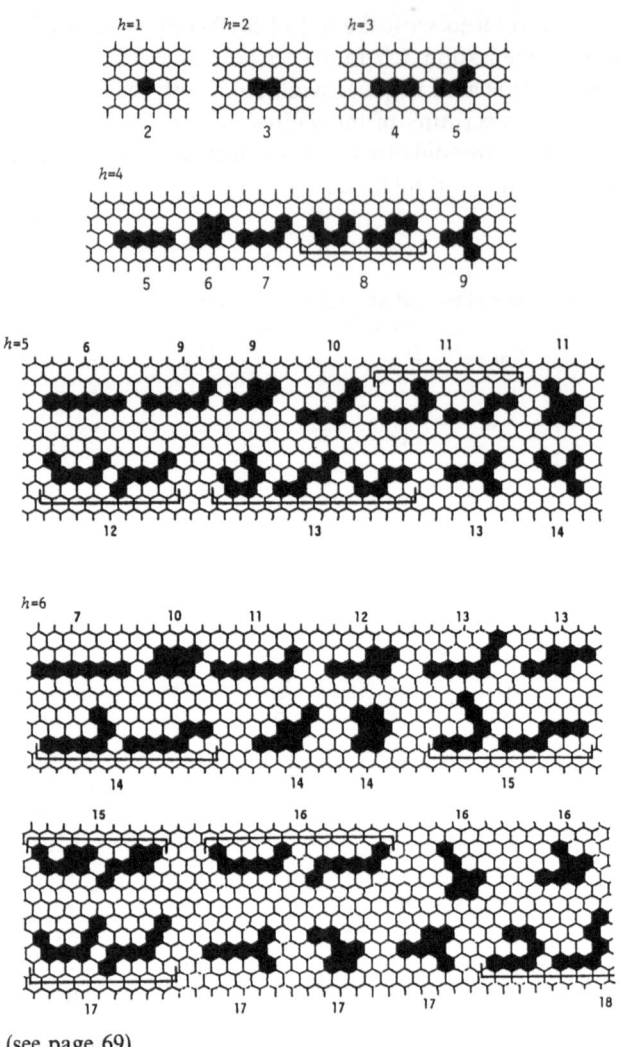

(see page 69)

# 9 Essentially Disconnected Benzenoids

Essentially disconnected benzenoids are, by definition, the Kekuléan benzenoids (with $K > 0$, $\Delta = 0$) which are not normal. All of them are pericondensed.

In Table 28 the numbers of essentially disconnected benzenoids are displayed with the distributions into symmetry groups. The documentations contain a work [105] in which the essentially disconnected benzenoids were recognized automatically by means of a computer program, based on Pauling bond orders. The same principles were also employed in the present supplementary computations. Extensions for the highest symmetries (hexagonal and trigonal) are accessible through some of the subsequent tables.

(*h*=6)

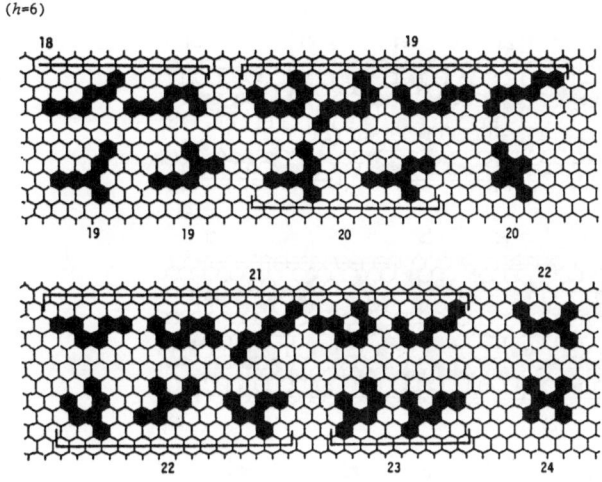

*h*=7

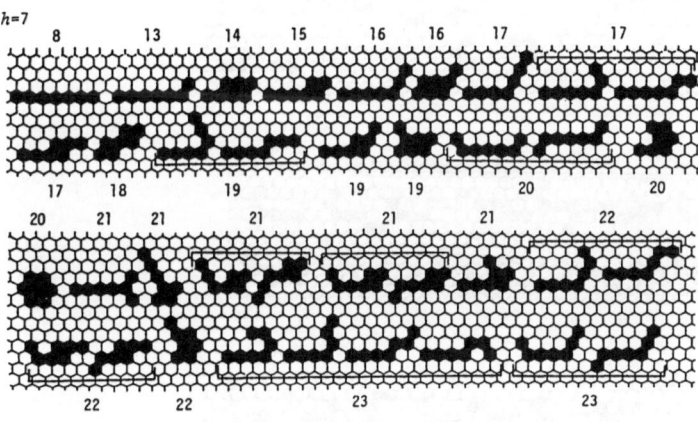

(see page 69)

Björg N. Cyvin, Jon Brunvoll, and Sven J. Cyvin

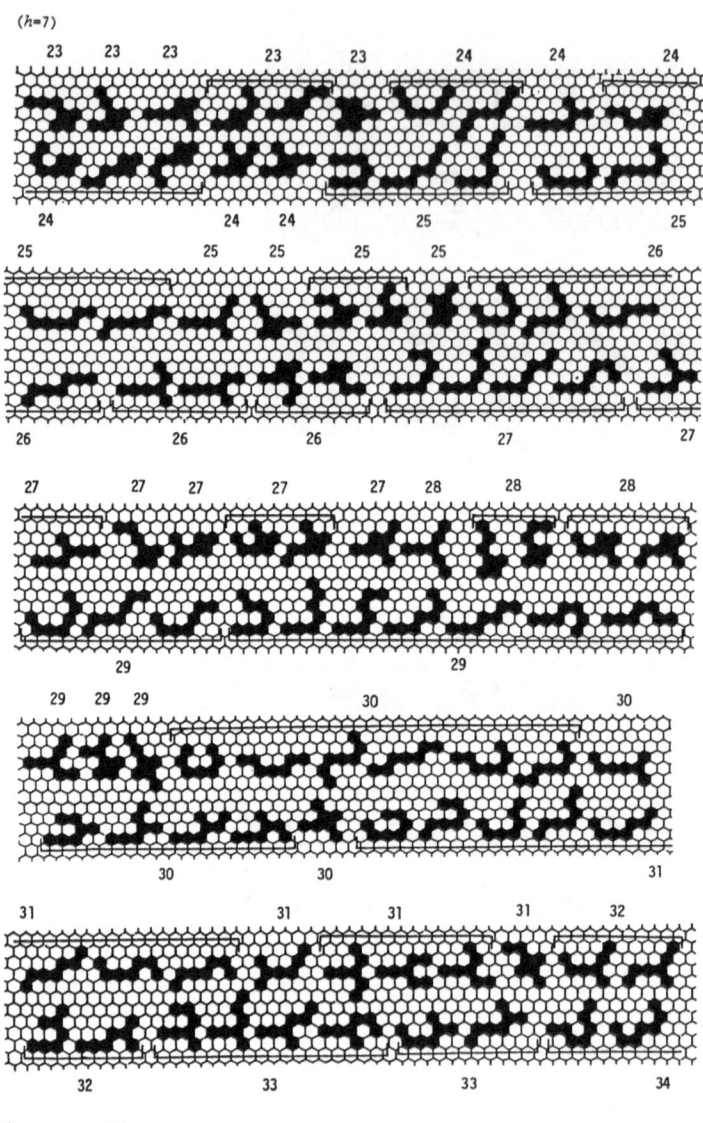

(see page 69)

($h=7$)

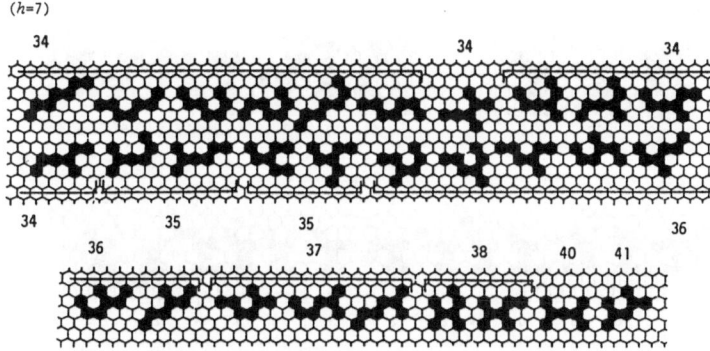

**Fig. 22.** All normal benzenoids with $h \leq 7$. $K$ numbers are given

**Table 28.** Numbers of essentially disconnected benzenoids classified according to symmetry [+]

| $h$ | $D_{3h}$ | $C_{3h}$ | $D_{2h}$ | $C_{2h}$ | $C_{2v}$ | $C_s$ |
|---|---|---|---|---|---|---|
| 5 | | | 1[a] | | | |
| 6 | | | 0 | 1[a] | | 2[a] |
| 7 | | | 0 | 3[a] | 6[a] | 14[a] |
| 8 | | | 2[a] | 7[a] | 2[a] | 110[a] |
| 9 | | | 2[a] | 16[a] | 29[a] | 645[a] |
| 10 | | | 1[a] | 53[a] | 31[a] | 3647[a] |
| 11 | | | 2[b] | 87[b] | 166[b] | 19705[b] |
| 12 | | | 5[b] | 306 | 202 | 104200 |
| 13 | | | 7[b] | 452 | 875 | 541928 |
| 14 | | | 7[b] | 1702 | 1199 | 2787150 |
| 15 | | | 9[b] | 2317 | 4577 | † |
| 16 | 1[c] | 2[c] | 19[b] | 9124 | 6651 | † |
| 17 | 0 | 0 | 27[b] | 11762 | † | † |
| 18 | 0 | 0 | 34[b] | † | † | † |
| 19 | 0 | 23[c] | 39[b] | † | † | † |
| 20 | 0 | 0 | 84[b] | † | † | † |

[+] The smallest essentially disconnected benzenoids with $D_{6h}$ and with $C_{6h}$ symmetries (one each) occur at $h = 25$ [71].
[a] Brunvoll, Cyvin and Cyvin (1987) [71]; [b] Brunvoll, Cyvin, Cyvin and Gutman (1988) [105];
[c] Cyvin, Brunvoll and Cyvin (1988) [104]; † Unknown

In Fig. 23 the forms of all essentially disconnected benzenoids up to $h = 8$ are displayed as black silhouettes. Such figures have been given previously for $h \leq 7$ [26] and $h \leq 8$ [105], for $h = 7$ also as dualists [55]. In one of these works [105] a number of (bizarre) forms of larger essentially disconnected benzenoids are included.

Björg N. Cyvin, Jon Brunvoll, and Sven J. Cyvin

**Fig. 23.** All essentially disconnected benzenoids with $h \leq 8$. $K$ numbers are given

# 10 Obvious Non-Kekuléan Benzenoids

## 10.1 Numbers and Forms

The benzenoids with $\Delta > 0$ are by definition the obvious non-Kekuléans. They have $K = 0$ and are pericondensed. Only four symmetry groups are possible for these systems, viz. $D_{3h}$, $C_{3h}$, $C_{2v}$ and $C_s$.

**Table 29.** Numbers of obvious non-Kekuléan benzenoids with different colour excess ($\Delta$ values), classified according to symmetry

| $h$ | $\Delta$ | $D_{3h}$ | $C_{3h}$ | $C_{2v}$ | $C_s$ |
|---|---|---|---|---|---|
| 3 | 1 | 1[a] | | | |
| 4 | 1 | 0 | | | 1[a] |
| 5 | 1 | 0 | | 3[a] | 4[a] |
| 6 | 1 | 0 | 1[a] | 1[a] | 26[a] |
| | 2 | 1[a] | 0 | 1[a] | 0 |
| 7 | 1 | 0 | 0 | 10[a] | 124[a] |
| | 2 | 0 | 0 | 1[a] | 6[a] |
| 8 | 1 | 0 | 0 | 5[a] | 614[a] |
| | 2 | 0 | 0 | 7[a] | 45[a] |
| 9 | 1 | 0 | 4[a] | 39[a] | 2914[a] |
| | 2 | 1[a] | 1[a] | 9[a] | 311[a] |
| | 3 | 0 | 0 | 1[a] | 2[a] |
| 10 | 1 | 0 | 0 | 20[a] | 14004[a] |
| | 2 | 0 | 0 | 38[a] | 1878[a] |
| | 3 | 1[a] | 1[a] | 0 | 37[a] |
| 11 | 1 | 0 | 0 | 156[b] | 66890[b] |
| | 2 | 0 | 0 | 52[b] | 10870[b] |
| | 3 | 0 | 0 | 13[b] | 361[b] |
| 12 | 1 | 3[b,c] | 13[b,c] | 80 | 320763 |
| | 2 | 0 | 7[b,c] | 176[b] | 60522[b] |
| | 3 | 0 | 0 | 5[b] | 2985[b] |
| | 4 | 0 | 1[b,c] | 6[b] | 7[b] |
| 13 | 1 | 0 | 0 | 652 | 1539522 |
| | 2 | 0 | 0 | 266 | 329972 |
| | 3 | 0 | 8[b,c] | 84[b] | 21583[b] |
| | 4 | 0 | 0 | 4[b] | 207[b] |
| 14 | 1 | 0 | 0 | 347 | 7408063 |
| | 2 | 0 | 0 | 853 | 1768772 |
| | 3 | 0 | 0 | 35 | 145473 |
| | 4 | 0 | 0 | 46[b] | 2542[b] |
| 15 | 1 | 0 | 58[b,c] | 2789 | † |
| | 2 | 2[b,c] | 29[b,c] | 1289 | † |
| | 3 | 0 | 0 | 486 | † |
| | 4 | 1[b,c] | 7[b,c] | 52 | † |
| | 5 | 0 | 1[b,c] | 4[b] | 43[b] |
| 16 | 1 | 0 | 0 | 1474 | † |
| | 2 | 0 | 0 | 4033 | † |
| | 3 | 1 | 43 | 214 | † |
| | 4 | 0 | 0 | 331 | † |
| | 5 | 0 | 0 | 2 | † |

[a] Brunvoll, Cyvin and Cyvin (1987) [71]; [b] Gutman and Cyvin (1988) [102]; [c] Cyvin, Brunvoll and Cyvin (1988) [104]; † Unknown

135

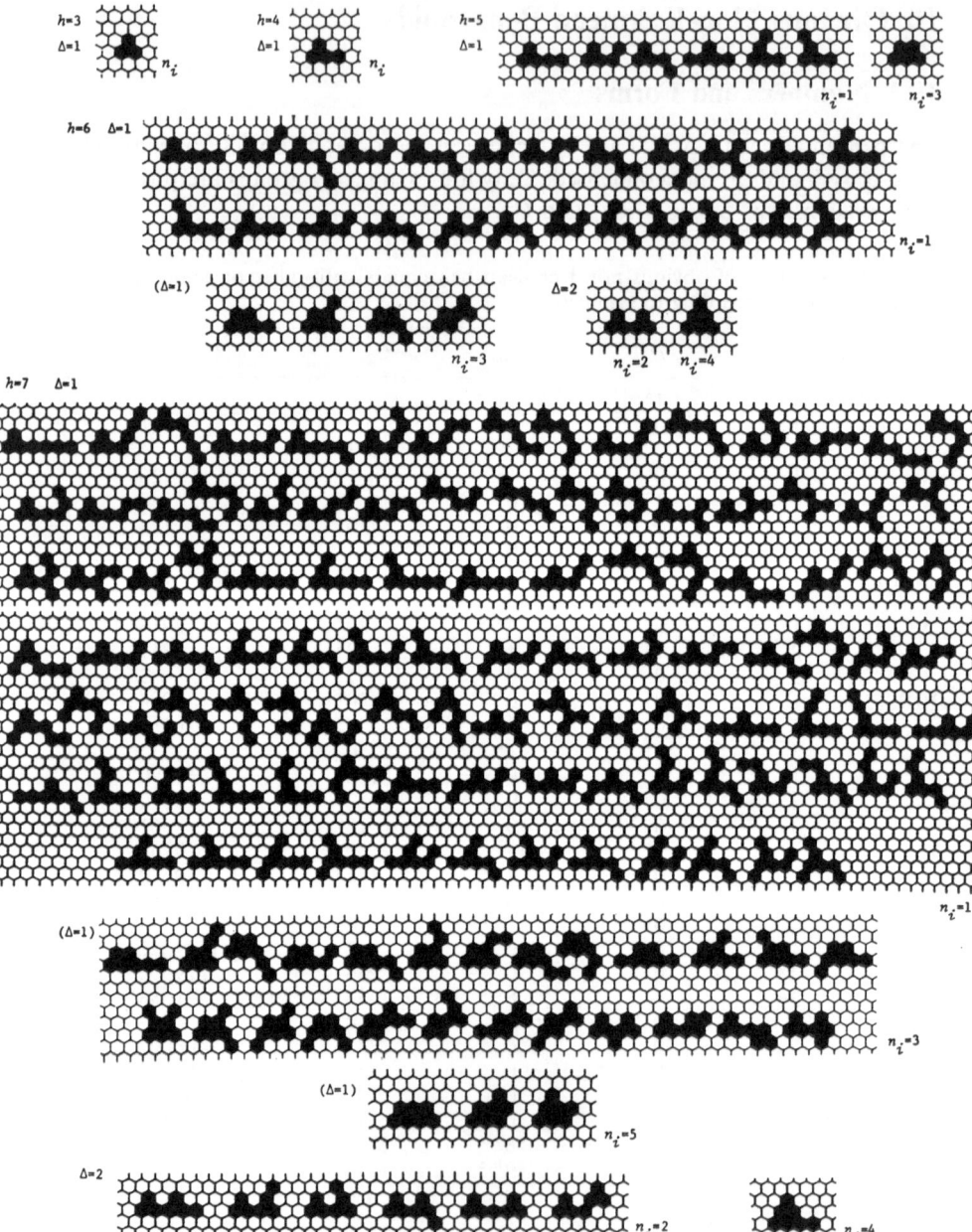

**Fig. 24.** All obvious non-Kekuléan benzenoids with $h \leq 7$. $\Delta$ values and numbers of internal vertices ($n_i$) are indicated

Table 29 gives a general survey of the numbers of obvious non-Kekuléans with given $\Delta$ values, including the distributions into symmetry groups. Extensions for the trigonal symmetries are accessible through subsequent tables.

In Fig. 24 the forms of the obvious non-Kekuléans up to $h = 7$ are shown as black silhouettes. They have been given previously as dualists [55].

*Erratum*

In Gutman and Cyvin [22], in the table which corresponds to Table 29 here, the entry for $h = 6$, $\Delta = 2$ (Total) is omitted by a minor misprint.

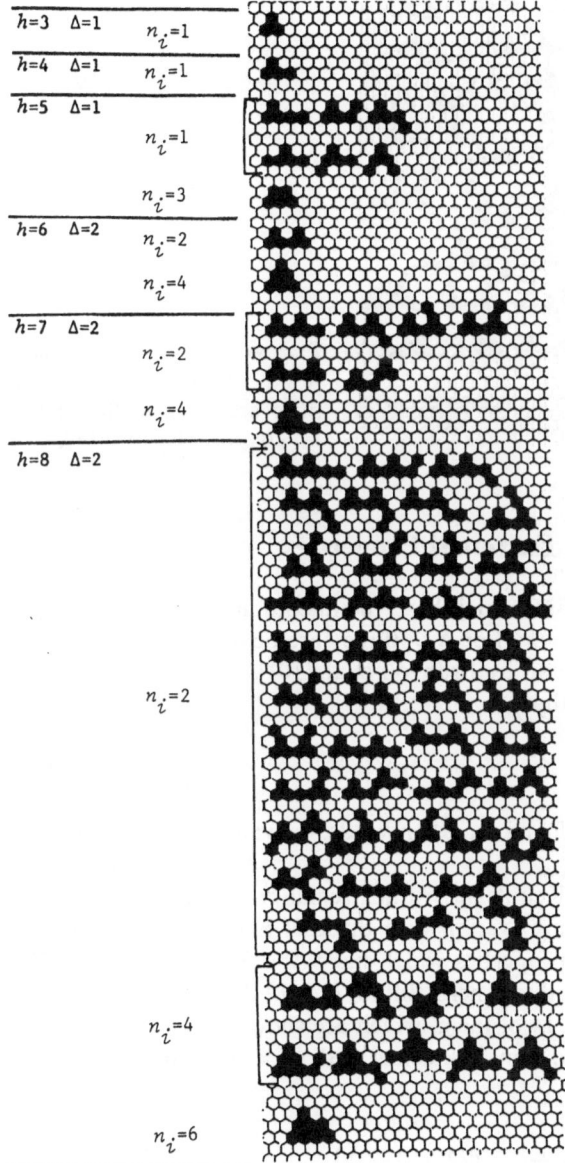

| $h=3$ | $\Delta=1$ | $n_i=1$ |
| $h=4$ | $\Delta=1$ | $n_i=1$ |
| $h=5$ | $\Delta=1$ | $n_i=1$ |
| | | $n_i=3$ |
| $h=6$ | $\Delta=2$ | $n_i=2$ |
| | | $n_i=4$ |
| $h=7$ | $\Delta=2$ | $n_i=2$ |
| | | $n_i=4$ |
| $h=8$ | $\Delta=2$ | |
| | | $n_i=2$ |
| | | $n_i=4$ |
| | | $n_i=6$ |

(see next page)

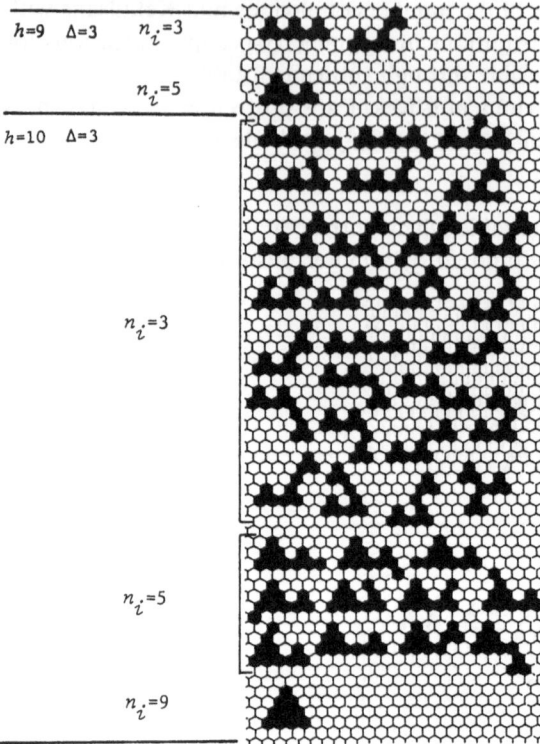

$h=9$  $\Delta=3$  $n_i=3$

$n_i=5$

$h=10$  $\Delta=3$

$n_i=3$

$n_i=5$

$n_i=9$

**Fig. 25.** All benzenoids with $\Delta = \Delta_{max}$ for $3 \leq h \leq 10$, classified according to their numbers of internal vertices $(n_i)$

## 10.2 Non-Kekuléans with Extremal Properties

The title class refers to benzenoids which have the maximum $\Delta$ value, $\Delta = \Delta_{max}$, for a given number of hexagons $(h)$. For $\Delta > 0$ they are obvious non-Kekuléans. These systems are treated in detail by Brunvoll et al. [101].

It has been shown [18, 101] that

$$\Delta_{max} = \lfloor h/3 \rfloor \tag{57}$$

where the floor function is employed. In other words, the value of $\Delta_{max}$ jumps one unit for every third $h$ value.

Figure 25 shows all the non-Kekuléans with extremal properties $(\Delta = \Delta_{max})$ to $h = 10$. Only the first 18 systems overlap with those of Fig. 24. For $h = 11$ the non-Kekuléans with extremal properties which have $\Delta = 3$, are too many (374 systems; cf. Table 25) to be reproduced here. But for $h = 12$ this type of benzenoids have $\Delta = 4$ and there are only 14 of them, which are shown in Fig. 26.

Especially interesting are the non-Kekuléans with $\Delta = \Delta_{max}$ for $h = 3, 6, 9, 12 ...$, i.e.

$$h = 3\Delta, \qquad \Delta = \Delta_{max} = h/3 \quad (\Delta > 0) \tag{58}$$

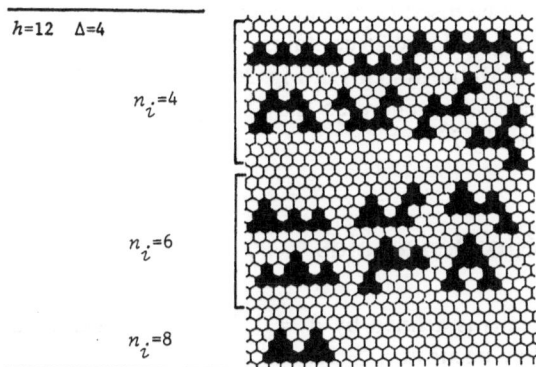

$h=12 \quad \Delta=4$

$n_i=4$

$n_i=6$

$n_i=8$

**Fig. 26.** The benzenoids (teepees) with $\Delta = \Delta_{max} = 4$ for $h = 12$, classified according to $n_i$

It has been shown [101] that these systems are characterized by some very restricted forms. They are called *teepees* (originally TP benzenoids, where T and P signify Triangulene and Phenalene, respectively). A teepee is defined as phenalene ($h = 3$) or triangulene ($h = 6$) or any number of these two units fused together so that the triangle apex of each unit points the same way (conventionally upwards). By

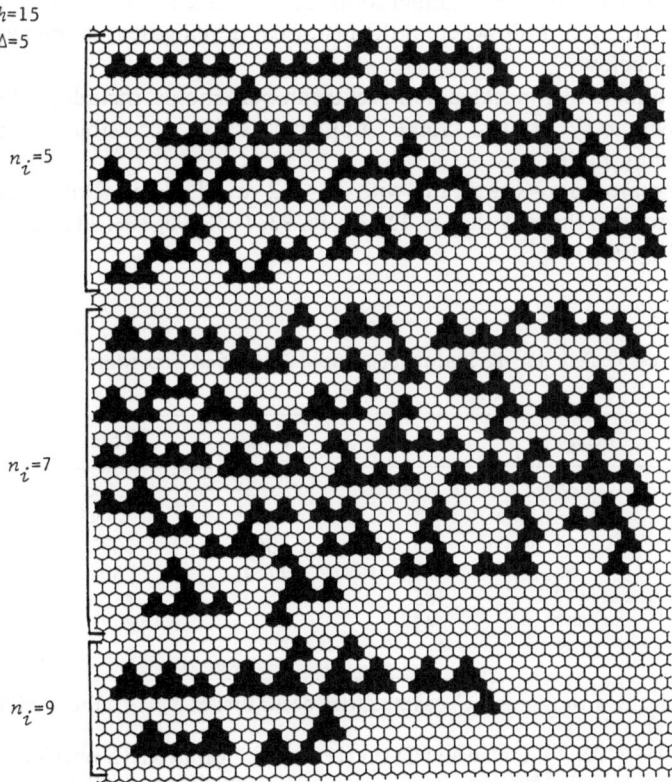

$h=15$
$\Delta=5$

$n_i=5$

$n_i=7$

$n_i=9$

**Fig. 27.** The benzenoids (teepees) with $\Delta = \Delta_{max} = 5$ for $h = 15$, classified according to $n_i$

139

definition, two benzenoids are said to be fused (in a restricted sense as used above) when they share one and only one edge. Figure 25 contains the teepees for: $h = 3$, $\Delta = 1$; $h = 6$, $\Delta = 2$; $h = 9$, $\Delta = 3$. All the systems in Fig. 26 (for $h = 12$, $\Delta = 4$) are teepees. Brunvoll et al. [101] have also generated the 48 teepees with $h = 15$ and $\Delta = 5$; they are reproduced in Fig. 27.

# 11 Concealed Non-Kekuléan Benzenoids

The class of concealed non-Kekuléans is defined by $\Delta = 0$, $K = 0$. All members of this class are pericondensed. They occur at $h \geq 11$.

A systematic search for concealed non-Kekuléans seems to have started in 1974 with Gutman [81], who inferred that no such systems with less than 11 hexagons can be constructed, and depicted two of those with $h = 11$. Not until thirteen years later was it proved by Brunvoll et al. [106] using computer generation that the number of concealed non-Kekuléans with $h = 11$ is 8. The corresponding analysis for $h = 12$ by He et al. [107] followed. Guo and Zhang [108], and independently Jiang and Chen [109], deduced analytically the numbers of concealed non-Kekuléans with $h = 12$ and $h = 13$. For $h = 14$, see *Comments and Errata* below. Two recent reviews on the enumeration of concealed non-Kekuléan benzenoids have appeared [93, 110], wherein the historical development is treated in particular; see also below.

Table 30 shows numbers of concealed non-Kekuléan benzenoids, including the distributions into symmetry groups. Some supplements are accessible from appropriate forthcoming tables.

In Fig. 28 the concealed non-Kekuléans with $h \leq 12$ are depicted. Both sets have been given previously: those with $h = 11$ [22, 26, 93, 103, 106, 108, 110–112] and with $h = 12$ [107, 108].

**Table 30.** Numbers of concealed non-Kekuléan benzenoids classified according to symmetry[+]

| $h$ | $D_{2h}$ | $C_{2h}$ | $C_{2v}$ | $C_s$ | Total |
|-----|----------|----------|----------|-------|-------|
| 11 | 1[a] | 2[b] | 1[b] | 4[b] | 8[c] |
| 12 | 0 | 5[b] | 0 | 93[b] | 98[d] |
| 13 | 0 | 23 | 23 | 1051 | 1097[e,f] |
| 14 | 1[a] | 58 | 11 | 9734 | 9804 |
| 15 | 3[a] | 177 | 185 | † | † |
| 16 | 3[a] | 502 | 145 | † | † |
| 17 | 2[a] | 1208 | † | † | † |

[+] Concealed non-Kekuléans with $D_{6h}$ and $C_{6h}$ symmetries occur first at $h = 43$, those with $D_{3h}$ at $h = 40$ and $C_{3h}$ at 34.
[a] Gutman and Cyvin (1988) [102]; [b] Gutman and Cyvin (1989) [22]; [c] Brunvoll, Cyvin, Cyvin, Gutman, He and He (1987) [106]; [d] He, He, Cyvin, Cyvin and Brunvoll (1988) [107]; [e] Guo and Zhang (1989) [108]; [f] Jiang and Chen (1989) [109]; † Unknown

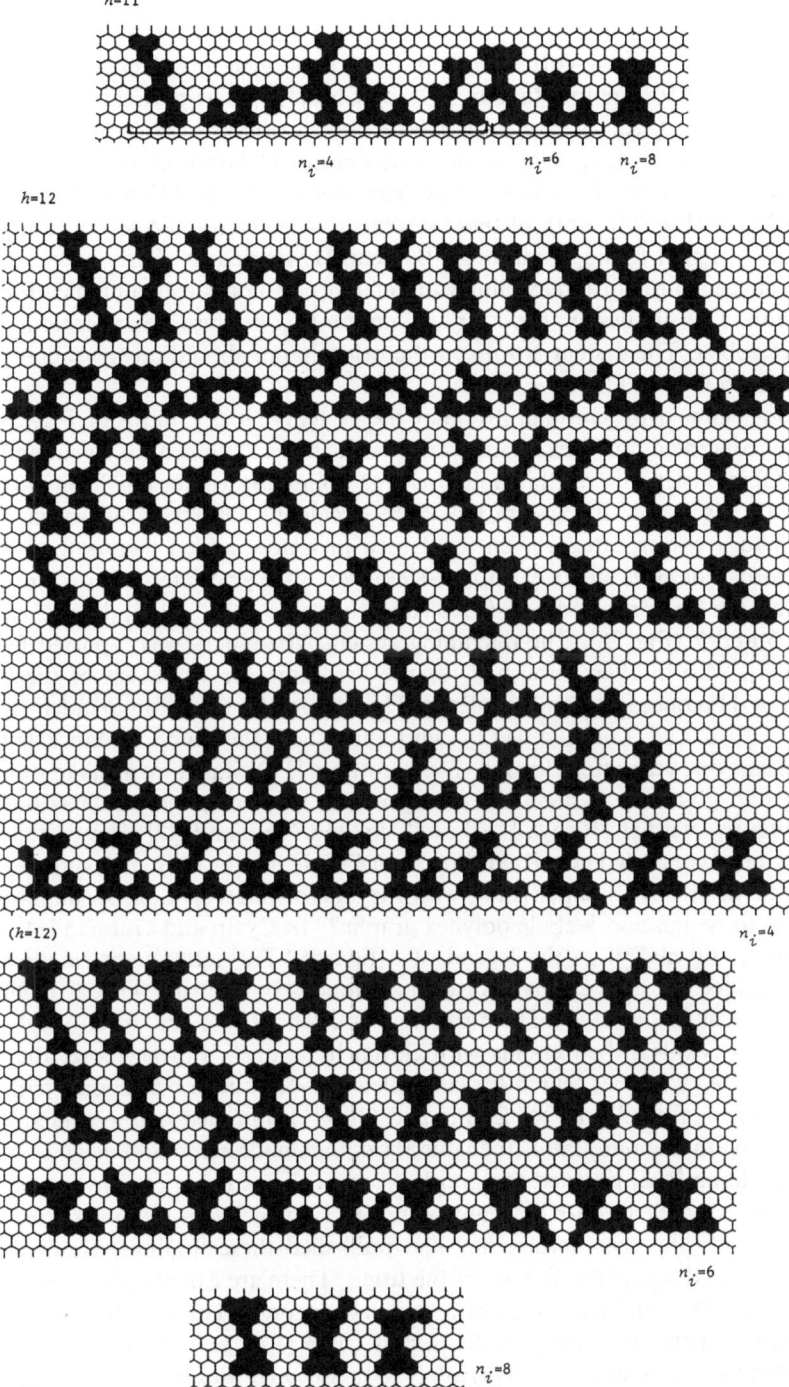

**Fig. 28.** All concealed non-Kekuléan benzenoids with $h \leq 12$, classified according to $n_i$

141

Björg N. Cyvin, Jon Brunvoll, and Sven J. Cyvin

*Comments and Errata*

A benzenoid hydrocarbon corresponding to a concealed non-Kekuléan was probably described for the first time in 1972 by Clar [113]. The particular system which he constructed, viz. the system No. 8 of Fig. 28 (the last one with $h = 11$) was referred to as the "Clar goblet" in the review entitled "The hunt for concealed non-Kekuléan polyhexes" [93]. The two systems constructed by Gutman [81] are No. 1 and No. 8 (Fig. 28). Both of these systems have been quoted several times as Gutman's original findings, although Balaban [114] tried to share the credit for No. 8 between Gutman and Mallion with reference to a private communication from Mallion. This discussion, however, is uninteresting because we are speaking of the Clar goblet. The system No. 1, on the other hand, should most probably be attributed to Gutman [81].

It was not claimed by Gutman [81] that the systems Nos. 1 and 8 of Fig. 28 are the only concealed non-Kekuléans with $h = 11$, but a statement in Cyvin and Gutman [83] is misleading on this point, as was pointed out later [103]. In 1981 Balaban [114] discovered the system No. 2 "by accident". Originally he assigned this system to a wrong category, but explained in a note added in proof, with different words, that it is a concealed non-Kekuléan. One year later, after a more conscious search for concealed non-Kekuléans, the same author [115] discovered Nos. 3, 4, 5 and 7 (Fig. 28). The remaining system, viz. No. 6, was reported by Hosoya [111], who in 1986 published for the first time the whole set of the 8 smallest ($h = 11$) concealed non-Kekuléans. The system No. 6 was found independently by Cyvin and Gutman [112], who published the same 8 concealed non-Kekuléans one year after Hosoya. These authors [112] stated by mistake that Dias [116] had given the system No. 6 before; the similar system of Dias has twelve hexagons. Neither Hosoya [111] or Cyvin and Gutman [112] claimed that the 8 constructed concealed non-Kekuléans with $h = 11$ are the only such systems. The very title of Hosoya's paper [111], with the sign of interrogation, speaks for itself: "How to design non-Kekulé polyhex graphs?" In Cyvin and Gutman [112] it is expressed still clearer in the legend of a figure: "The (all?) eight smallest possible concealed non-Kekuléans". As was mentioned above, Brunvoll et al. [106] resolved this problem in 1987. Their paper is entitled "There are exactly eight concealed non-Kekuléan benzenoids with eleven hexagons". Furthermore, they stated clearly: "Any further search for concealed non-Kekuléan benzenoids with eleven hexagons is futile." In 1988 Zhang and Guo [117] proved mathematically (by graph theory) a theorem stating the same about the smallest non-Kekuléans.

Guo and Zhang [108] pursued the graph-theoretical analysis and proved that the number of concealed non-Kekuléans with $h = 12$, which can be constructed, is 98. But already the year before they had published these findings, He et al. [107] had published a paper with part of the title: "There are exactly ninety eight concealed non-Kekuléan benzenoids with twelve hexagons." This conclusion had been reached by computer analysis. Jiang and Chen [109] referred to a wrong number instead of 98 communicated privately to them by He and He. This (wrong) number was published as a preliminary communication in He and He [35], but the error was detected and corrected before the final publication [107]. Herein,

142

by the way, in the depiction of the $h = 12$ concealed non-Kekuléans one black hexagon is missing.

The mathematical deductions did not always come after the computer-aided analysis. In the mentioned work of Guo and Zhang [108] also the number of concealed non-Kekuléans with $h = 13$ (see Table 30) was reported. An analytical deduction of the same number was achieved by Jiang and Chen [109], who extended their mathematical analysis to attain at the corresponding number for $h = 14$. Cyvin et al. [93] offered the following comment. "We wish to emphasize that these numbers (viz. 1097 and 9781) were obtained by mathematical analyses without computer aid. Brilliant achievements!" The former number (viz. 1097 for $h = 13$) was confirmed by a computer analysis of the present work. For $h = 14$, however, a very recent computer result of the present work deviates from the Jiang and Chen number. The new number (viz. 9804) is supposed to be correct and is therefore entered in Table 30.

# 12 Benzenoids with Specific Symmetries

## 12.1 Hexagonal Symmetry: Snowflakes

*Topological Properties*

A benzenoid of hexagonal symmetry belongs to one of the symmetry groups $D_{6h}$ and $C_{6h}$. For obvious reasons these systems are called *snowflakes*. Sometimes it is distinguished between the $D_{6h}$ and $C_{6h}$ groups by means of the terms proper and improper snowflakes, respectively. The proper snowflakes are also said to have regular hexagonal symmetry. Snowflakes, both of $D_{6h}$ and $C_{6h}$, occur for

$$h = 6\eta + 1; \qquad \eta = 0, 1, 2, \ldots \qquad (59)$$

but $\eta > 2$ for $C_{6h}$.

Every snowflake has a hexagon at its centre; it is the central hexagon. Furthermore, for $\eta > 0$ every snowflake has a coronene configuration at its centre; it is sometimes referred to as the core. It may also be natural to consider larger units as cores, such as circumcoronene.

From the above considerations it is clear that all snowflakes are pericondensed except for benzene, which corresponds to $\eta = 0$.

All snowflakes have vanishing color excess; $\Delta = 0$. Therefore they can be either normal, essentially disconnected or concealed non-Kekuléan (but not obvious non-Kekuléan).

*Numbers and Forms*

Table 31 shows the numbers of snowflakes, which supplement the few, small numbers found in Table 26. Tables 32 and 33 show the numbers of proper and *improper snowflakes*, respectively, classified into the neo categories. The data were

143

**Table 31.** Numbers of benzenoids with hexagonal symmetry*

| $h$ | $D_{6h}$ | $C_{6h}$ | Total $(D_{6h} + C_{6h})$ |
|---|---|---|---|
| 1 | 1[a] | | 1[a] |
| 7 | 1[b,c] | | 1[b,c] |
| 13 | 2[b,c] | | 2[b,c] |
| 19 | 2[b,c] | 2[b,c] | 4[b,c] |
| 25 | 3[b] | 8[b] | 11[b] |
| 31 | 5[b] | 32[b] | 37[b] |
| 37 | 8[b] | 128[b] | 136[b] |
| 43 | 13[b] | 527[b] | 540[b] |
| 49 | 20[b] | 2209[b] | 2229[b] |
| 55 | 35[b] | 9470[b] | 9505[b] |
| 61 | 60[d] | † | † |
| 67 | 104[d] | † | † |
| 73 | 183[d] | † | † |

* Contains supplements to Table 26.
[a] Rouvray (1973) [56]; [b] Brunvoll, Cyvin and Cyvin (1987) [71]; [c] Balaban, Brunvoll, Cioslowski, Cyvin, Cyvin, Gutman, He, He, Knop, Kovačević, Müller, Szymanski, Tošić and Trinajstić (1987) [18]; [d] He, He, Wang, Brunvoll and Cyvin (1988) [19]; † Unknown

**Table 32.** Numbers of classified benzenoids with regular hexagonal symmetry, $D_{6h}$ (proper snowflakes)

| $h$ | Kekuléan* | | | Concealed non-Kekuléan |
|---|---|---|---|---|
| | n | e | Total Kek. | |
| 1 | 1[a] | | 1[a] | |
| 7 | 1[a] | | 1[a] | |
| 13 | 2[a] | | 2[a] | |
| 19 | 2[a] | | 2[a] | |
| 25 | 2[a] | 1[a] | 3[a] | |
| 31 | 5[a] | 0 | 5[a] | |
| 37 | 7[a] | 1[a] | 8[a] | |
| 43 | 11[a] | 1[a] | 12[a] | 1[a] |
| 49 | 17[b,c] | 3[b,c] | 20[b,c] | 0 |
| 55 | 30[b,c] | 4[b,c] | 34[b,c] | 1[d] |
| 61 | 51 | 8 | 59[b,c] | 1[b,c] |
| 67 | 87 | 13 | 100[b,c] | 4[b,c] |
| 73 | 150 | 26 | 176[b,c] | 7[b,c] |

* Abbreviations: e essentially disconnected; n normal.
[a] Brunvoll, Cyvin and Cyvin (1987) [71]; [b] Cyvin, Brunvoll and Cyvin (1989) [103]; [c] Gutman and Cyvin (1989) [22]; [d] Cyvin, Brunvoll and Cyvin (1988) [118]

produced by specific generations of the systems with hexagonal symmetry [71, 103, 118].

The first specific generation of snowflakes [71] contains computer-generated figures in the form of mini-hexagons for all the 191 such systems with $7 \leq h \leq 37$ and selected systems for $h = 43$. The five smallest ($h = 43$) concealed non-

**Table 33.** Numbers of classified benzenoids with $C_{6h}$ symmetry (improper snowflakes)

| $h$ | Kekuléan* | | | Concealed non-Kekuléan |
|---|---|---|---|---|
| | n | e | Total Kek. | |
| 19 | 2[a] | | 2[a] | |
| 25 | 7[a] | 1[a] | 8[a] | |
| 31 | 24[a] | 8[a] | 32 | |
| 37 | 84[a] | 44[a] | 128 | |
| 43 | 310[a] | 213[a] | 523 | 4[a] |
| 49 | † | † | 2167 | 42[b,c] |
| 55 | † | † | 9158 | 312[b] |

* Abbreviations: See footnote to Table 32.
[a] Brunvoll, Cyvin and Cyvin (1987) [71]; [b] Cyvin, Brunvoll and Cyvin (1988) [118]; [c] He, He, Cyvin, Cyvin and Brunvoll (1988) [107]; † Unknown

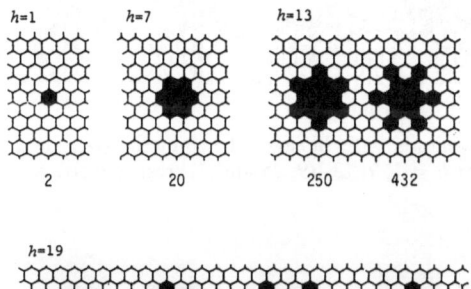

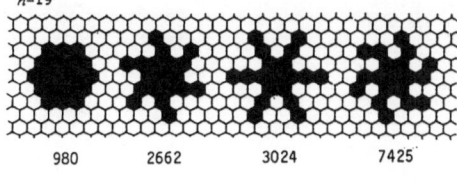

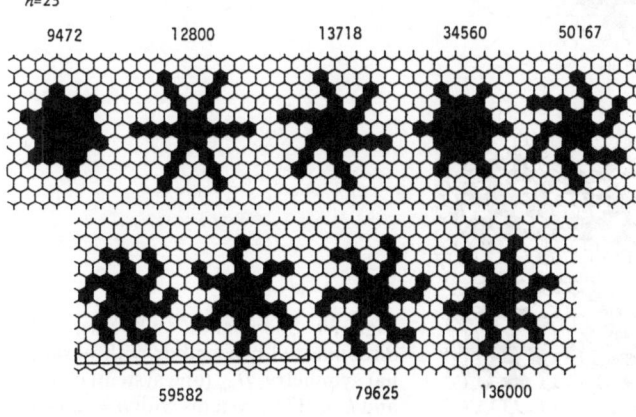

(see next page)

145

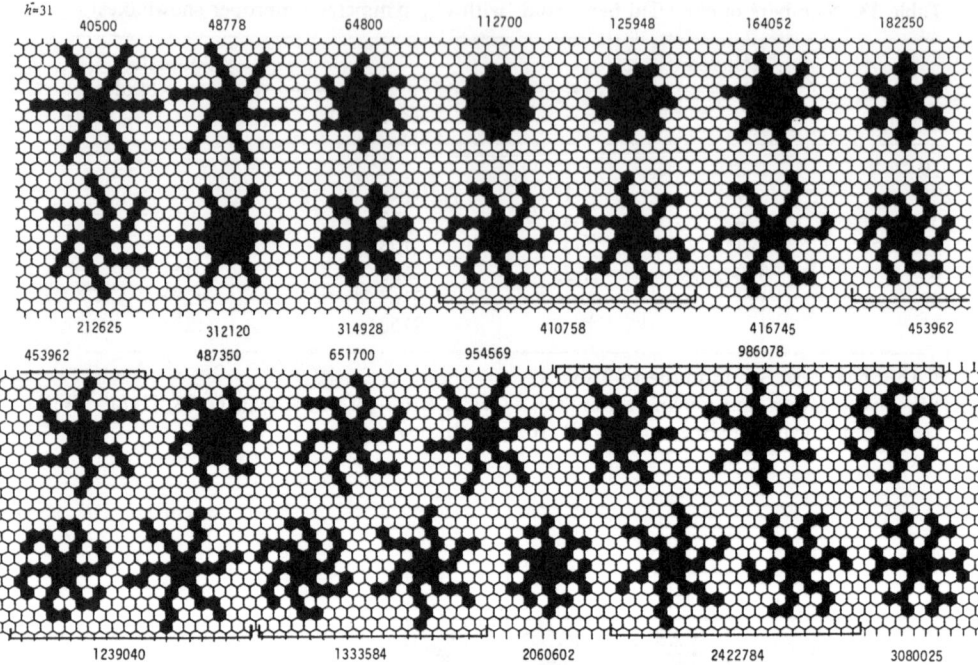

**Fig. 29.** Snowflakes: all normal benzenoids with hexagonal symmetry ($D_{6h}$ or $C_{6h}$) and $h < 37$: 1, 1, 2, 4, 9 and 29 systems for $h = 1, 7, 13, 19, 25$ and 31, respectively. $K$ numbers are given

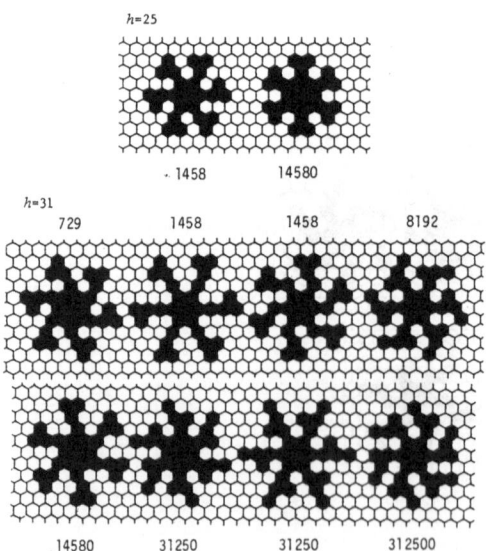

**Fig. 30.** Snowflakes: all essentially disconnected benzenoids with hexagonal symmetry, $D_{6h}$ (one system) or $C_{6h}$, and $h < 37$: 2 systems with $h = 25$ and 8 with $h = 31$. $K$ numbers are given

Kekuléan snowflakes have been depicted or reproduced several times [71, 103, 107, 110]. The $D_{6h}$ system out of these was actually identified and depicted for the first time by Hosoya [119], while three out of the four $C_{6h}$ systems were depicted by Cyvin et al. [120]. The 42 concealed non-Kekuléan (improper) snowflakes with $h = 49$ have also been depicted before [107]. The 313 concealed non-Kekuléan snowflakes with $h = 55$ have been described, and selected representatives of them have been depicted [118]. In this set there is 1 proper snowflake, which also has been depicted together with all such systems for $h \leq 73$ [103]; the 7 largest of these systems (for $h = 73$) are reproduced elsewhere [93]. One of these papers [103] shows the computer-generated pictures of all proper snowflakes with $h \leq 55$.

Here we give a re-edited selection of the forms of snowflakes. Figures 29, 30 and 31 display the forms of normal, essentially disconnected and concealed non-Kekuléan snowflakes, respectively, both proper $(D_{6h})$ and improper $(C_{6h})$. Similarly, the forms of proper snowflakes in particular are displayed in Figs. 32, 33 and 34, pertaining to the normal, essentially disconnected and concealed non-Kekuléan systems, respectively.

$h=43$

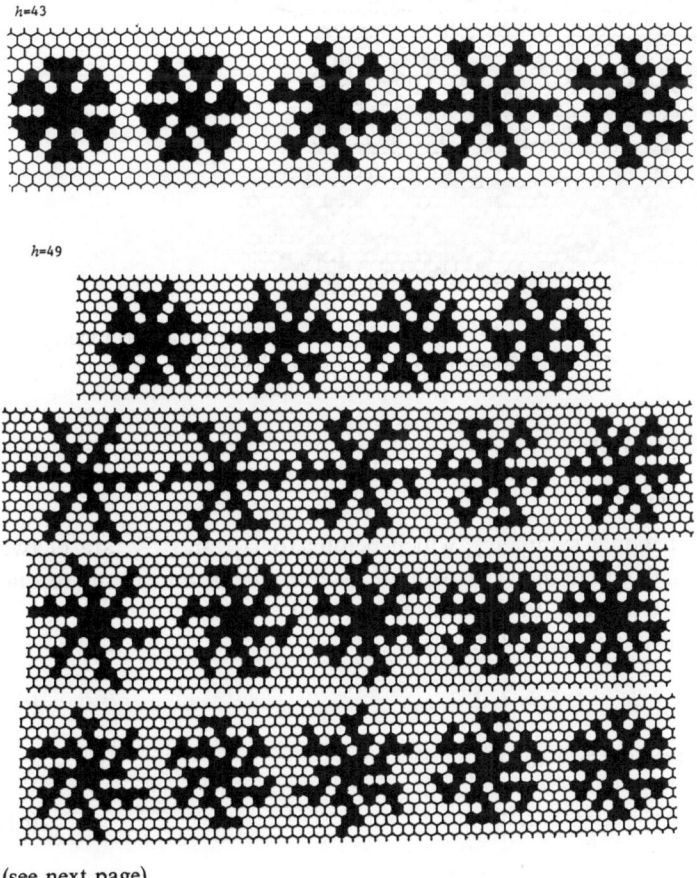

$h=49$

(see next page)

147

*(h=49)*

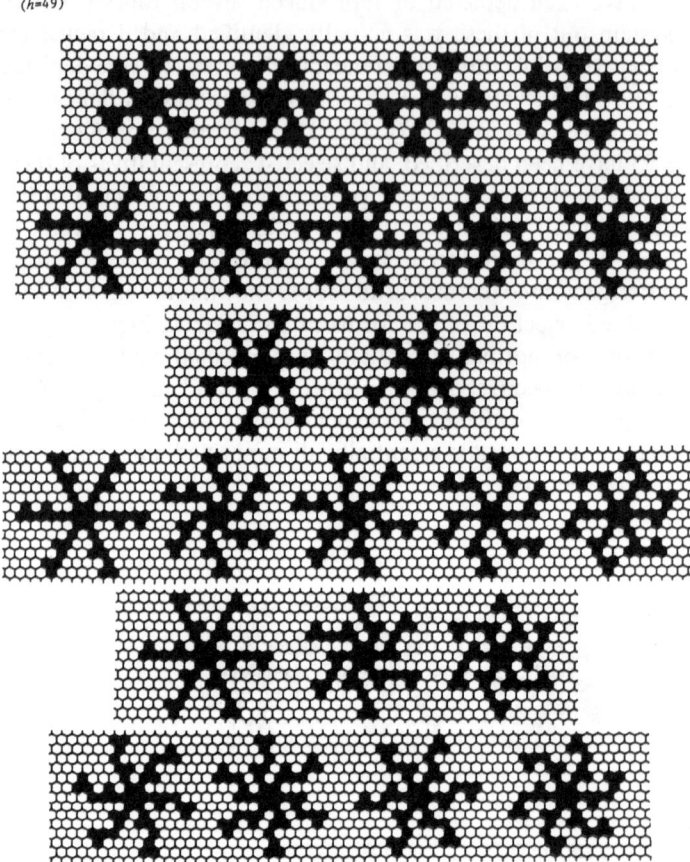

**Fig. 31.** Snowflakes: all concealed non-Kekuléan benzenoids with hexagonal symmetry, $D_{6h}$ (one system) or $C_{6h}$, and $h < 55$: 5 systems with $h = 43$ and 42 with $h = 49$

## 12.2 Trigonal Symmetry

*Topological Properties*

The benzenoids of trigonal symmetry, which belong to $D_{3h}$ and $C_{3h}$ are of two kinds: (i) the first kind, where the systems have a central hexagon; (ii) the second kind, where they have a vertex in the centre, a central vertex.

The numbers of hexagons are restricted to:

$$h = 3\xi + 1; \qquad \xi = 1, 2, 3, \dots \tag{60}$$

for the first kind (i), but $\xi > 1$ for $C_{3h}$;

$$h = 3\xi; \qquad \xi = 1, 2, 3, \dots \tag{61}$$

for the second kind (ii), but again $\xi > 1$ for $C_{3h}$.

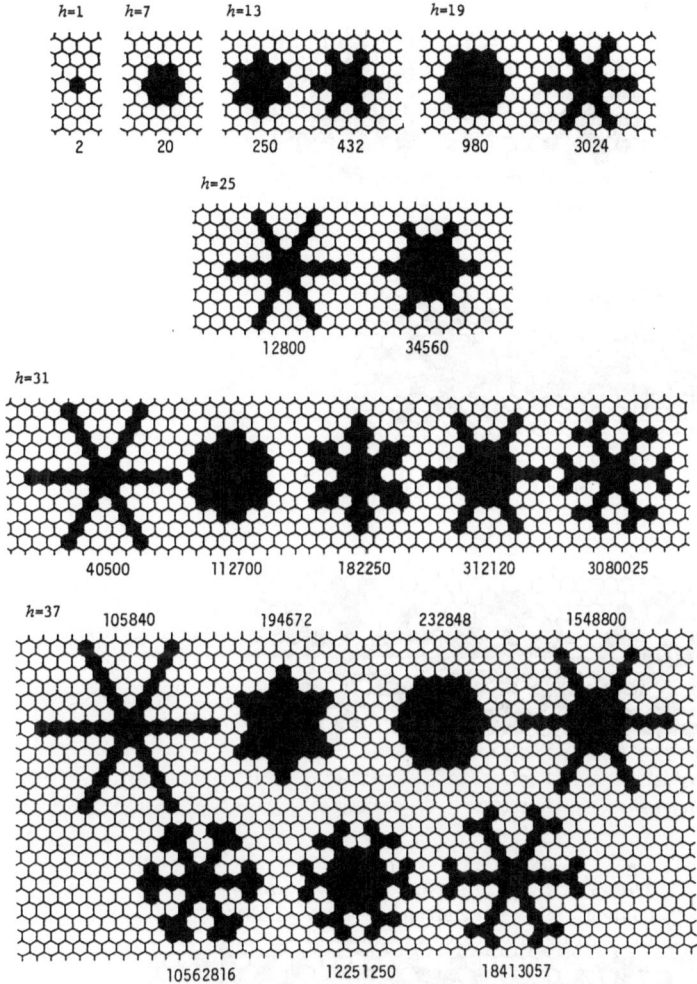

**Fig. 32.** Proper snowflakes: all normal benzenoids with $D_{6h}$ symmetry and $h < 43$. $K$ numbers are given

The benzenoids of regular trigonal symmetry, viz. $D_{3h}$, and of the first kind, viz. $D_{3h}(i)$, are subdivided into: (a) those where the two-fold symmetry axes cut edges perpendicularly; (b) those where the two-fold symmetry axes go through edges (and vertices). Under the adopted convention to draw a benzenoid with some of its edges vertical, the systems $D_{3h}(ia)$ and $D_{3h}(ib)$ will possess a horizontal or a vertical two-fold symmetry axis, respectively. In the systems of the second kind, $D_{3h}(ii)$, the two-fold symmetry axes invariably go through edges, and therefore a vertical two-fold symmetry axis is found in such a system under the adopted convention.

There are restrictions on the color excess. For the $D_{3h}(i)$ and $C_{3h}(i)$ systems one has $\Delta = 3\eta$, where $\eta = 0, 1, 2, \ldots$; those of $D_{3h}(ia)$ can only assume the value

Björg N. Cyvin, Jon Brunvoll, and Sven J. Cyvin

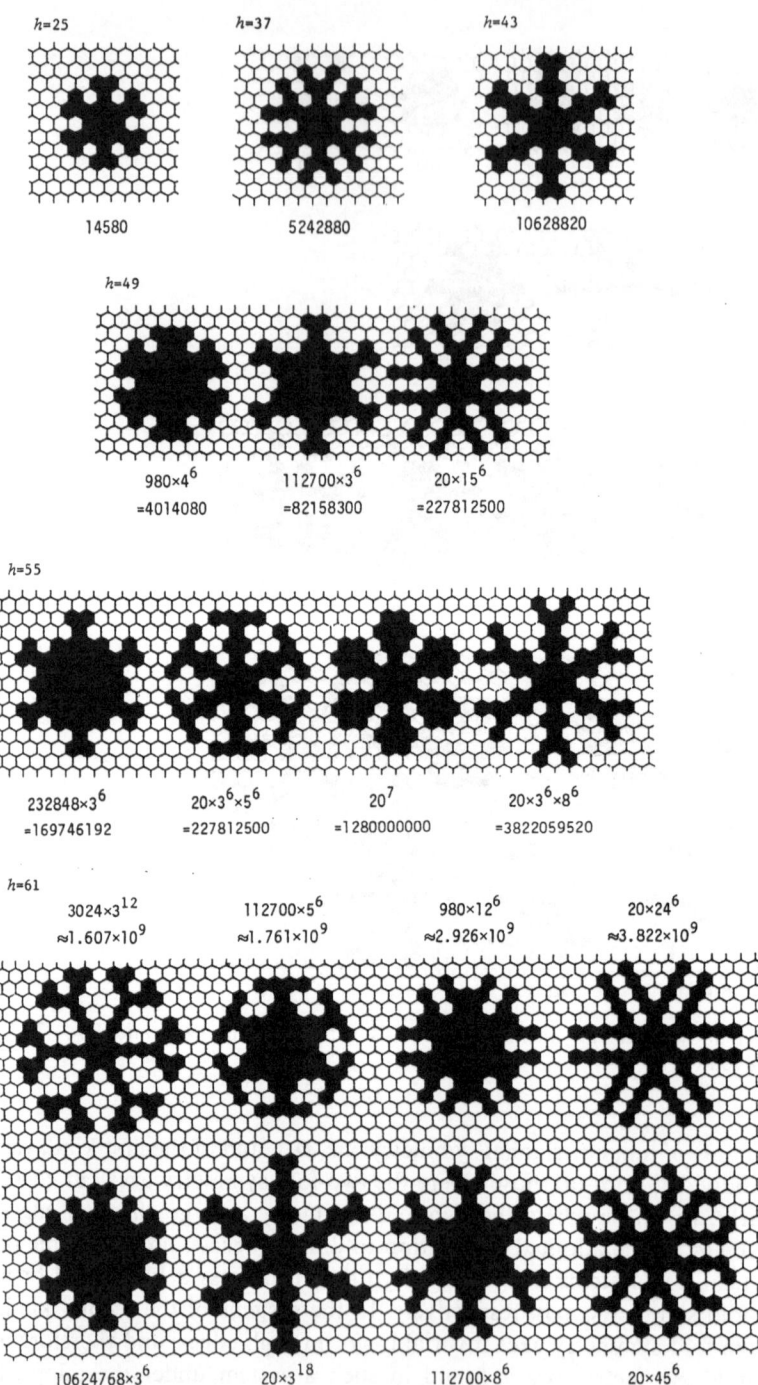

$h=25$     $h=37$     $h=43$

14580     5242880     10628820

$h=49$

$980 \times 4^6$     $112700 \times 3^6$     $20 \times 15^6$
$=4014080$     $=82158300$     $=227812500$

$h=55$

$232848 \times 3^6$     $20 \times 3^6 \times 5^6$     $20^7$     $20 \times 3^6 \times 8^6$
$=169746192$     $=227812500$     $=1280000000$     $=3822059520$

$h=61$

$3024 \times 3^{12}$     $112700 \times 5^6$     $980 \times 12^6$     $20 \times 24^6$
$\approx 1.607 \times 10^9$     $\approx 1.761 \times 10^9$     $\approx 2.926 \times 10^9$     $\approx 3.822 \times 10^9$

$10624768 \times 3^6$     $20 \times 3^{18}$     $112700 \times 8^6$     $20 \times 45^6$
$\approx 7.745 \times 10^9$     $\approx 7.748 \times 10^9$     $\approx 2.954 \times 10^{10}$     $\approx 1.661 \times 10^{11}$

**Fig. 33.** Proper snowflakes: all essentially disconnected benzenoids with $D_{6h}$ symmetry and $h < 66$. $K$ numbers are given

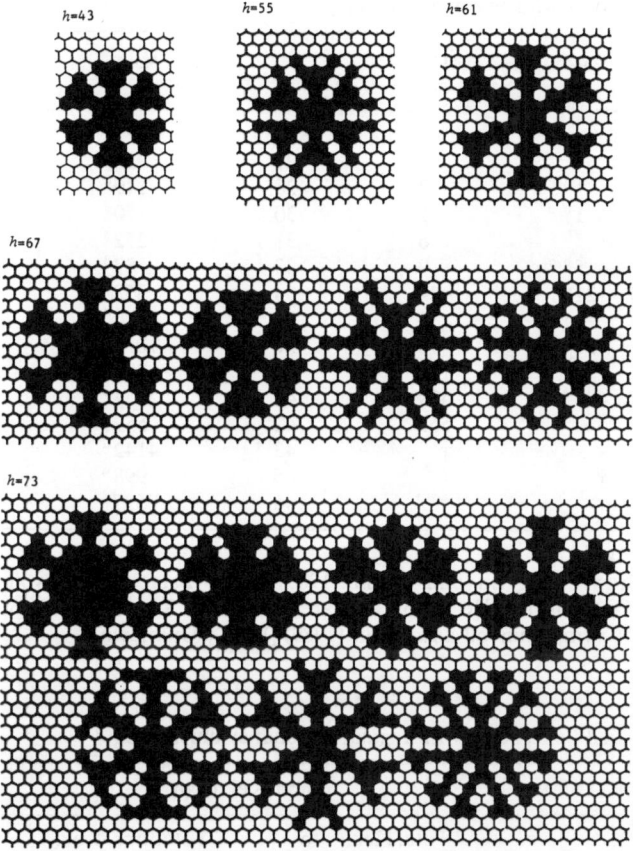

**Fig. 34.** Proper snowflakes: all concealed non-Kekuléan benzenoids with $D_{6h}$ symmetry and $h < 79$

$\Delta = 0$. In the case of $D_{3h}$(ii) and $C_{3h}$(ii) the allowed values for the color excess are exactly those which are forbidden in the former case (i); the allowed values are $\Delta = 3\eta + 1, 3\eta + 2; \eta = 0, 1, 2, \ldots$ . In consequence, all benzenoids of trigonal symmetry belonging to the second kind (ii) are obvious non-Kekuléans.

*Numbers and Forms*

The numbers of catacondensed benzenoids of trigonal symmetry are listed in Table 17 with a continuation in Table 19. A listing for all benzenoids of trigonal symmetry (catacondensed + pericondensed) is given in Table 34, which supplements Table 26. Tables 35 and 36 take into account various divisions into subclasses for the $D_{3h}$ and $C_{3h}$ systems, respectively. The total numbers of $D_{3h}$(ia) and of $D_{3h}$(ib) systems have been given elsewhere [110] and are consistent with those of Table 35.

Björg N. Cyvin, Jon Brunvoll, and Sven J. Cyvin

**Table 34.** Numbers of benzenoids with trigonal symmetry*

| $h$ | $D_{3h}$ | $C_{3h}$ | Total $(D_{3h} + C_{3h})$ | $h$ | $D_{3h}$ |
|---|---|---|---|---|---|
| 3 | 1[a] | | 1 | 25 | 72[d] |
| 4 | 1[a] | | 1 | 27 | 28[d] |
| 6 | 1[b,c] | 1[b,c] | 2 | 28 | 149[d] |
| 7 | 1[b,c] | 1[b,c] | 2 | 30 | 50[d] |
| 9 | 1[b,c] | 5[b,c] | 6 | 31 | 272[d] |
| 10 | 4[b,c] | 5[b,c] | 9 | 33 | 87[d] |
| 12 | 3[c] | 21[c] | 24 | 34 | 557[d] |
| 13 | 4[c] | 26[c] | 30 | 36 | 164[d] |
| 15 | 3[c] | 95[c] | 98 | 37 | 1050[d] |
| 16 | 12[c] | 118[c] | 130 | 39 | 286[d] |
| 18 | 6[c] | 423[c] | 429 | 40 | 2154[d] |
| 19 | 19[c] | 543[c] | 562 | 42 | 557[d] |
| 21 | 10[d] | † | † | 43 | 4142[d] |
| 22 | 41[d] | † | † | 45 | 998[d] |
| 24 | 16[d] | † | † | 46 | 8537[d] |

* Contains supplements to Table 26.
[a] Rouvray (1973) [56]; [b] Brunvoll, Cyvin and Cyvin (1987) [71]; [c] Balaban, Brunvoll, Cioslowski, Cyvin, Cyvin, Gutman, He, He, Knop, Kovačević, Müller, Szymanski, Tošić and Trinajstić (1987) [18]; [d] Cyvin, Brunvoll and Cyvin (1989) [110]; † Unknown

**Table 35.** Numbers of classified benzenoids with regular trigonal symmetry, $D_{3h}$*

| $h$ | (Kind/ type) | $\Delta$ | Kekuléan | | | non-Kekuléan | |
|---|---|---|---|---|---|---|---|
| | | | n | e | Total Kek. | o | Total non-Kek. |
| 3 | (ii) | 1 | | | | 1[a,b] | 1 |
| 4 | (ia) | 0 | 1[a] | | 1 | 0 | 0 |
| 6 | (ii) | 2 | 0 | | 0 | 1[a,b] | 1 |
| 7 | (ia) | 0 | 1[a] | | 1 | 0 | 0 |
| 9 | (ii) | 2 | 0 | | 0 | 1[a,b] | 1 |
| 10 | (ia) | 0 | 3[a] | | } 3 | 0 | } 1 |
| | (ib) | 3 | 0 | | | 1[a,b] | |
| 12 | (ii) | 1 | 0 | | 0 | 3[a,b] | 3 |
| 13 | (ia) | 0 | 4[a] | | 4 | 0 | 0 |
| 15 | (ii) | 2 | 0 | | } 0 | 2[a,b] | } 3 |
| | | 4 | 0 | | | 1[a,b] | |
| 16 | (ia) | 0 | 10[a] | | | 0 | |
| | (ib) | 0 | 0 | 1[a] | } 11 | 0 | } 1 |
| | | 3 | 0 | 0 | | 1[a,b] | |
| 18 | (ii) | 1 | 0 | 0 | | 3[a,b] | |
| | | 2 | 0 | 0 | } 0 | 2[a,b] | } 6 |
| | | 5 | 0 | 0 | | 1[a,b] | |
| 19 | (ia) | 0 | 17[a] | 0 | } 17 | 0 | } 2 |
| | (ib) | 3 | 0 | 0 | | 2[a,b] | |

152

**Table 35.** (Continued)

| $h$ | (Kind/ type) | $\Delta$ | Kekuléan | | | non-Kekuléan | |
|---|---|---|---|---|---|---|---|
| | | | n | e | Total Kek. | o | Total non-Kek. |
| 21 | (ii) | 1 | 0 | 0 | | 4 | |
| | | 2 | 0 | 0 | | 4 | |
| | | 4 | 0 | 0 | 0 | 1 | 10 |
| | | 5 | 0 | 0 | | 1 | |
| 22 | (ia) | 0 | 35 | 1 | | 0 | |
| | (ib) | 0 | 2 | 2 | 40 | 0 | 1 |
| | | 3 | 0 | 0 | | 1 | |
| 24 | (ii) | 1 | 0 | 0 | | 5 | |
| | | 2 | 0 | 0 | | 6 | |
| | | 4 | 0 | 0 | 0 | 2 | 16 |
| | | 5 | 0 | 0 | | 3 | |
| 25 | (ia) | 0 | 62 | 2 | | 0 | |
| | (ib) | 0 | 1 | 1 | 66 | 0 | 6 |
| | | 3 | 0 | 0 | | 6 | |
| 27 | (ii) | 1 | 0 | 0 | | 13 | |
| | | 2 | 0 | 0 | | 11 | |
| | | 4 | 0 | 0 | 0 | 3 | 28 |
| | | 5 | 0 | 0 | | 1 | |
| 28 | (ia) | 0 | 124 | 9 | | 0 | |
| | (ib) | 0 | 2 | 4 | | 0 | |
| | | 3 | 0 | 0 | 139 | 6 | 10 |
| | | 6 | 0 | 0 | | 4 | |
| 30 | (ii) | 1 | 0 | 0 | | 19 | |
| | | 2 | 0 | 0 | | 15 | |
| | | 4 | 0 | 0 | | 6 | |
| | | 5 | 0 | 0 | 0 | 8 | 50 |
| | | 7 | 0 | 0 | | 1 | |
| | | 8 | 0 | 0 | | 1 | |
| 31 | (ia) | 0 | 227 | 20 | | 0 | |
| | (ib) | 0 | 4 | 6 | 257 | 0 | 15 |
| | | 3 | 0 | 0 | | 14 | |
| | | 6 | 0 | 0 | | 1 | |
| 33 | (ii) | 1 | 0 | 0 | | 29 | |
| | | 2 | 0 | 0 | | 34 | |
| | | 4 | 0 | 0 | | 13 | |
| | | 5 | 0 | 0 | 0 | 8 | 87 |
| | | 7 | 0 | 0 | | 2 | |
| | | 8 | 0 | 0 | | 1 | |
| 34 | (ia) | 0 | 446 | 61 | | 0 | |
| | (ib) | 0 | 7 | 14 | | 0 | |
| | | 3 | 0 | 0 | 528 | 20 | 29 |
| | | 6 | 0 | 0 | | 8 | |
| | | 9 | 0 | 0 | | 1 | |
| 36 | (ii) | 1 | 0 | 0 | | 61 | |
| | | 2 | 0 | 0 | | 42 | |
| | | 4 | 0 | 0 | | 21 | |
| | | 5 | 0 | 0 | 0 | 28 | 164 |
| | | 7 | 0 | 0 | | 6 | |
| | | 8 | 0 | 0 | | 5 | |
| | | 11 | 0 | 0 | | 1 | |

**Table 35.** (Continued)

| h | (Kind/ type) | Δ | Kekuléan n | e | Total Kek. | non-Kekuléan o | Total non-Kek. |
|---|---|---|---|---|---|---|---|
| 37 | (ia) | 0 | 829 | 136 | | 0 | |
| | (ib) | 0 | 15 | 15 | | 0 | |
| | | 3 | 0 | 0 | 995 | 43 | 55 |
| | | 6 | 0 | 0 | | 11 | |
| | | 9 | 0 | 0 | | 1 | |
| 39 | (ii) | 1 | 0 | 0 | | 86 | |
| | | 2 | 0 | 0 | | 103 | |
| | | 4 | 0 | 0 | | 46 | |
| | | 5 | 0 | 0 | 0 | 30 | 286 |
| | | 7 | 0 | 0 | | 10 | |
| | | 8 | 0 | 0 | | 10 | |
| | | 10 | 0 | 0 | | 1 | |
| 40 | (ia) | 0 | 1635 | 350 | | 3[c] | |
| | (ib) | 0 | 19 | 45 | | 0 | |
| | | 3 | 0 | 0 | 2049 | 64 | 105 |
| | | 6 | 0 | 0 | | 33 | |
| | | 9 | 0 | 0 | | 5 | |
| 42 | (ii) | 1 | 0 | 0 | | 192 | |
| | | 2 | 0 | 0 | | 142 | |
| | | 4 | 0 | 0 | | 71 | |
| | | 5 | 0 | 0 | 0 | 99 | 557 |
| | | 7 | 0 | 0 | | 28 | |
| | | 8 | 0 | 0 | | 18 | |
| | | 10 | 0 | 0 | | 2 | |
| | | 11 | 0 | 0 | | 5 | |
| 43 | (ia) | 0 | 3081 | 759 | | 6[c] | |
| | (ib) | 0 | 47 | 50 | | 1[c] | |
| | | 3 | 0 | 0 | 3937 | 140 | 205 |
| | | 6 | 0 | 0 | | 48 | |
| | | 9 | 0 | 0 | | 10 | |
| 45 | (ii) | 1 | 0 | 0 | | 275 | |
| | | 2 | 0 | 0 | | 332 | |
| | | 4 | 0 | 0 | | 164 | |
| | | 5 | 0 | 0 | 0 | 119 | 998 |
| | | 7 | 0 | 0 | | 49 | |
| | | 8 | 0 | 0 | | 49 | |
| | | 10 | 0 | 0 | | 7 | |
| | | 11 | 0 | 0 | | 3 | |
| 46 | (ia) | 0 | 6059 | 1865 | | 34[c] | |
| | (ib) | 0 | 56 | 156 | | 0 | |
| | | 3 | 0 | 0 | 8136 | 214 | 401 |
| | | 6 | 0 | 0 | | 126 | |
| | | 9 | 0 | 0 | | 22 | |
| | | 12 | 0 | 0 | | 5 | |

* Abbreviations: e essentially disconnected; n normal; o non-Kekuléan.
[a] Cyvin, Brunvoll and Cyvin (1988) [78]; [b] Gutman and Cyvin (1988) [102]; [c] Cyvin, Brunvoll and Cyvin (1989) [110]

**Table 36.** Numbers of classified benzenoids with $C_{3h}$ symmetry*

| $h$ | (Kind) | $\Delta$ | Kekuléan | | | non-Kekuléan | |
|---|---|---|---|---|---|---|---|
| | | | $n$ | $e$ | Total Kek. | $o$ | Total non-Kek. |
| 6 | (ii) | 1 | | | | $1^{a,b}$ | 1 |
| 7 | (i) | 0 | $1^a$ | | 1 | | 0 |
| 9 | (ii) | 1 | 0 | | $\left.\right\}$ 0 | $4^{a,b}$ | $\left.\right\}$ 5 |
| | | 2 | 0 | | | $1^{a,b}$ | |
| 10 | (i) | 0 | $4^a$ | | $\left.\right\}$ 4 | 0 | $\left.\right\}$ 1 |
| | | 3 | 0 | | | $1^{a,b}$ | |
| 12 | (ii) | 1 | 0 | | | $13^{a,b}$ | |
| | | 2 | 0 | | $\left.\right\}$ 0 | $7^{a,b}$ | $\left.\right\}$ 21 |
| | | 4 | 0 | | | $1^{a,b}$ | |
| 13 | (i) | 0 | $18^a$ | | $\left.\right\}$ 18 | 0 | $\left.\right\}$ 8 |
| | | 3 | 0 | | | $8^{a,b}$ | |
| 15 | (ii) | 1 | 0 | | | $58^{a,b}$ | |
| | | 2 | 0 | | $\left.\right\}$ 0 | $29^{a,b}$ | $\left.\right\}$ 95 |
| | | 4 | 0 | | | $7^{a,b}$ | |
| | | 5 | 0 | | | $1^{a,b}$ | |
| 16 | (i) | 0 | $73^a$ | $2^a$ | $\left.\right\}$ 75 | 0 | $\left.\right\}$ 43 |
| | | 3 | 0 | 0 | | $43^{a,b}$ | |
| 18 | (ii) | 1 | 0 | 0 | | $234^{a,b}$ | |
| | | 2 | 0 | 0 | $\left.\right\}$ 0 | $136^{a,b}$ | $\left.\right\}$ 423 |
| | | 4 | 0 | 0 | | $45^{a,b}$ | |
| | | 5 | 0 | 0 | | $8^{a,b}$ | |
| 19 | (i) | 0 | $298^a$ | $23^a$ | | 0 | |
| | | 3 | 0 | 0 | $\left.\right\}$ 321 | $217^{a,b}$ | $\left.\right\}$ 222 |
| | | 6 | 0 | 0 | | $5^{a,b}$ | |

* Abbreviations: See footnote to Table 35.
[a] Cyvin, Brunvoll and Cyvin (1988) [78]; [b] Gutman and Cyvin (1988) [102]


In Figs. 35, 36 and 37 the smallest benzenoids with trigonal symmetry are illustrated; the figures pertain to the normal, essentially disconnected and non-Kekuléan systems, repectively. The same collection of forms has been displayed elsewhere [78].


*Erratum*

In the work of Cyvin et al. [78] the two first (silhouette) drawings of essentially disconnected benzenoids should be switched in order to match the given $K$ numbers.

The ten smallest concealed non-Kekuléan benzenoids with regular trigonal symmetry ($D_{3h}$) are depicted in Fig. 38 and taken from Cyvin et al. [110]. The reader is referred to the below comments about the search for concealed non-Kekuléans with $C_{3h}$ symmetry.

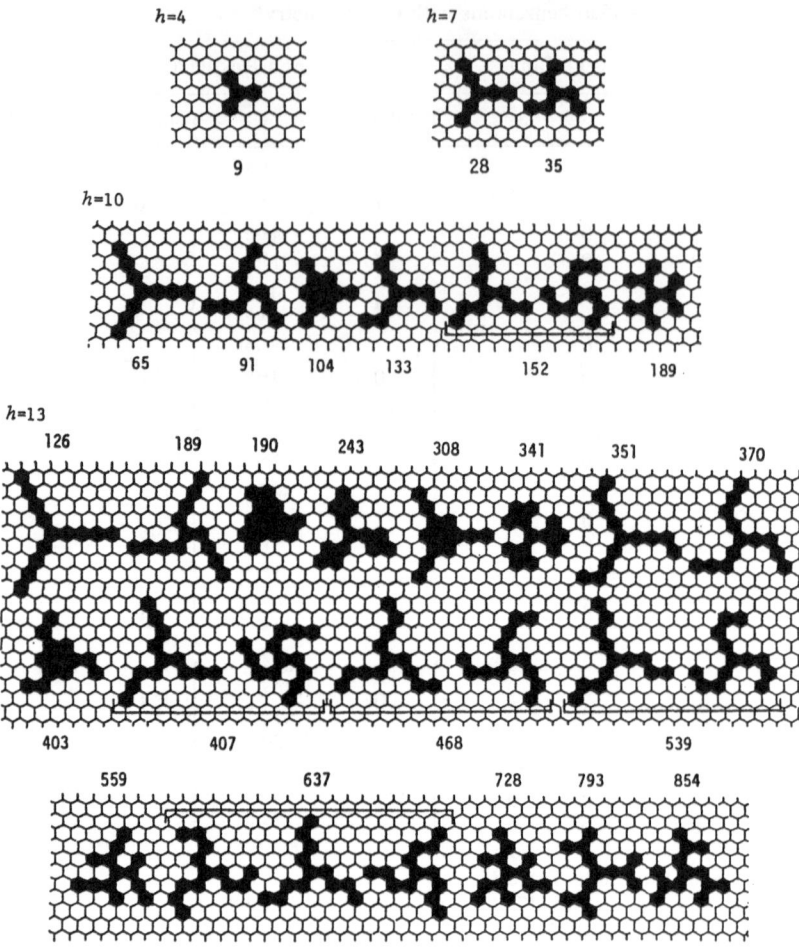

**Fig. 35.** All normal benzenoids with trigonal symmetry ($D_{3h}$ or $C_{3h}$) and $h < 16$: 1, 2, 7 and 22 systems for $h = 4$, 7, 10 and 13, respectively; all of them are of the first kind. $K$ numbers are given

## Comments

There are no concealed non-Kekuléan benzenoids with $C_{3h}$ symmetry among the enumerated systems ($h \leq 19$; cf. Table 36), and it has been pointed out by Cyvin et al. [78] that it is not easy to construct small concealed non-Kekuléans with trigonal symmetry. Nevertheless, in the mentioned work [78] two such systems with $h = 34$ are presented and are supposed to be among the smallest. Later Cyvin et al. [110] depicted twelve concealed non-Kekuléans with 34 hexagons each and presented convincing arguments to the effect that they really are among the smallest such systems. However, the authors emphasized that they are "... well aware of the fact that the list is not complete."

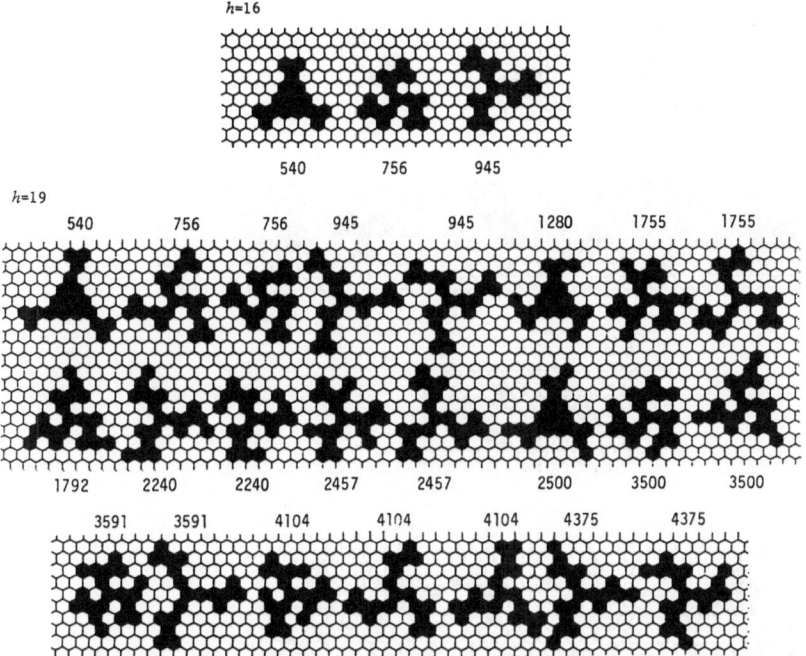

**Fig. 36.** All essentially disconnected benzenoids with trigonal symmetry, $D_{3h}$ (one system) and $C_{3h}$, and $h < 22$: 3 and 23 systems with $h = 16$ and 19, respectively; the additions to triphenylene are of the first kind, the additions to triangulene (four systems) of the second kind. $K$ numbers are given

## 12.3 Dihedral Symmetry and Centrosymmetry

*Topological Properties*

Benzenoids of both dihedral symmetry $(D_{2h})$ and centrosymmetry $(C_{2h})$ are divided into two kinds: (i) the first kind, where the systems have a central hexagon; (ii) the second kind, where they have an edge in the centre, a central edge.

The numbers of hexagons are restricted to:

$$h = 2\xi + 1; \qquad \xi = 1, 2, 3, \ldots \tag{62}$$

for the first kind (i), but $\xi > 1$ for $C_{2h}$;

$$h = 2\xi; \qquad \xi = 1, 2, 3, \ldots \tag{63}$$

for the second kind, but $\xi > 1$ for $C_{2h}$.

All benzenoids belonging to the $D_{2h}$ and $C_{2h}$ symmetries have $\Delta = 0$.

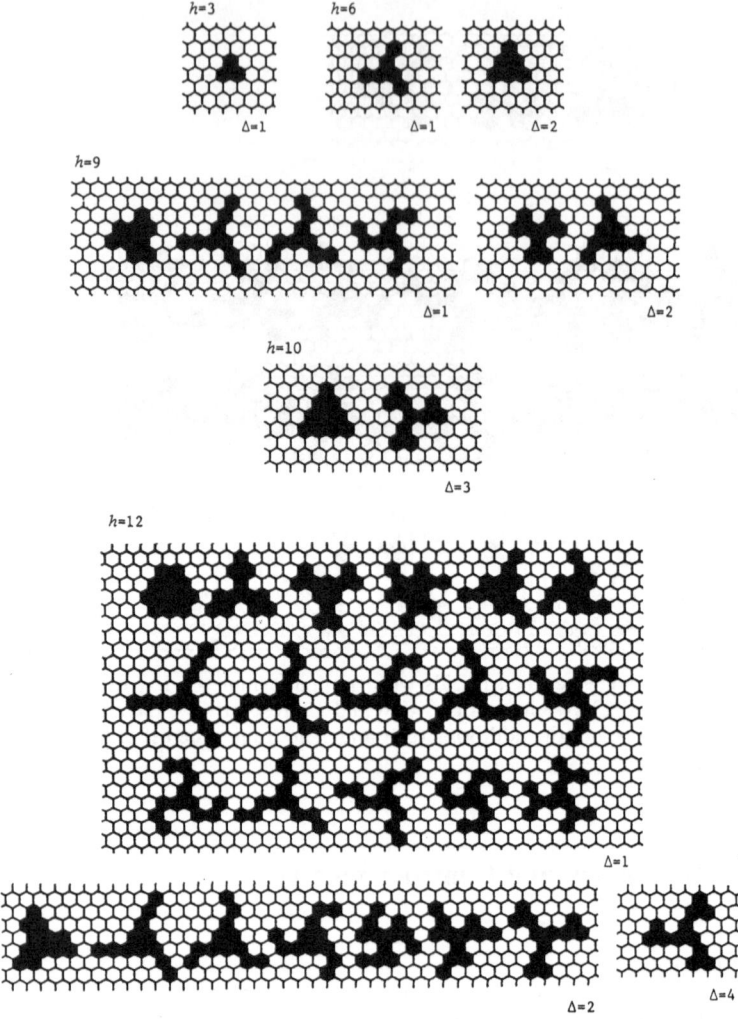

**Fig. 37.** All (obvious) non-Kekuléan benzenoids with trigonal symmetry ($D_{3h}$ or $C_{3h}$) and $h < 13$: 1, 2, 6, 2 and 24 systems with $h = 3, 6, 9, 10$ and $12$, respectively; the two systems with $h = 10$ are of the first kind, all the other of the second kind. $\Delta$ values are indicated

### Numbers and Forms

The numbers of catacondensed benzenoids belonging to the symmetries $D_{2h}$ and $C_{2h}$ are listed in Tables 14, 15, 17 and 20. When taking the catacondensed and pericondensed benzenoids of these symmetries together, most of the relevant information on the numbers of dihedral ($D_{2h}$) and centrosymmetrical ($C_{2h}$) benzenoids, which is available so far, is already contained above; cf. Tables 26, 27, 28 and 30. Collections of the specific data for the $D_{2h}$ and $C_{2h}$ systems are

$h=40$

$h=43$

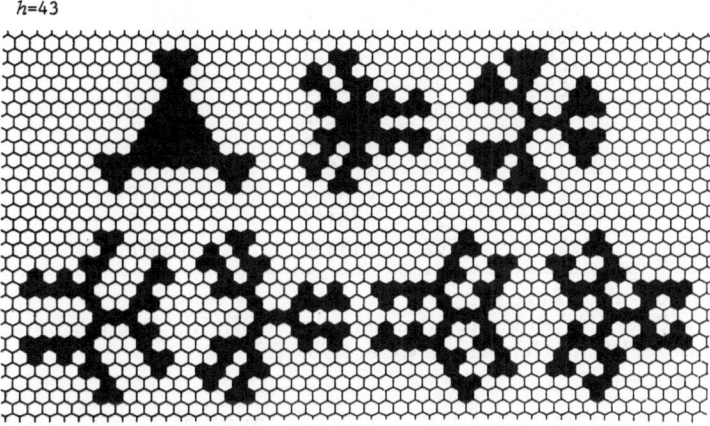

**Fig. 38.** All concealed non-Kekuléan benzenoids with $D_{3h}$ symmetry and $h < 46$: 3 and 7 systems with $h = 40$ and 43, respectively. The first system at $h = 43$ belongs to the class $D_{3h}$ (ib), all the others to $D_{3h}$ (ia)

**Table 37.** Numbers of classified benzenoids with dihedral symmetry $(D_{2h})$

| h | (Kind) | Kekuléan* | | | Concealed non-Kekuléan |
|---|---|---|---|---|---|
| | | n | e | Total Kek. | |
| 2 | (ii) | 1[a] | | 1 | |
| 3 | (i) | 1[a] | | 1 | |
| 4 | (ii) | 2[a] | | 2 | |
| 5 | (i) | 1[a] | 1[a] | 2 | |
| 6 | (ii) | 3[a] | 0 | 3 | |
| 7 | (i) | 3[a] | 0 | 3 | |
| 8 | (ii) | 4[a] | 2[a] | 6 | |
| 9 | (i) | 5[a] | 2[a] | 7 | |
| 10 | (ii) | 10[a] | 1[a] | 11 | |
| 11 | (i) | 11 | 2[b] | 13 | 1[c] |
| 12 | (ii) | 16 | 5[b] | 21 | 0 |
| 13 | (i) | 16 | 7[b] | 23 | 0 |
| 14 | (ii) | 33 | 7[b] | 40 | 1[c] |
| 15 | (i) | 38 | 9[b] | 47 | 3[c] |
| 16 | (ii) | 58 | 19[b] | 77 | 3[c] |
| 17 | (i) | 65 | 27[b] | 92 | 2[c] |
| 18 | (ii) | 117 | 34[b] | 151 | 5[c] |
| 19 | (i) | 136 | 39[b] | 175 | 14[c] |
| 20 | (ii) | 211 | 84[b] | 295 | 15[c] |

* Abbreviations: See footnote to Table 32.
[a] Brunvoll, Cyvin and Cyvin (1987) [71]; [b] Brunvoll, Cyvin, Cyvin and Gutman (1988) [105]; [c] Gutman and Cyvin (1988) [102]

Björg N. Cyvin, Jon Brunvoll, and Sven J. Cyvin

**Table 38.** Numbers of classified benzenoids with centrosymmetry ($C_{2h}$)

| $h$ | (Kind) | Kekuléan* | | | Concealed non-Kekuléan |
| | | n | e | Total Kek. | |
|---|---|---|---|---|---|
| 4 | (ii) | 1[a] | | 1 | |
| 5 | (i) | 1[a] | | 1 | |
| 6 | (ii) | 6[b] | 1[b] | 7 | |
| 7 | (i) | 4[b] | 3[b] | 7 | |
| 8 | (ii) | 28[b] | 7[b] | 35 | |
| 9 | (i) | 20[b] | 16[b] | 36 | |
| 10 | (ii) | 116[b] | 53[b] | 169 | |
| 11 | (i) | 88[b] | 87[c] | 175 | 2[d] |
| 12 | (ii) | 496 | 306 | 802 | 5[d] |
| 13 | (i) | 384 | 452 | 836 | 23 |
| 14 | (ii) | 2104 | 1702 | 3806 | 58 |
| 15 | (i) | 1651 | 2317 | 3968 | 177 |
| 16 | (ii) | 8990 | 9124 | 18114 | 502 |
| 17 | (i) | 7128 | 11762 | 18890 | 1208 |

* Abbreviations: See footnote to Table 32.
[a] Balaban and Harary (1968) [13]; [b] Brunvoll, Cyvin and Cyvin (1987) [71]; [c] Brunvoll, Cyvin, Cyvin and Gutman (1988) [105]; [d] Gutman and Cyvin (1989) [22]

presented in Tables 37 and 38, respectively. Here only Table 37 includes some supplementary numbers for $D_{2h}$ concealed non-Kekuléans in continuation of Table 30.

The 15 smallest concealed non-Kekuléans with $D_{2h}$ symmetry have been depicted [93] and are also reproduced in Fig. 39. Otherwise the forms of dihedral and

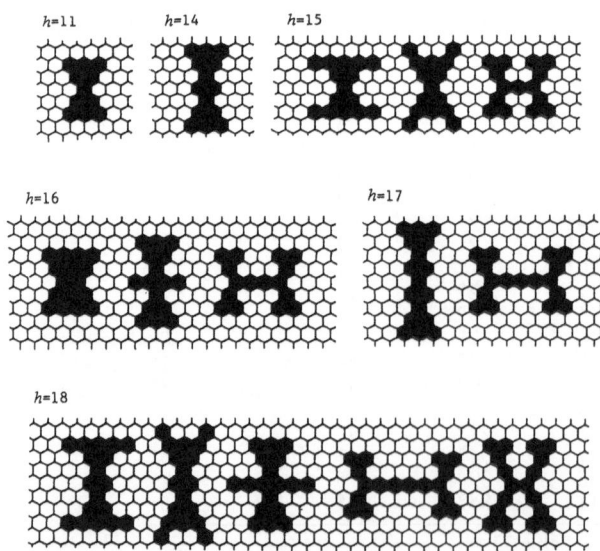

$h=11$    $h=14$    $h=15$

$h=16$    $h=17$

$h=18$

**Fig. 39.** All concealed non-Kekuléan benzenoids with $D_{2h}$ symmetry and $h \leq 18$

centrosymmetrical benzenoids have not been treated in particular, except for the catacondensed $D_{2h}$ systems (cf. Fig. 19).

*Erratum*

One of the $D_{2h}$ concealed non-Kekuléans with $h = 14$ is depicted by Gutman and Cyvin [102] under a wrong indication of its number of hexagons.

## 12.4 Mirror Symmetry

*Topological Properties*

Mirror-symmetrical (symmetry $C_{2v}$) benzenoids occur for all $h \geq 3$.

Two types of $C_{2v}$ benzenoids are distinguished: (a) those where the (two-fold) symmetry axis cuts edges perpendicularly; (b) those where the symmetry axis goes through vertices (of one or more edges). Adherring to the convention that some of the edges should be vertical, a system of $C_{2v}$(a), resp. $C_{2v}$(b), can be drawn so that the symmetry axis is horizontal, resp. vertical.

The $C_{2v}$(a) systems are restricted to $\Delta = 0$, while all $\Delta$ values are allowed for $C_{2v}$(b).

*Numbers*

The systematic investigations of benzenoids with specific symmetries started naturally with the highest symmetries (hexagonal, trigonal, ...). With regard to the mirror symmetry ($C_{2v}$) there has still not been very much done. We have supplemented the existing data by means of a specific generation of the $C_{2v}$ benzenoids achieved for the first time; cf. Table 39. But still there is more which can be done in this area.

**Table 39.** Numbers of classified benzenoids with mirror symmetry, $C_{2v}$*

| $h$ | (Type) | $\Delta$ | Kekuléan | | | non-Kekuléan | |
|---|---|---|---|---|---|---|---|
| | | | n | e | Total Kek. | o | Total non-Kek. |
| 3 | (a) | 0 | 1 | | 1 | | |
| 4 | (b) | 0 | 1 | | 1 | | |
| 5 | (a) | 0 | 5 | | | | |
| | (b) | 0 | 1 | | 6 | | 3 |
| | | 1 | 0 | | | 3 | |
| 6 | (a) | 0 | 7 | | | 0 | |
| | (b) | 0 | 3 | | | 0 | |
| | | 1 | 0 | | 10 | 1 | 2 |
| | | 2 | 0 | | | 1 | |
| 7 | (a) | 0 | 18 | 4 | | 0 | |
| | (b) | 0 | 4 | 2 | | 0 | |
| | | 1 | 0 | 0 | 28 | 10 | 11 |
| | | 2 | 0 | 0 | | 1 | |

161

**Table 39.** (Continued)

| h | (Type) | Δ | Kekuléan | | | non-Kekuléan | |
|---|---|---|---|---|---|---|---|
| | | | n | e | Total Kek. | o | Total non-Kek. |
| 8 | (a) | 0 | 32 | 2 | | 0 | |
| | (b) | 0 | 15 | 0 | 49 | 0 | 12 |
| | | 1 | 0 | 0 | | 5 | |
| | | 2 | 0 | 0 | | 7 | |
| 9 | (a) | 0 | 82 | 23 | | 0 | |
| | (b) | 0 | 18 | 6 | | 0 | |
| | | 1 | 0 | 0 | 129 | 39 | 49 |
| | | 2 | 0 | 0 | | 9 | |
| | | 3 | 0 | 0 | | 1 | |
| 10 | (a) | 0 | 131 | 24 | | 0 | |
| | (b) | 0 | 54 | 7 | 216 | 0 | 58 |
| | | 1 | 0 | 0 | | 20 | |
| | | 2 | 0 | 0 | | 38 | |
| 11 | (a) | 0 | 334 | 139 | | 1 | |
| | (b) | 0 | 74 | 27 | 574 | 0 | 222 |
| | | 1 | 0 | 0 | | 156 | |
| | | 2 | 0 | 0 | | 52 | |
| | | 3 | 0 | 0 | | 13 | |
| 12 | (a) | 0 | 560 | 164 | | 0 | |
| | (b) | 0 | 222 | 38 | | 0 | |
| | | 1 | 0 | 0 | 984 | 80 | 267 |
| | | 2 | 0 | 0 | | 176 | |
| | | 3 | 0 | 0 | | 5 | |
| | | 4 | 0 | 0 | | 6 | |
| 13 | (a) | 0 | 1377 | 755 | | 21 | |
| | (b) | 0 | 297 | 120 | | 2 | |
| | | 1 | 0 | 0 | 2549 | 652 | 1029 |
| | | 2 | 0 | 0 | | 266 | |
| | | 3 | 0 | 0 | | 84 | |
| | | 4 | 0 | 0 | | 4 | |
| 14 | (a) | 0 | 2322 | 990 | | 11 | |
| | (b) | 0 | 879 | 209 | | 0 | |
| | | 1 | 0 | 0 | 4440 | 347 | 1292 |
| | | 2 | 0 | 0 | | 853 | |
| | | 3 | 0 | 0 | | 35 | |
| | | 4 | 0 | 0 | | 46 | |
| 15 | (a) | 0 | 5703 | 4018 | | 176 | |
| | (b) | 0 | 1205 | 559 | | 9 | |
| | | 1 | 0 | 0 | | 2789 | |
| | | 2 | 0 | 0 | 11485 | 1289 | 4805 |
| | | 3 | 0 | 0 | | 486 | |
| | | 4 | 0 | 0 | | 52 | |
| | | 5 | 0 | 0 | | 4 | |
| 16 | (a) | 0 | 9657 | 5547 | | 141 | |
| | (b) | 0 | 3562 | 1104 | | 4 | |
| | | 1 | 0 | 0 | | 1474 | |
| | | 2 | 0 | 0 | 19870 | 4033 | 6199 |
| | | 3 | 0 | 0 | | 214 | |
| | | 4 | 0 | 0 | | 331 | |
| | | 5 | 0 | 0 | | 2 | |

\* Abbreviations: See footnote to Table 35

# 13 All-Benzenoids

## 13.1 Some Topological Properties

The class of benzenoids called all-benzenoids (or fully benzenoids) is a subclass of the normal benzenoids. Hence each all-benzenoid is Kekuléan and has $\Delta = 0$.

The class of all-benzenoids is also a subclass of 2-factorable benzenoids. A 2-factorable benzenoid is also 1-factorable, where a 1-factorable benzenoid is synonymous with a Kekuléan.

An all-benzenoid may be catacondensed, but only for every third $h$ value; specifically $h = 1, 4, 7, 10, \dots$ . Pericondensed all-benzenoids occur for $h = 6$ and $h \geq 8$.

## 13.2 Catacondensed and Pericondensed All-Benzenoids

The enumeration of all-benzenoids was foreshadowed by Dias [121–123], who discussed 2-factorable benzenoids in the frame of the enumeration of benzenoid isomers (according to the chemical formulas $C_nH_s$). In these works Dias depicted some all-benzenoids as examples. It was Knop et al. [91] who presented the first list of the numbers of all-benzenoids according to the number of hexagons ($h$); see Table 40. In later works, Dias [25, 124, 125] enumerated some all-benzenoid

**Table 40.** Numbers of all-benzenoids, including their subdivision into catacondensed and pericondensed systems

| $h$ | Catacondensed | Pericondensed | Total |
|----|----|----|----|
| 1  | 1[a]   |        | 1[a]   |
| 4  | 1[a]   |        | 1[a]   |
| 6  | 0      | 1[a]   | 1[a]   |
| 7  | 2[a]   | 0      | 2[a]   |
| 8  | 0      | 1[a]   | 1[a]   |
| 9  | 0      | 3[a]   | 3[a]   |
| 10 | 6[a]   | 3[a]   | 9[a]   |
| 11 | 0      | 10[b]  | 10[b]  |
| 12 | 0      | 29[b]  | 29[b]  |
| 13 | 32[b]  | 25[b]  | 57[b]  |
| 14 | 0      | 102[b] | 102[b] |
| 15 | 0      | 259[b] | 259[b] |
| 16 | 172[b] | 354[b] | 526[b] |
| 17 | 0      | 1136[b]| 1136[b]|
| 18 | 0      | 2713[b]| 2713[b]|
| 19 | 1139[b]| †      | †      |
| 20 | 0      | †      | †      |
| 21 | 0      | †      | †      |
| 22 | 7661[b]| †      | †      |

[a] Knop, Müller, Szymanski and Trinajstić (1986) [91]; [b] Cyvin, Brunvoll, Cyvin and Gutman (1988) [127]; † Unknown

Björg N. Cyvin, Jon Brunvoll, and Sven J. Cyvin

isomers with emphasis on the strain-free systems. These enumerations are
summarized in a recent review [126]. The list of Knop et al. [91] was extended
substantially by Cyvin et al. [127]; cf. Table 40.

Figure 40 shows the forms of the all-benzenoids up to $h = 13$ reproduced from
Cyvin et al. [127].

Tables 41 and 42 give an account of the classification according to symmetry
for the catacondensed and pericondensed benzenoids, respectively. Most of the
data are from Cyvin et al. [127]: cf. also Gutman and Cyvin [22]. For the
catacondensed all-benzenoids with $D_{2h}$ symmetry, we have also generated by hand,
in continuation of Table 41, 3 systems with $h = 31$, and 13 systems with $h = 43$,
claiming that this covers all such systems for $h \leq 43$.

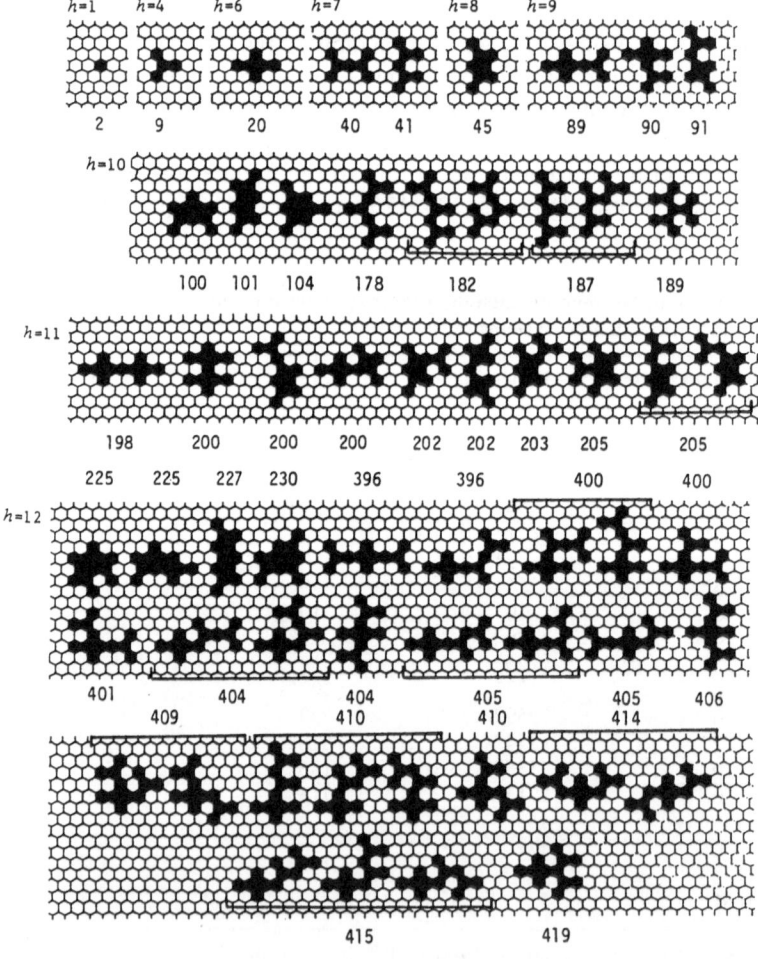

(see next page)

164

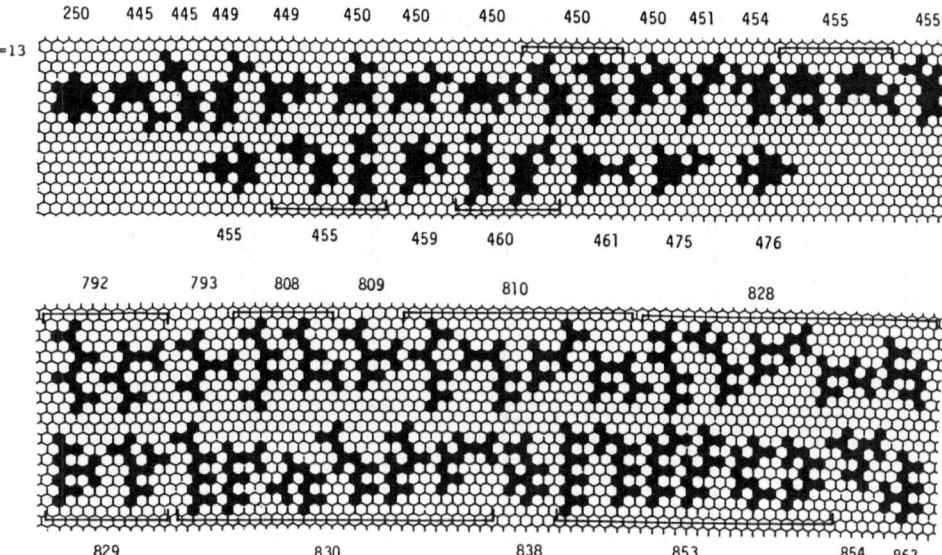

**Fig. 40.** The all-benzenoids for $h \leq 13$. $K$ numbers are given

Dias [25, 124] defined the strain-free all-benzenoids by the absence of fjords; a cove never occurs in an all-benzenoid. A fjord gives a zero-carbon gap between vertices of degree two (corresponding to secondary carbon atoms), thus causing steric hindrance for the hydrogen atoms. This proximity of carbon atoms could also be achieved, if the hydrocarbons were planar, without fjords. The smallest examples of this kind are represented by the below systems with $h = 16$ (left) and $h = 18$ (right). Primarily this kind of systems are included here among the

**Table 41.** Numbers of catacondensed all-benzenoids, classified according to symmetry*

| $h$ | $D_{3h}$ | $C_{3h}$ | $D_{2h}$ | $C_{2h}$ | $C_{2v}$ | $C_s$ |
|-----|----------|----------|----------|----------|----------|-------|
| 4   | 1[a]     |          |          |          |          |       |
| 7   | 0        |          | 1[a]     |          | 1[a]     |       |
| 10  | 1[a]     |          | 0        |          | 2[a]     | 3[a]  |
| 13  | 1[a]     | 1[a]     | 0        | 3[a]     | 7[a]     | 20[a] |
| 16  | 0        | 0        | 0        | 0        | 14[a]    | 158[a] |
| 19  | 0        | 2[a]     | 1[a]     | 19[a]    | 41[a]    | 1076[a] |
| 22  | 0        | 8[a]     | 0        | 0        | 79[a]    | 7574[a] |
| 25  | 0        | 0        | 0        | †        | †        | †     |
| 28  | 0        | 11       | 0        | †        | †        | †     |

* Benzene ($h = 1$, not included in this table) is the only (trivial) catacondensed all-benzenoid of $D_{6h}$ symmetry. There are no catacondensed all-benzenoids of $C_{6h}$ symmetry. For $D_{2h}$, see also the text.
[a] Cyvin, Brunvoll, Cyvin and Gutman (1988) [127]; † Unknown

165

Björg N. Cyvin, Jon Brunvoll, and Sven J. Cyvin

**Table 42.** Numbers of pericondensed all-benzenoids, classified according to symmetry*

| $h$ | $D_{6h}$ | $D_{3h}$ | $C_{3h}$ | $D_{2h}$ | $C_{2h}$ | $C_{2v}$ | $C_s$ |
|----|------|------|------|------|------|------|------|
| 6  |      |      |      | 1[a] |      |      |      |
| 8  |      |      |      | 0    |      | 1[a] |      |
| 9  |      |      |      | 0    |      | 1[a] | 2[a] |
| 10 |      | 1[a] |      | 0    | 1[a] | 1[a] | 0    |
| 11 |      | 0    |      | 2[a] | 0    | 2[a] | 6[a] |
| 12 |      | 0    |      | 1[a] | 2[a] | 5[a] | 21[a] |
| 13 | 1[a] | 0    |      | 0    | 0    | 3[a] | 21[a] |
| 14 | 0    | 0    |      | 1[a] | 2[a] | 12[a] | 87[a] |
| 15 | 0    | 0    |      | 0    | 2[a] | 14[a] | 243[a] |
| 16 | 0    | 1[a] | 1[a] | 2[a] | 9[a] | 18[a] | 323[a] |
| 17 | 0    | 0    | 0    | 2[a] | 11[a] | 38[a] | 1085[a] |
| 18 | 0    | 0    | 0    | 1[a] | 22[a] | 58[a] | 2632[a] |
| 19 | 0    | 4    | 3    | 2    | †    | †    | †    |
| 20 | 0    | 0    | 0    | 3    | †    | †    | †    |
| 21 | 0    | 0    | 0    | 3    | †    | †    | †    |
| 22 | 0    | 1    | 4    | 3    | †    | †    | †    |
| 23 | 0    | 0    | 0    | 5    | †    | †    | †    |

* All-benzenoids with $C_{6h}$ symmetry occur at $h \geq 31$.
[a] Cyvin, Brunvoll, Cyvin and Gutman (1988) [127]; † Unknown

strain-free all-benzenoids, adherring to the absence of fjords as the decisive criterion. A list of the numbers of strain-free all-benzenoids (for $h > 1$) is presented in Table 43. However, it is also of interest to distinguish the systems with

$C_{66}H_{36}$          $C_{72}H_{38}$

zero-carbon gaps in spite of the absence of fjords (like the examples of $C_{66}H_{36}$ and $C_{72}H_{38}$ above). Therefore, in Table 43, the numbers of "absolutely strain-free" all-benzenoids, where such systems are excluded, are given in parentheses. Figure 41 displays the forms of the strain-free all-benzenoids up to $h = 16$.

*Comments and Errata*

In Knop et al. [91], by an obvious misprint, there is a missing (full) hexagon in one of the depictions for $h = 10$.

**Table 43.** Numbers of strain-free all-benzenoids, including their sub-division into catacondensed and pericondensed systems*

| $h$ | Catacondensed | Pericondensed | Total |
|---|---|---|---|
| 4 | 1 [a, b] | | 1 [a, b] |
| 6 | 0 | 1 [a−c] | 1 [a−c] |
| 7 | 1 [a, b] | 0 | 1 [a, b] |
| 8 | 0 | 1 [b] | 1 [b] |
| 9 | 0 | 1 [a, b] | 1 [a, b] |
| 10 | 1 [d, e] | 3 [d, e] | 4 [d, e] |
| 11 | 0 | 2 [d, e] | 2 [d, e] |
| 12 | 0 | 5 [f] | 5 [f] |
| 13 | 3 [d, e] | 5 [d, e] | 8 [d, e] |
| 14 | 0 | 13 [f] | 13 [f] |
| 15 | 0 | 16 [f] | 16 [f] |
| 16 | 4 (3) | 25 [f] | 29 (28) |
| 17 | 0 | 42 | 42 |
| 18 | 0 | 73 (72) | 73 (72) |
| 19 | 11 (8) | 110 (108) | 121 (116) |
| 20 | 0 | 187 (184) | 187 (184) |
| 21 | 0 | 321 (305) | 321 (305) |
| 22 | 23 (12) | 501 (485) | 524 (497) |
| 23 | 0 | 886 (834) | 886 (834) |
| 24 | 0 | 1477 (1370) | 1477 (1370) |
| 25 | 62 (25) | 2447 (2276) | 2509 (2301) |

* Numbers of "absolutely strain-free" systems (without any zero-carbon gap) are given in parentheses.
[a] Dias (1985) [121]; [b] Dias (1985) [123]; [c] Dias (1985) [122]; [d] Dias (1987) [124]; [e] Dias (1987) [25]; [f] Dias (1989) [125]

Four all-benzenoid systems were omitted in a report by Dias [124], but enclosed as erratum with the reprints. The material without correction was reproduced in the book of Dias [25], but an erratum appeared later [128]. The four systems in question are also reproduced elsewhere [125]; herein the numbers of strain-free all-benzenoids with $h = 17$, $n_i$ (number of internal vertices) = 10 ($C_{60}H_{28}$), and with $h = 18$, $n_i = 14$ ($C_{60}H_{26}$) are in error. In both cases we have located two isomorphic systems among the depictions.

## 13.3 Hexagonal Symmetry: All-Flakes

All-benzenoids with hexagonal ($D_{6h}$ or $C_{6h}$) symmetry have been referred to as *all-flakes* [129]. In other words, an all-flake is an all-benzenoid snowflake. We may also speak about proper ($D_{6h}$) and improper ($C_{6h}$) all-flakes as subclasses of the proper and improper snowflakes, respectively.

The known numbers of all-flakes are given in Table 44 as a continuation of a part of Table 42.

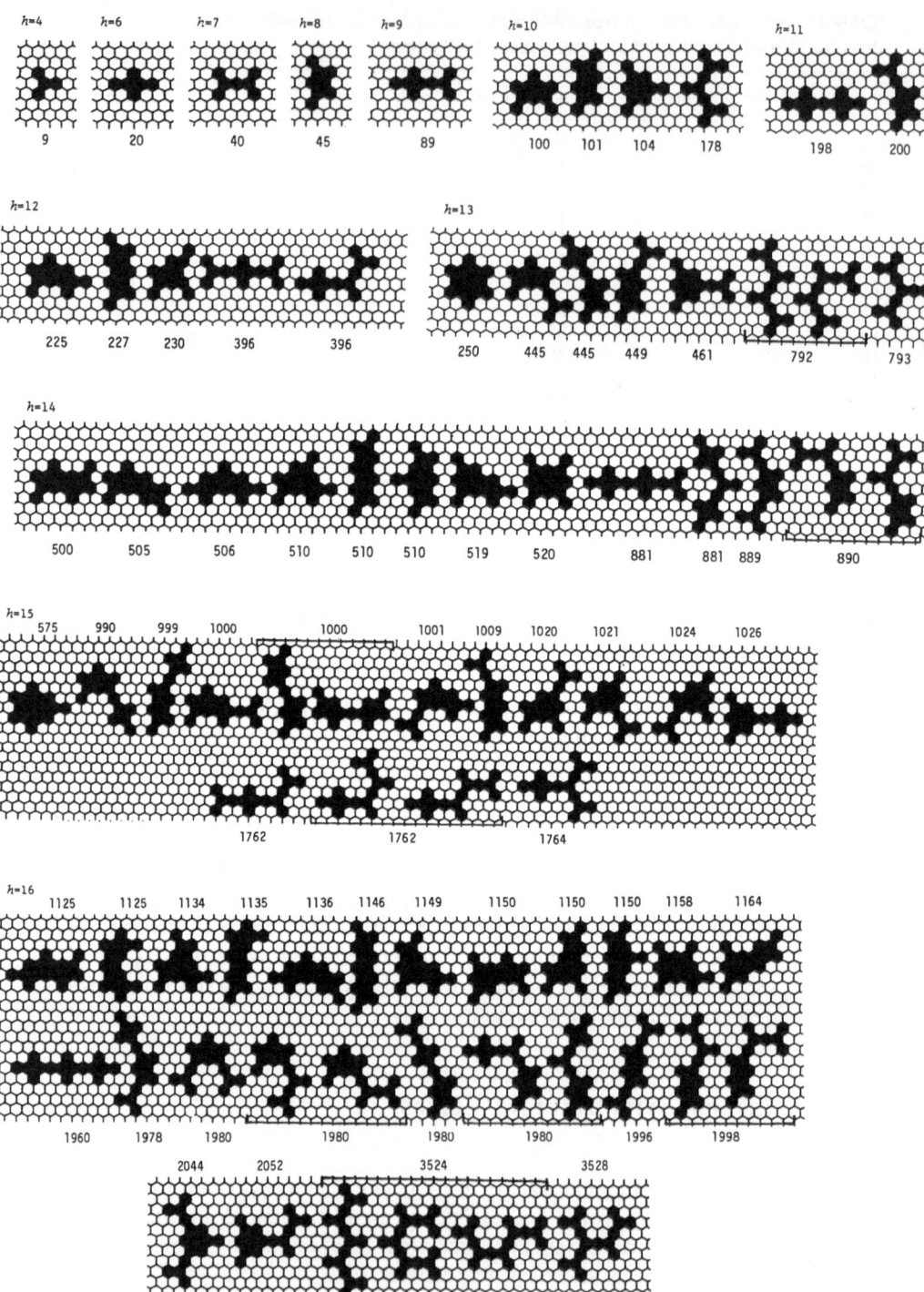

**Fig. 41.** The strain-free all-benzenoids for $4 \le h \le 16$. $K$ numbers are given

**Table 44.** Numbers of all-benzenoids with hexagonal symmetry*

| $h$ | $D_{6h}$ | $C_{6h}$ | Total ($D_{6h} + C_{6h}$) | $h$ | $D_{6h}$ |
|---|---|---|---|---|---|
| 25 | 1 | | 1 | 61 | 1 |
| 31 | 0 | 1 | 1 | 67 | 1 |
| 37 | 0 | 1 | 1 | 73 | 3 |
| 43 | 2 | 2 | 4 | 79 | 1 |
| 49 | 0 | 4 | 4 | 85 | 3 |
| 55 | 1 | 8 | 9 | 91 | 3 |

* Supplements Table 42. For $h = 24$ there are no benzenoids with hexagonal symmetry. Data from: Cyvin, Cyvin and Brunvoll (1989) [129]

In Fig. 42 the forms of the snowflakes for $h \leq 55$ are shown. Those for $h = 43$ and $h = 49$ are also found in Cyvin et al. [129]. The proper all-flakes up to $h = 85$ are depicted in Fig. 43; cf. also Cyvin et al. [129].

*Erratum*

Among the proper all-flakes of Cyvin et al. [129] the depiction for $h = 67$ is a wrong system.

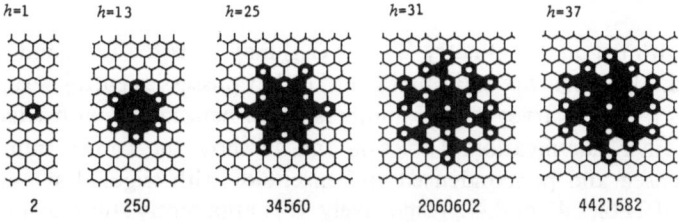

| $h=1$ | $h=13$ | $h=25$ | $h=31$ | $h=37$ |
|---|---|---|---|---|
| 2 | 250 | 34560 | 2060602 | 4421582 |

$h=43$

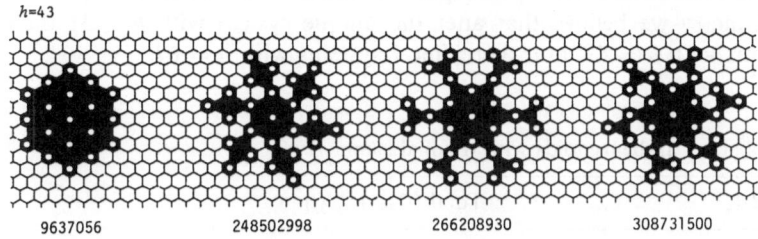

9637056     248502998     266208930     308731500

$h=49$

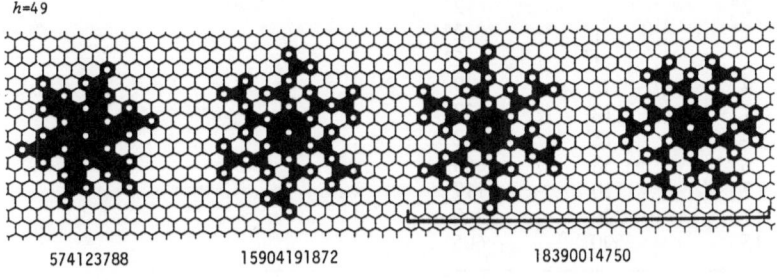

574123788     15904191872     18390014750

(see next page)

$h=55$

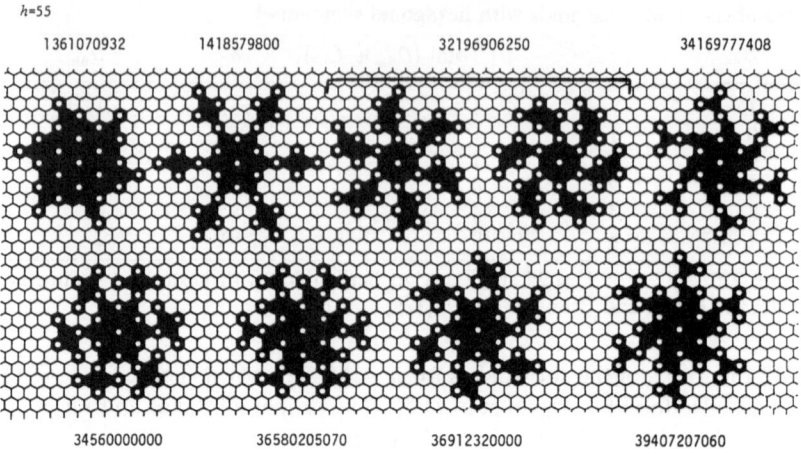

Fig. 42. All-flakes: the all-benzenoids with hexagonal symmetry ($D_{6h}$ or $C_{6h}$) and $h < 61$. $K$ numbers are given

## 13.4 Trigonal Symmetry

All-benzenoids with trigonal symmetry ($D_{3h}$ or $C_{3h}$) may be either catacondensed or pericondensed. All of them are of the first kind, i.e. they possess a central hexagon.

Table 45 gives a gross survey of the numbers of all-benzenoids with trigonal symmetry. As a finer classification, Tables 46 and 47 show the known numbers of the catacondensed and pericondensed all-benzenoids with trigonal symmetry in continuation of Tables 41 and 42, respectively. It is apparently still much work which could be done in this area. With regard to the catacondensed all-benzenoids with $D_{3h}$ symmetry we believe that after the unique system with $h = 31$ there come 2 systems with $h = 49$, and thereafter 7 systems with $h = 67$.

Table 45. Numbers of all-benzenoids with trigonal symmetry

| $h$ | $D_{3h}$ | $C_{3h}$ | Total ($D_{3h} + C_{3h}$) |
|---|---|---|---|
| 4 | 1[a] | | 1[a] |
| 10 | 2[a] | | 2[a] |
| 13 | 1[a] | 1[a] | 2[a] |
| 16 | 1[a] | 1[a] | 2[a] |
| 19 | 4 | 5 | 9 |
| 22 | 1 | 12 | 13 |
| 25 | 4 | 18 | 22 |
| 28 | 6 | 58 | 64 |
| 31 | 7 | 110 | 117 |
| 34 | 11 | 229 | 240 |
| 37 | 15 | 590 | 605 |

[a] Cyvin, Brunvoll, Cyvin and Gutman (1988) [127]

**Table 46.** Numbers of catacondensed all-benzenoids with trigonal symmetry*

| $h$ | $D_{3h}$ | $C_{3h}$ | Total ($D_{3h}$ + $C_{3h}$) |
|-----|----------|----------|------------------------------|
| 31  | 1        | 47       | 48                           |
| 34  | 0        | 0        | 0                            |
| 37  | 0        | 68       | 68                           |

* Supplements Table 41; see also the text. For $h = 29$ and $h = 30$ there are no benzenoids with trigonal symmetry of the first kind

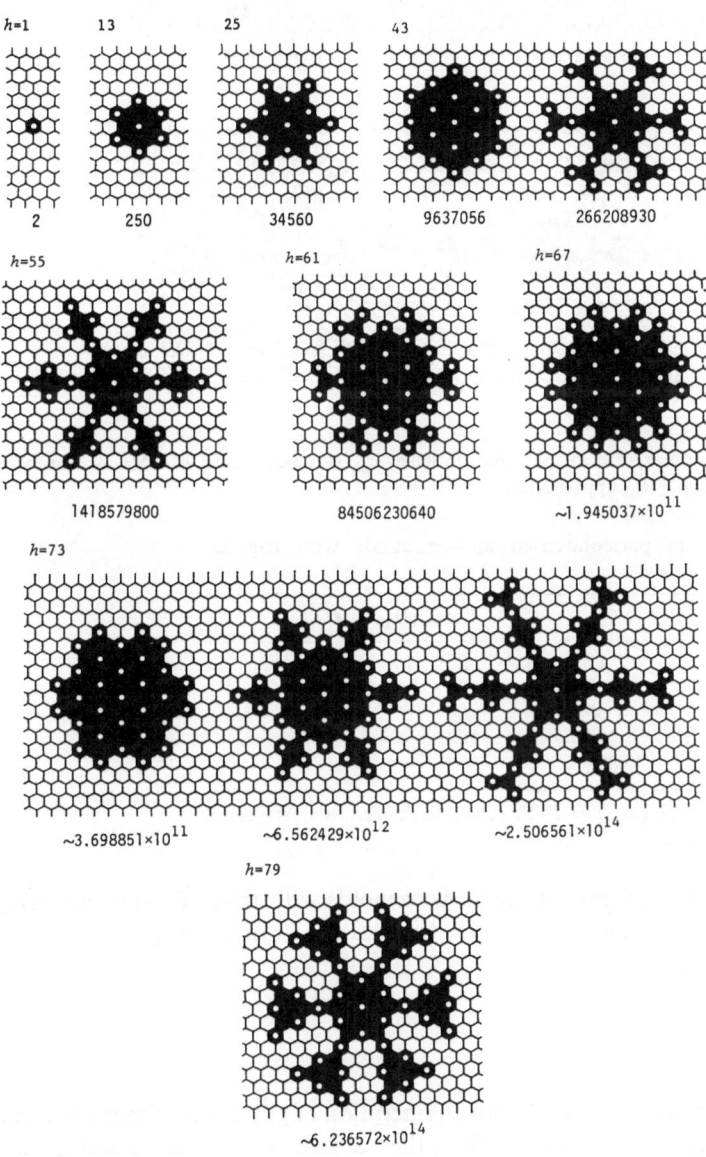

(see next page)

$h=85$

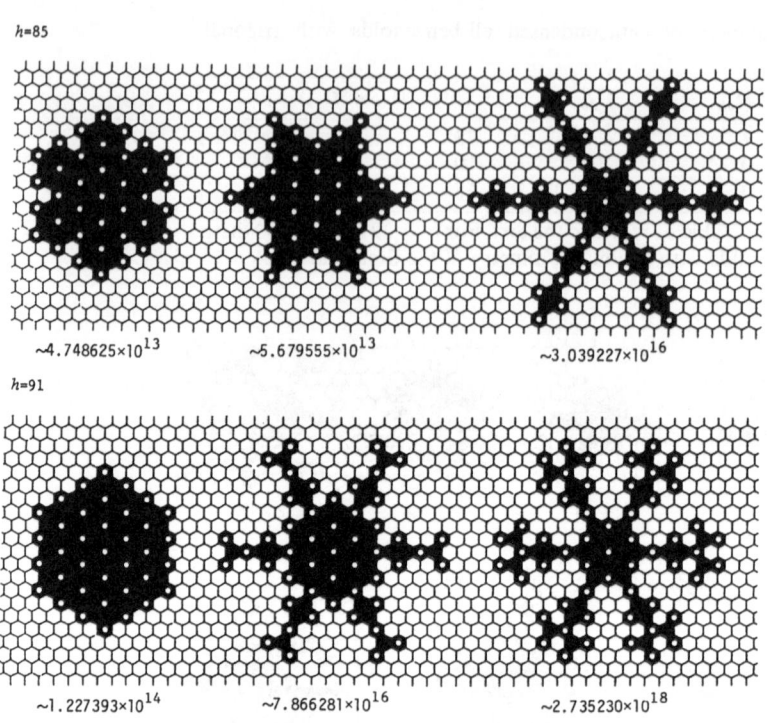

~$4.748625\times10^{13}$     ~$5.679555\times10^{13}$          ~$3.039227\times10^{16}$

$h=91$

~$1.227393\times10^{14}$          ~$7.866281\times10^{16}$          ~$2.735230\times10^{18}$

**Fig. 43.** Proper all-flakes: the all-benzenoids with $D_{6h}$ symmetry and $h < 97$. K numbers are given; those for $h \geq 67$ approximate

**Table 47.** Numbers of pericondensed all-benzenoids with trigonal symmetry*

| $h$ | $D_{3h}$ | $C_{3h}$ | Total $(D_{3h} + C_{3h})$ |
|---|---|---|---|
| 25 | 4 | 18 | 22 |
| 28 | 6 | 47 | 53 |
| 31 | 6 | 63 | 69 |
| 34 | 11 | 229 | 240 |
| 37 | 15 | 522 | 537 |

* Supplements Table 42. For $h = 24$ there are no benzenoids of trigonal symmetry of the first kind

Figure 44 shows the forms of the all-benzenoids with trigonal symmetry ($D_{3h}$ and $C_{3h}$) and $h \leq 25$. Those with regular trigonal ($D_{3h}$) symmetry and $h \leq 37$ are displayed in Fig. 45.

# 14 Conclusion

It is not intended to close research into enumeration of polyhexes with the present review. This should be clear, not only from the last sentence of Sect. 12 and similar statements, but also from the spirit shown throughout the whole chapter, and not least in several of the comments. On the contrary, it is rather intended to inspire researchers in the field to further achievements in this realm and provide them with a comprehensive survey on the work which has already been done.

Balaban and Artemi wrote very recently (1990) [63]. "The enumeration of polycyclic benzenoid hydrocarbons (polyhexes, or benzenoids) continues to be a challenging problem." The field has been flourishing, not least during the very last few years, as documented by the present list of references. It contains 16 relevant publications from 1989, which can be supplemented by a few more

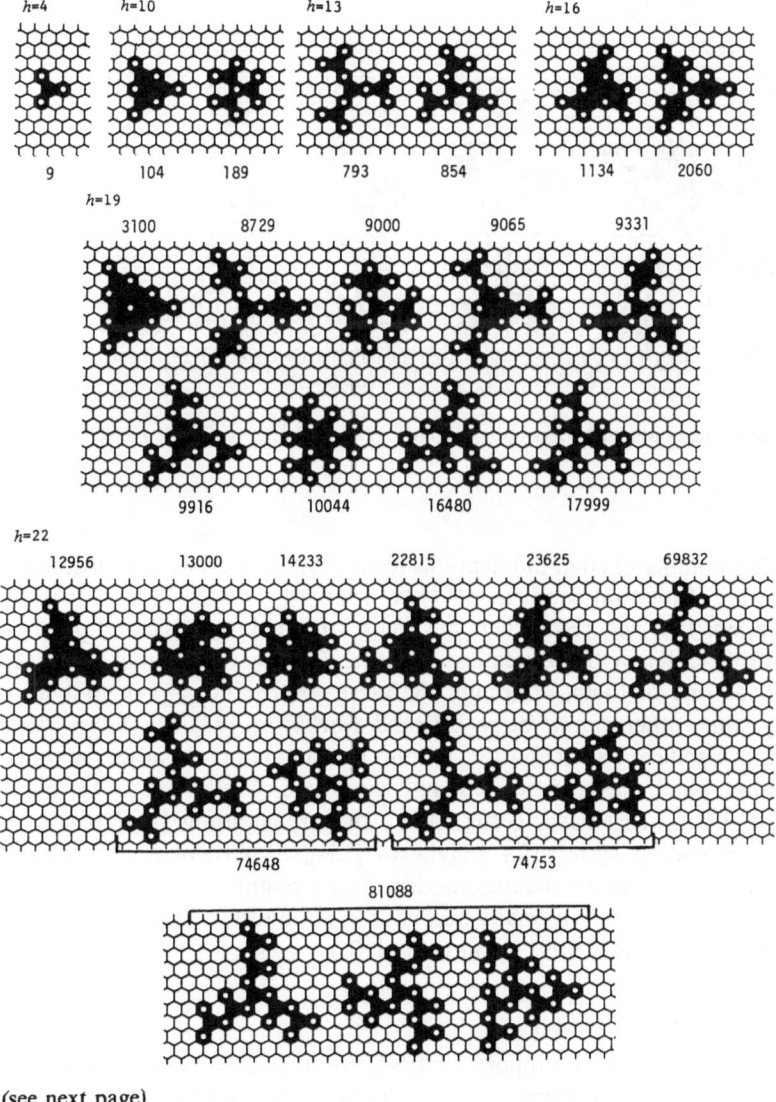

(see next page)

Björg N. Cyvin, Jon Brunvoll, and Sven J. Cyvin

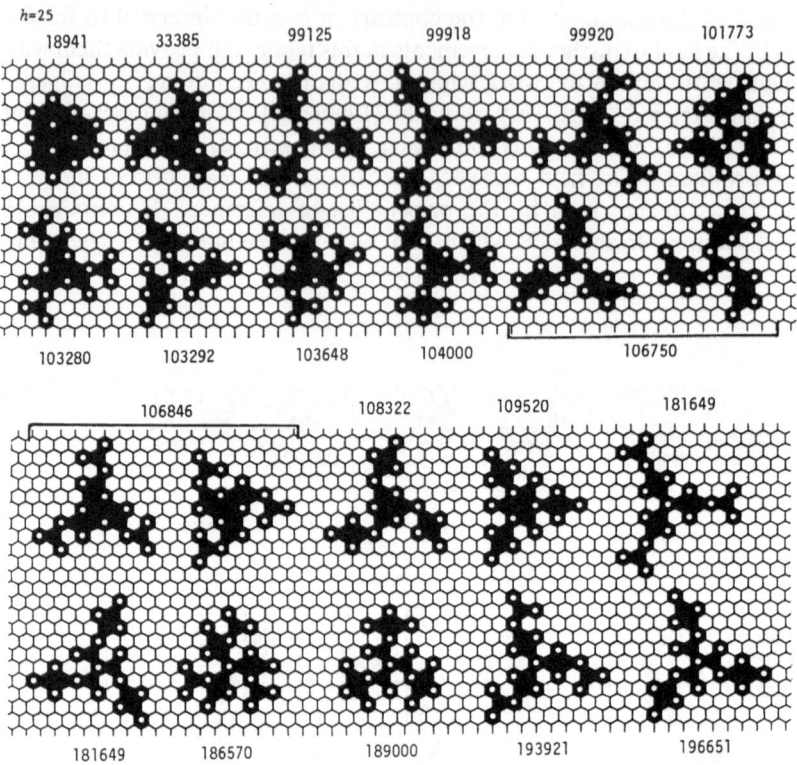

**Fig. 44.** The all-benzenoids with trigonal symmetry ($D_{3h}$ and $C_{3h}$) and $h < 28$. $K$ numbers are given

[130–133]. The number of cited publications from 1990/91 is 20 with a substantial number (viz. 19) of relevant supplements [134–152].

It also happens that new researchers are being attracted to the field of polyhex enumerations. E. C. Kirby (Resource Use Institute, Pitlochry, Scotland, UK) may be reckoned as one of them, although his latest contributions [141, 143] were preceded by some other enumeration-oriented works [85, 153]. His latest work [141] makes a significant contribution to our understanding of all-benzenoids. Another name, which has recently entered the arena, is William C. Herndon (University of Texas at El Paso, Texas, USA). His enumerations [138], based on a computerized coding system for polyhexes [154], is particularly chemistry-oriented inasmuch as it takes stereoisomerism into account.

A monograph which contains many new enumeration results for coronoids has been announced as an early 1991 publication [155].

*Acknowledgement:* Financial support to BNC from The Norwegian Research Council for Science and the Humanities is gratefully acknowledged.

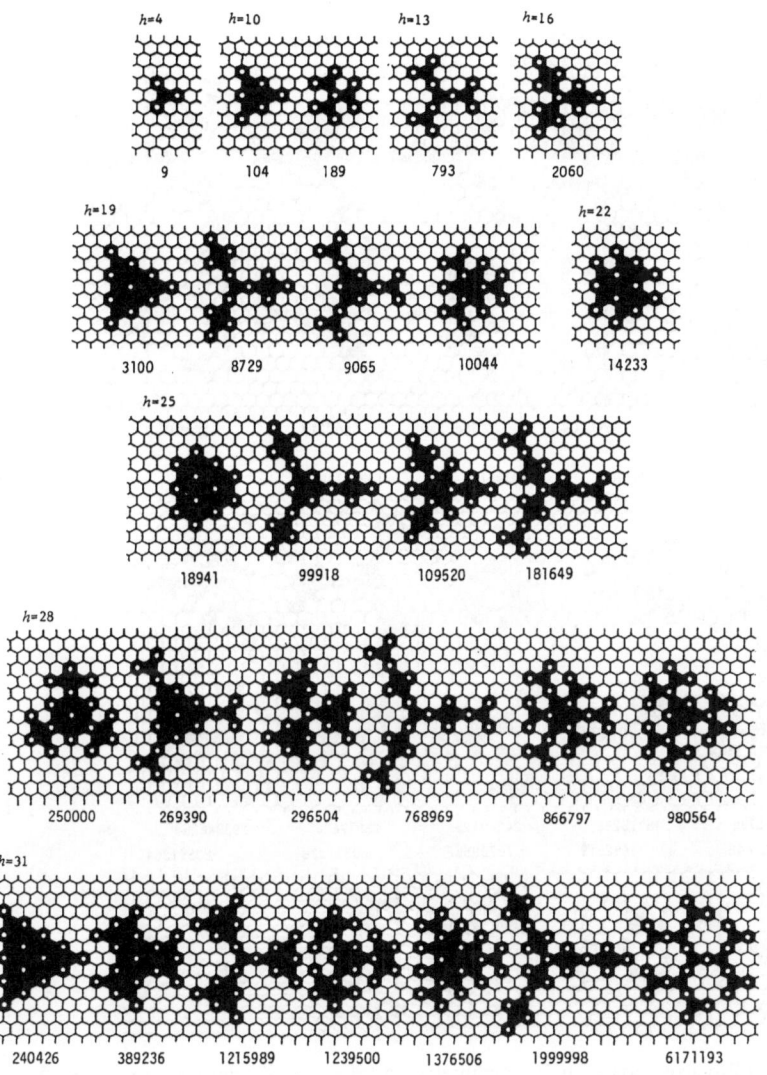

(see next page)

175

Björg N. Cyvin, Jon Brunvoll, and Sven J. Cyvin

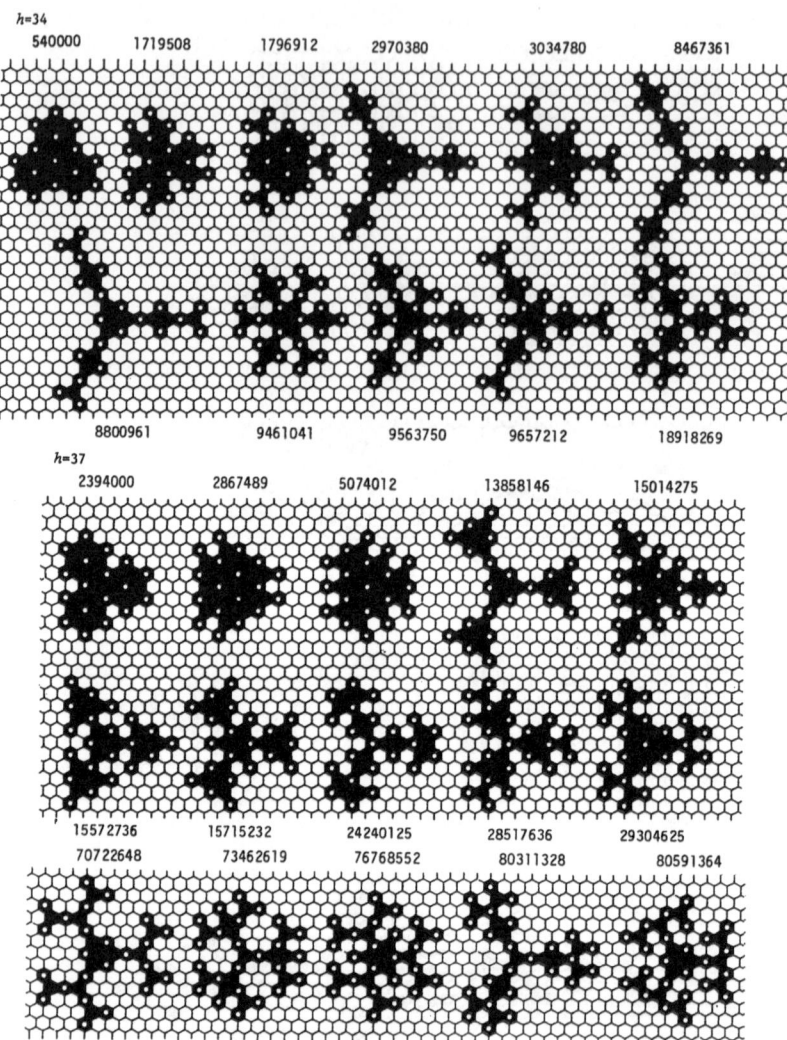

**Fig. 45.** The all-benzenoids with $D_{3h}$ symmetry and $h < 40$. $K$ numbers are given

# 15 References

1. Klarner DA (1967) Can J Math 19: 851
2. Harary F (1967) In: Harary F (ed) Graph theory and theoretical physics, Academic, London, p 1
3. Palmer EM (1972) In: Alavi Y, Lick DR, White AT (eds) Graph theory and applications, Proceedings of the conference at Western Michigan University, May 10–13, 1972 (Lecture Notes in Mathematics 303); Springer, Berlin Heidelberg New York, p 215
4. Harary F, Palmer EM (1973) Graphical enumeration, Academic, New York
5. Harary F, Palmer EM, Read RC (1975) Discrete Math 11: 371
6. Harary F, Harborth H (1976) J Combinat Inf System Sci 1: 1

7. Klarner DA (1965) Fibonacci Quarterly 3: 9
8. Golomb SW (1954) Am Math Monthly 61: 675
9. Harary F, Read RC (1970) Proc. Edinburgh Math Soc 17 (Ser II): 1
10. Lunnon WF (1972) In: Read RC (ed) Graph theory and computing, Academic, London, p 87
11. Gutman I (1983) Croat Chem Acta 56: 365
12. Harary F (1968) In: Sachs H, Voss HJ, Walther H (eds) Beiträge zur Graphentheorie, Teubner, Leipzig, p 49
13. Balaban AT, Harary F (1968) Tetrahedron 24: 2505
14. Balasubramanian K, Kaufman JJ, Koski WS, Balaban AT (1980) J Comput Chem 1: 149
15. Knop JV, Szymanski K, Jeričević Ž, Trinajstić N (1983) J Comput Chem 4: 23
16. Knop JV, Müller WR, Szymanski K, Trinajstić N (1985) Computer generation of certain classes of molecules, SKTH/Kemija u industriji (Association of Chemists and Technologists of Croatia), Zagreb
17. Knop JV, Szymanski K, Jeričević Ž, Trinajstić N (1984) Match 16: 119
18. Balaban AT, Brunvoll J, Cioslowski J, Cyvin BN, Cyvin SJ, Gutman I, He WC, He WJ, Knop JV, Kovačević M, Müller WR, Szymanski K, Tošić R, Trinajstić N (1987) Z. Naturforsch 42a: 863
19. He WJ, He WC, Wang QX, Brunvoll J and Cyvin SJ (1988) Z. Naturforsch. 43a: 693
20. Balaban AT (1976) In: Balaban AT (ed) Chemical applications of graph theory, Academic, London, p 63
21. Balaban AT (1989) Carbon and its nets, Computers Math Applic 17: 397; reprinted in: Hargittai I (ed) (1989) Symmetry 2 Unifying human understanding, Pergamon, Oxford
22. Gutman I, Cyvin SJ (1989) Introduction to the theory of benzenoid hydrocarbons, Springer, Berlin Heidelberg New York
23. Trinajstić N (1983) Chemical graph theory, Vols I, II, CRC Press, Boca Raton, FL
24. Gutman I, Polansky OE (1986) Mathematical concepts in organic chemistry, Springer, Berlin Heidelberg New York
25. Dias JR (1987) Handbook of polycyclic hydrocarbons — Part A — Benzenoid hydrocarbons, Elsevier, Amsterdam
26. Cyvin SJ, Gutman I (1988) Kekulé structures in benzenoid hydrocarbons (Lecture Notes in Chemistry 46), Springer, Berlin Heidelberg New York
27. Gutman I (1982) Bull Soc Chim Beograd 47: 453
28. Ege G, Vogler H (1972) Theor Chim Acta 26: 55
29. Ege G, Vogler H (1972) Z Naturforsch 27b: 918
30. Balaban AT (1969) Tetrahedron 25: 2949
31. Knop JV, Müller WR, Szymanski K, Nikolić S, Trinajstić N (1990) In: Rouvray DH (ed) Computational graph theory, Nova, New York, p 9
32. He WC, He WJ (1985) Theor Chim Acta 68: 301
33. He WC, He WJ (1986) Tetrahedron 19: 5291
34. Cioslowski J (1987) J Comput Chem 8: 906
35. He WJ, He WC (1987) In: King RB, Rouvray DH (eds) Graph theory and topology in chemistry, A collection of papers presented at an international conference held at the University of Georgia, Athens, Georgia, USA, 16–20 March 1987 (Studies in Physical and Theoretical Chemistry 51), Elsevier, Amsterdam, p 476
36. Müller WR, Szymanski K, Knop JV, Nikolić S, Trinajstić N (1990) J Comput Chem 11: 223
37. Knop JV, Müller WR, Szymanski K, Trinajstić N (1990) J Chem Inf Comput Sci 30: 159
38. Dias JR (1982) J Chem Inf Comput Sci 22: 15
39. Knop JV, Müller WR, Szymanski K, Trinajstić N (1986) Match 20: 197
40. Brunvoll J, Cyvin BN, Cyvin SJ (1987) J Chem Inf Comput Sci 27: 14
41. Knop JV, Müller WR, Szymanski K, Trinajstić N (1990) J Mol Stuct (Theochem) 205: 361
42. Brunvoll J, Cyvin BN, Cyvin SJ, Knop JV, Müller WR, Szymanski K, Trinajstić N (1990) J Mol Struct (Theochem) 207: 131

43. Cyvin SJ, Brunvoll J, Cyvin BN (1990) J Chem Inf Comput Sci 30: 210
44. Brunvoll J, Cyvin BN, Cyvin SJ (1990) Croat Chem Acta 63: 585
45. Cyvin SJ, Brunvoll J, Cyvin BN, Bergan JL, Brendsdal E (1991) Struct Chem.: 2: 555
46. Stojmenović I, Tošić R, Doroslovački R (1986) In: Tošić R, Acketa D, Petrović V (eds) Graph theory, Proceedings of the sixth Yugoslav seminar on graph theory, Dubrovnik April 18–19, 1985, University of Novi Sad, Novi Sad, p 189
47. Cyvin SJ, Brunvoll J (1990) Chem Phys Letters 170: 364
48. Müller WR, Szymanski K, Knop JV, Nikolić S, Trinajstić N (1989) Croat Chem Acta 62: 481
49. Nikolić S, Trinajstić N, Knop JV, Müller WR, Szymanski K (1990) J Math Chem 4: 357
50. Knop JV, Müller WR, Szymanski K, Trinajstić N (1990) Reports on Molecular Theory 1: 95
51. Cyvin SJ, Brunvoll J (1989) Chem Phys Letters 164: 635
52. Dias JR (1984) Can J Chem 62: 2914
53. Knop JV, Szymanski K, Klasinc L, Trinajstić N (1984) Computers & Chemistry 8: 107
54. Brunvoll J, Cyvin SJ, Cyvin BN (1987) J Comput Chem 8: 189
55. Cyvin SJ, Brunvoll J, Cyvin BN (1989) J Chem Inf Comput Sci 29: 79
56. Rouvray DH (1973) J South African Chem Inst 26: 141
57. Trinajstić N (1990) Reports on Molecular Theory 1: 185
58. Balaban AT (1976) Match 2: 51
59. Brunvoll J, Tošić R, Kovačević M, Balaban AT, Gutman I, Cyvin SJ (1990) Rev. Roumaine Chim 35: 85
60. Balaban AT, Brunvoll J, Cyvin BN, Cyvin SJ (1988) Tetrahedron 44: 221
61. Gutman I (1987) J Serb Chem Soc 52: 611
62. Balaban AT (1989) Match 24: 29; erratum (1990) ibid 25: 275
63. Balaban AT, Artemi C (1990) Polycyclic Aromatic Compounds 1: 171
64. Harary F, Schwenk AJ (1973) Discrete Mathematics 6: 359
65. Cyvin SJ, Brunvoll J, Cyvin BN (1989) In: Graovac A (ed) MATH/CHEM/COMP 1988, Proceedings of an international conference on the interfaces between mathematics, chemistry and computer science, Dubrovnik, Yugoslavia, 20–25 June 1988 (Studies in Physical and Theoretical Chemistry 63), Elsevier, Amsterdam, p 127
66. Cyvin SJ, Brunvoll J, Gutman I (1990) Rev Roumaine Chim 35: 985
67. Gutman I (1986) Z Naturforsch 41 a: 1089
68. Cyvin SJ, Brunvoll J, Cyvin BN (1986) Z Naturforsch 41a: 1429
69. Aboav D, Gutman I (1988) Chem Phys Letters 148: 90
70. Aboav D, Gutman I (1989) J Serb Chem Soc 54: 249
71. Brunvoll J, Cyvin BN, Cyvin SJ (1987) J Chem Inf Comput Sci 27: 171
72. Tošić R, Kovačević M (1988) J Chem Inf Comput Sci 28: 29
73. Balaban AT (1970) Rev Roumaine Chim 15: 1243
74. Džonova-Jerman-Blažič B, Trinajstić N (1982) Croat Chem Acta 55: 347
75. Gutman I (1982) Coll Sci Papers Fac Sci Kragujevac 3: 43
76. Trinajstić N, Jeričević Ž, Knop JV, Müller WR, Szymanski K (1983) Pure & Appl Chem 55: 379
77. El-Basil S (1984) Croat Chem Acta 57: 21
78. Cyvin SJ, Brunvoll J, Cyvin BN (1988) J. Mol. Struct. (Theochem) 180: 329
79. Tošić R, Budimac Z, Brunvoll J, Cyvin SJ (1990) J Mol Struct (Theochem) 209: 289
80. Tošić R, Budimac Z, Cyvin SJ, Brunvoll J (1991) J Mol Struct 247: 129
81. Gutman I (1974) Croat Chem Acta 46: 209
82. Gutman I (1983) Coll Sci Papers Fac Sci Kragujevac 4: 189
83. Cyvin SJ, Gutman I (1986) Computers Math Applic 12B: 859; reprinted in: Hargittai I (ed) (1986) Symmetry unifying human understanding, Pergamon, New York
84. Elk SB (1985) Match 17: 255
85. Kirby EC (1989) J Mol Struct (Theochem) 185: 39
86. Rouvray DH (1974) J South African Chem Inst 27: 20
87. Elk SB (1980) Match 8: 121

88. Cyvin SJ, Gutman I (1986) Z. Naturforsch 41a: 1079
89. Yamaguchi T, Suzuki M, Hosoya H (1975) Natural Science Report, Ochanomizu University 26: 39
90. Bonchev D, Balaban AT (1981) J Chem Inf Comput Sci 21: 223
91. Knop JV, Müller WR, Szymanski K, Trinajstić N (1986) J Comput Chem 7: 547
92. Pólya G (1936) Zeitschr f Kristallographie 93: 415
93. Cyvin SJ, Brunvoll J, Cyvin BN (1990) J Math Chem 4: 47
94. Cyvin SJ (1986) Match 20: 165
95. Cyvin BN, Brunvoll J, Cyvin SJ, Gutman I (1986) Match 21: 301
96. Brunvoll J, Cyvin SJ, Cyvin BN, Gutman I (1989) Match 24: 51
97. Brunvoll J, Cyvin BN, Cyvin SJ (1991) Match 26: 3
98. Brunvoll J, Cyvin SJ, Cyvin BN (1988) Match 23: 239
99. Hosoya H, Yamaguchi T (1975) Tetrahedron 52: 4659
100. Ohkami N, Hosoya H (1984) Natural Science Report, Ochanomizu University 35: 71
101. Brunvoll J, Cyvin BN, Cyvin SJ, Gutman I (1988) Z Naturforsch 43a: 889
102. Gutman I, Cyvin SJ (1988) J Serb Chem Soc 53: 391
103. Cyvin SJ, Brunvoll J, Cyvin BN (1989) Computers Math Applic 17: 355; reprinted in: Hargittai I (ed) (1989) Symmetry 2 unifying human understanding, Pergamon, Oxford
104. Cyvin SJ, Brunvoll J, Cyvin BN (1988) J Mol Struct (Theochem) 180: 329
105. Brunvoll J, Cyvin BN, Cyvin SJ, Gutman I (1988) Match 23: 209
106. Brunvoll J, Cyvin SJ, Cyvin BN, Gutman I, He WJ, He WC (1987) Match 22: 105
107. He WC, He WJ, Cyvin BN, Cyvin SJ, Brunvoll J (1988) Match 23: 201
108. Guo XF, Zhang FJ (1989) Match 24: 85
109. Jiang Y, Chen GY (1989) In: Graovac A (ed) MATH/CHEM/COMP 1988, Proceedings of an international conference on the interfaces between mathematics, chemistry and computer science, Dubrovnik, Yugoslavia, 20–25 June 1988 (Studies in Physical and Theoretical Chemistry 63), Elsevier, Amsterdam, p 107
110. Cyvin SJ, Brunvoll J, Cyvin BN (1989) J Chem Inf Comput Sci 29: 236
111. Hosoya H (1986) Croat Chem Acta 59: 583
112. Cyvin SJ, Gutman I (1987) J Mol Struct (Theochem) 150: 157
113. Clar E (1972) The aromatic sextet, Wiley, London
114. Balaban AT (1981) Rev Roumaine Chim 26: 407
115. Balaban AT (1982) Pure & Appl Chem 54: 1075
116. Dias JR (1986) J Mol Struct (Theochem) 137: 9
117. Zhang FJ, Guo XF (1988) Match 23: 229
118. Cyvin SJ, Brunvoll J, Cyvin BN (1988) Match 23: 189
119. Hosoya H (1986) Computers Math Applic 12B: 271; reprinted in: Hargittai I (ed) (1986) Symmetry unifying human understanding, Pergamon, New York
120. Cyvin SJ, Bergan JL, Cyvin BN (1987) Acta Chim Hung 124: 691
121. Dias JR (1985) Accounts Chem Res 18: 241
122. Dias JR (1985) J Macromol Sci-Chem A22: 335
123. Dias JR (1985) Nouv J Chim 9: 125
124. Dias JR (1987) Thermochim Acta 122: 313
125. Dias JR (1989) J Mol Struct (Theochem) 185: 57; erratum (1990) ibid 207: 141
126. Dias JR (1990) In: Gutman I, Cyvin SJ (eds) Advances in the theory of benzenoid hydrocarbons (Topics in Current Chemistry 153), Springer, Berlin Heidelberg New York, p 123
127. Cyvin BN, Brunvoll J, Cyvin SJ, Gutman I (1988) Match 23: 163
128. Dias JR (1988) Handbook of polycyclic hydrocarbons − Part B − Polycyclic isomers and heteroatom analogs of benzenoids, Elsevier, Amsterdam
129. Cyvin SJ, Cyvin BN, Brunvoll J (1989) J Mol Struct 198: 31
130. Brunvoll J, Cyvin BN, Gutman I, Tošić R, Kovačević M (1989) J Mol Struct (Theochem) 184: 165
131. Cyvin SJ, Brunvoll J, Cyvin BN, Tošić R, Kovačević M (1989) J Mol Struct (Theochem) 200: 261
132. Cyvin SJ (1989) Monatsh Chem 120: 243

133. Dias JR (1989) Z Naturforsch 44a: 765
134. Cyvin SJ (1991) Acta Chim Hung 127: 849
135. Cyvin SJ, Brunvoll J (1990) Chem Phys Letters 170: 364
136. Dias JR (1991) Chem Phys Letters 176: 559
137. Cyvin SJ (1991) Coll Sci Papers Fac Sci Kragujevac 12: 95
138. Herndon WC (1990) J Am Chem Soc 112: 4546
139. Dias JR (1990) J Chem Inf Comput Sci 30: 61
140. Dias JR (1990) J Chem Inf Comput Sci 30: 159
141. Kirby EC (1990) J Chem Soc Faraday Trans 86: 447
142. Dias JR (1990) J Math Chem 4: 17
143. Kirby EC (1990) J Math Chem 4: 31
144. Tošić R, Stojmenović I (1990) J Mol Struct (Theochem) 207: 285
145. Cyvin SJ (1990) J Mol Struct (Theochem) 208: 173
146. Nikolić S, Trinajstić N, Knop JV, Müller WR, Szymanski K (1991) J Mol Struct (Theochem) 231: 219
147. Balaban AT, Brunvoll J, Cyvin SJ (1991) Rev Roumaine Chim: in press
148. Cyvin SJ, Brunvoll J, Cyvin BN (1990) Struct Chem 1: 429
149. Cyvin SJ, Balaban AT (1991) Struct Chem 2: 485
150. Jiang YS, Chen GY (1990) Theor Chim Acta 76: 437
151. Dias JR (1990) Theor Chim Acta 77: 143
152. Brunvoll J, Cyvin SJ (1990) Z Naturforsch 45a: 69
153. Kirby EC (1987) In: King RB, Rouvray DH (eds) Graph theory and topology in chemistry, A collection of papers presented at an international conference held at the University of Georgia, Athens, Georgia, USA, 16–20 March 1987 (Studies in Physical and Theoretical Chemistry 51), Elsevier, Amsterdam, p 529
154. Herndon WC, Bruce AJ (1987) In: King RB, Rouvray DH (eds) Graph theory and topology in chemistry, A collection of papers presented at an international conference held at the University of Georgia, Athens, Georgia, USA, 16–20 March 1987 (Studies in Physical and Theoretical Chemistry 51), Elsevier, Amsterdam, p 491
155. Cyvin SJ, Brunvoll J, Cyvin BN (1991) Theory of coronoid hydrocarbons (Lecture Notes in Chemistry 54), Springer, Berlin Heidelberg New York

# Benzenoid Chemical Isomers and Their Enumeration

Jon Brunvoll, Björg N. Cyvin, and Sven J. Cyvin

Division of Physical Chemistry, The University of Trondheim,
N-7034 Trondheim-NTH, Norway

## Table of Contents

1 Foreword . . . . . . . . . . . . . . . . . . . . . . . . . . 183

2 Introduction . . . . . . . . . . . . . . . . . . . . . . . . 183

3 Description and Precise Definition of the Problem . . . . . . . . . 184

4 Invariants and Classes of Benzenoids . . . . . . . . . . . . . 185
  4.1 Hexagons and Vertices . . . . . . . . . . . . . . . . . 185
  4.2 The Dias Parameter . . . . . . . . . . . . . . . . . . 185
  4.3 The "neo" Classification . . . . . . . . . . . . . . . 186
  4.4 Color Excess . . . . . . . . . . . . . . . . . . . . . 186

5 First Enumerations of Benzenoid Isomers . . . . . . . . . . . 187
  5.1 Catacondensed Benzenoids . . . . . . . . . . . . . . . 187
  5.2 Classification According to the Number of Internal Vertices . . . 193
  5.3 Classification According to the Perimeter Length . . . . . . . 193

6 Complete Data for Some Benzenoid Isomers . . . . . . . . . . . 193
  6.1 Arrangement of Tables . . . . . . . . . . . . . . . . . 193
  6.2 Preliminary Account on Extremal Benzenoids . . . . . . . . 194

7 Periodic Table for Benzenoid Hydrocarbons . . . . . . . . . . . 195
  7.1 Description of the Table . . . . . . . . . . . . . . . . 195
  7.2 How to Find the Place of a Formula in the Periodic Table? . . . 197
  7.3 The Position of Benzene . . . . . . . . . . . . . . . . 197
  7.4 Shape of the Staircase-Like Boundary . . . . . . . . . . . 198

8 Detailed Analysis of the Formulas . . . . . . . . . . . . . . . 198
  8.1 Notation and Circumscribing . . . . . . . . . . . . . . 198
  8.2 More Classes of Benzenoids . . . . . . . . . . . . . . . 199

9 Strictly Pericondensed Benzenoid and Excised Internal Structure . . . 200

Topics in Current Chemistry, Vol. 162
© Springer-Verlag Berlin Heidelberg 1992

Jon Brunvoll, Björg N. Cyvin, and Sven J. Cyvin

10  Incomplete Data for Some Benzenoid Isomers . . . . . . . . . . 202

11  Benzenoid Isomers and Number of Edges . . . . . . . . . . . 210

12  Forms of Some Benzenoid Isomers . . . . . . . . . . . . . . 214

13  Conclusion . . . . . . . . . . . . . . . . . . . . . . . . . 218

14  References . . . . . . . . . . . . . . . . . . . . . . . . . 220

Benzenoid (chemical) isomers are, in a strict sense, the benzenoid systems compatible with a formula $C_nH_s$. Several invariants, including the Dias parameter, are treated and relations between them are given. Many of the relations involve upper and lower bounds. The periodic table for benzenoid hydrocarbons is revisited and new aspects of it are pointed out. In this connection some new classes of benzenoids are defined: extreme-left, protrusive and circular. Extensive tables of enumeration data for benzenoid isomers are presented. Some of their forms are displayed in figures.

# 1 Foreword

The topic of the present chapter falls under "Enumeration of Benzenoid Systems", which is part of the title of the preceding chapter. The numbers in the two chapters are strongly inter-related in the same way as many of the numbers within the preceding chapter. In particular, the sums of the numbers in each column for a given number of hexagons ($h$) in the first four tables of the present chapter can be checked against appropriate numbers in the preceding chapter.

However, for the individual numbers of the two chapters there is practically no overlap. It was avoided by omitting a classification of the benzenoids with a given $h$ according to their numbers of internal vertices ($n_i$) in the preceding chapter.

Nevertheless, the two chapters meet at the enumeration of catacondensed benzenoids ($n_i = 0$).

# 2 Introduction

The enumeration of chemical isomers has engaged mathematicians and chemists for more than one hundred years. Perhaps the most familiar example is the enumeration of alkanes, $C_N H_{2N+2}$. Some of the key references to this story should include Cayley from 1875 [1], Herrmann from 1880 [2], Henze and Blair from 1931 [3], and finally the more recent computer works of Davis et al. [4] and Knop et al. [5]. The latter authors [5] have given a vivid description of details of this story, as also to be found in a monograph of Trinajstić [6].

In the light of these long traditions, extensive enumerations of the isomers of benzenoid hydrocarbons is a very new area. A systematic investigation can be dated to 1982 with the first paper of Dias [7] (but see also below). He published an article series in ten parts [7–16] entitled "A Periodic Table for Polycyclic Aromatic Hydrocarbons" and more recent works [17, 18]. With the invention of the periodic table, Dias created orderness in the chaotic myriads of chemical formulas for benzenoid hydrocarbons, which may be written. He has also written a monograph [19] with relevance to this topic and some other reviews [20–22]. Two years before Dias, Elk [23] published a paper on benzenoids, which contains explicitly the enumeration of isomers up to $h = 5$. It seems that the work of Elk has largely been overlooked in the context of benzenoid isomer enumeration.

Knop et al. [5] summarized the pertinent work of Dias, his periodic table for benzenoid hydrocarbons (see also below) and enumeration of isomers. Hereby these authors [5] pointed out some erroneous omissions in the material of Dias. Many other numerical errors were later detected by Cyvin [24], Brunvoll et al. [25] and Cyvin et al. [26]. These works [24–26] provide a considerable amount of supplements to the enumeration data of Dias. Further supplements are found in the present work.

For a general background and basic definitions the reader is referred to the mentioned monographs [5, 6, 19] in addition to some others [27–29].

# 3 Description and Precise Definition of the Problem

The present work deals with benzenoid systems (benzenoids) or polyhexes without holes. We are using the definition [28, 29] which allows for Kekuléan or non-Kekuléan benzenoids, depending on whether they possess or do not possess Kekulé structures. Kekuléan benzenoids correspond to conjugated closed-shell (polycyclic aromatic) hydrocarbons with six-membered (benzenoid) rings only, either known or unknown as existing molecules in organic chemistry. Non-Kekuléan benzenoids correspond to hypothetic (so far never synthesized) radicals. Benzenoids have also been characterized as simply connected and planar, thus excluding coronoids or polyhexes with holes and also excluding helicenic systems.

A benzenoid isomer is defined by a pair of invariants $(n, s)$ and usually written as the chemical formula $C_n H_s$. Here $n$ is the total number of vertices, corresponding to the number of carbon (C) atoms, while $s$ is the number of vertices of degree two (on the perimeter), corresponding to the number of secondary carbon atoms (hence the symbol $s$). This number ($s$) is also the number of hydrogens (H).

The problem of enumerating benzenoid isomers ($C_n H_s$) consists of finding the number of non-isomorphic benzenoids for a given pair of the invariants $n$ and $s$. It is also of interest to classify the set of isomers into Kekuléan and non-Kekuléan systems and occasionally further into more subclasses.

In most of the enumerations of benzenoid systems the number of hexagons, $h$, has been used as a leading parameter. This is to say that the numbers of non-isomorphic benzenoids with a given $h$ have been determined, and this set has occasionally been subdivided into different classes; see e.g. a consolidated report [30] with supplements [31]. Also when special classes of benzenoids have been generated specifically the numbers of benzenoids were produced as a function of $h$.

It has been pointed out [25] that the two problems, enumeration of benzenoids with $h$ hexagons and the enumeration of $C_n H_s$ isomers, are not so much contrasted to each other as it may seem. One may get this (false) impression, for instance, from the statement of Dias [7]: "In this paper, the scope and framework for achieving this goal [systematically enumerate all possible polycyclic aromatic hydrocarbons] is defined. The basis for this framework is the molecular formula in contrast to the number of hexagonal rings [32]." As a matter of fact, all benzenoid isomers with a given molecular formula ($C_n H_s$) have the same number of hexagonal rings ($h$). Therefore the classes of benzenoid isomers form subclasses under the sets of benzenoids with the same $h$ values. For instance, benzenoids with $h = 5$ comprise exactly the isomers of $C_{19}H_{11}$, $C_{20}H_{12}$, $C_{21}H_{13}$ and $C_{22}H_{14}$. The members of these four sets are distinguished by having 3, 2, 1 and 0 internal vertices, respectively, simultaneously with 16, 18, 20 and 22 external vertices, respectively. In general, an isomer $C_n H_s$ is fully characterized by the pair of invariants $(h, n_i)$ or alternatively $(h, n_e)$, where $n_i$ and $n_e$ denote the numbers of internal and external vertices, respectively; $n_e$ is also the perimeter length in terms of the number of its edges. The explicit connections between different invariants are treated in the next section.

# 4 Invariants and Classes of Benzenoids

## 4.1 Hexagons and Vertices

The number of hexagons, $h$, and the number of internal vertices, $n_i$, are often taken as a pair of independent invariants $(h, n_i)$ of a benzenoid. Then the invariants $n$ and $s$ of the formula $C_n H_s$ are given by

$$n = 4h - n_i + 2, \qquad s = 2h - n_i + 4. \tag{1}$$

Similarly, in terms of the pair $(h, n_e)$ the same invariants read

$$n = 2h + (n_e/2) + 1, \qquad s = (n_e/2) + 3. \tag{2}$$

It is interesting that $s$ is a function of $n_e$ independent of $h$. The reverse relation reads

$$n_e = 2s - 6. \tag{3}$$

Now let $n_e$ be written as

$$n_e = s + t \tag{4}$$

where $t$ denotes the number of external vertices of degree three. They correspond to the tertiary carbon atoms (hence the symbol $t$) on the perimeter. The relation

$$t = s - 6 \tag{5}$$

emerges immediately on combining (3) and (4).

## 4.2 The Dias Parameter

The Dias parameter, $d_s$ [7], is an invariant for benzenoid systems and defined in terms of other invariants by

$$d_s = h - n_i - 2 = (n_e/2) - h - 3. \tag{6}$$

Dias [7] interpreted the invariant $d_s$ as the number of tree disconnections of internal edges. Figure 1 shows some examples. In the top row the three $h = 5$ benzenoids have zero, one and two disconnections, respectively; hence $d_s = 0, 1$ and 2, respectively. In general, when the internal edges form a tree, $d_s = 0$. The two catacondensed benzenoids with $h = 5$ in the middle row (Fig. 1) have both $d_s = 3$. In general, it is clear that $d_s = h - 2$ for catacondensed benzenoids, a result which is consistent with Eq. (6) on inserting $n_i = 0$. Negative values of $d_s$ indicate tree connections. This phenomenon occurs for the first time (at $h = 7$) for coronene, where six of the internal edges form a cycle; see the bottom row of Fig. 1. The parameter $d_s$ actually indicates the net number between disconnections and connections [20]; cf. the case of benzo[a]coronene depicted as the bottom-right system of Fig. 1. With regard to the interpretation of $d_s$ it should finally be noted that it does not hold for benzene, which is the only benzenoid without any internal edge. According to Eq. (6) it has $d_s = -1$.

Jon Brunvoll, Björg N. Cyvin, and Sven J. Cyvin

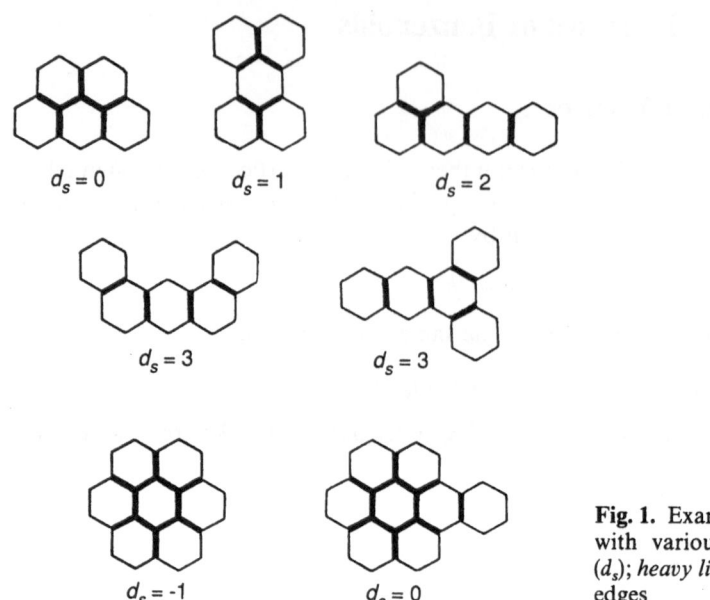

$d_s = 0$    $d_s = 1$    $d_s = 2$

$d_s = 3$    $d_s = 3$

$d_s = -1$    $d_s = 0$

**Fig. 1.** Examples of benzenoids with various Dias parameters ($d_s$); *heavy lines* indicate internal edges

The pair of invariants ($d_s$, $n_i$) plays an important role in connection with the periodic table for benzenoid hydrocarbons (see below). Therefore we give some relations in terms of these invariants. Firstly, the coefficients of $C_nH_s$ read:

$$n = 4d_s + 3n_i + 10, \qquad s = 2d_s + n_i + 8.$$ (7)

Secondly, it is useful to keep track of the number of hexagons, which is

$$h = d_s + n_i + 2.$$ (8)

Finally we give the number of external vertices or the perimeter length;

$$n_e = 4d_s + 2n_i + 10.$$ (9)

## 4.3 The "neo" Classification

The classification referred to as **neo** takes into account all benzenoids. They are either normal (**n**), essentially disconnected (**e**) or non-Kekuléan (**o**); cf., e.g., the multi-author report of Balaban et al. [30] and references cited therein. Among the Kekuléan systems (**n** + **e**) the Kekulé structures possess fixed bonds in the case of essentially disconnected benzenoid, while those of the normal benzenoids do not.

## 4.4 Color Excess

Another important classification of benzenoids follows the $\Delta$ values [30]. Here $\Delta$ is the color excess, defined as the absolute magnitude of the difference between

186

the numbers of black and white (or starred and unstarred) vertices. It is referred to the coloring (or starring) of vertices in benzenoids, which are known to correspond to alternant hydrocarbons. It is also known that the $\Delta$ value is the absolute magnitude of the difference between the numbers of valleys and peaks. The $\Delta$ value is another invariant for the benzenoid systems.

It is clear that $\Delta = 0$ holds for all Kekuléan ($\mathbf{n} + \mathbf{e}$) benzenoids. Hence, if $\Delta > 0$, the system is non-Kekuléan ($\mathbf{o}$); then it is called an obvious non-Kekuléan benzenoid. But also non-Kekuléan systems with $\Delta = 0$ can be constructed; they are called concealed non-Kekuléan benzenoids.

Depending on the number of hexagons ($h$) the $\Delta$ values occur in the range [30, 33]

$$0 \leq \Delta \leq \lfloor h/3 \rfloor \tag{10}$$

where both the upper and lower bound are realized. Here the special brackets are used in the sense that $\lfloor x \rfloor$ means the largest integer smaller than or equal to $x$. In consequence, the benzenoids with $\Delta = \Delta_{max}$ occur for: $h = 1$ and 2 when $\Delta_{max} = 0$; $h = 3\,\Delta_{max}$, $3\,\Delta_{max} + 1$ and $3\,\Delta_{max} + 2$ when $\Delta_{max} > 0$.

There are (obvious) connections between the $\Delta$ values and numbers of internal vertices ($n_i$) of benzenoid systems. If $n_i = 0$, then $\Delta = 0$. If $n_i$ is an even number, then $\Delta$ may only assume an even number or zero. If $n_i$ is odd, then $\Delta$ must be odd. The lower and upper bounds for $\Delta$ are given by

$$(1/2)\,[1 - (-1)^{n_i}] \leq \Delta \leq n_i. \tag{11}$$

All $\Delta$ values within the specified restrictions are realized. Examples: for $n_i = 6$, $\Delta = 6, 4, 2$ or 0; for $n_i = 7$, $\Delta = 7, 5, 3$ or 1.

# 5 First Enumerations of Benzenoid Isomers

## 5.1 Catacondensed Benzenoids

The catacondensed benzenoids (defined by $n_i = 0$) with a given $h$ actually form a class of isomers, viz. $C_{4h+2}H_{2h+4}$; cf. Eq. (1). Their number has sometimes been identified by the symbol $C_h$.

In the first enumerations of catacondensed polyhexes [32, 34, 35] the helicenic systems are included. The smallest helicenic system, viz. hexahelicene, occurs for $h = 6$. However, from the forms of generated catacondensed benzenoids depicted in some of the early works cited above [32, 34] and others [23, 36–41] the numbers of catacondensed benzenoids (without helicenes) for $h$ up to 7 ($C_7$) are easily extracted. Specific documentations are found in Tables 1 and 2.

A list of numbers of catacondensed benzenoids was produced for the first time in 1983 by Knop and Trinajstić with collaborators (the Düsseldorf-Zagreb group) [5, 42–44]. These investigators generated and enumerated all benzenoids up to $h = 10$. Somewhat later also $C_{11}$ [45] and $C_{12}$ [31, 46] were computed; cf. Table 3 and 4, respectively. Table 4 also includes the very recent values of $C_{13}$ and $C_{14}$ [47].

Jon Brunvoll, Björg N. Cyvin, and Sven J. Cyvin

**Table 1.** Numbers of benzenoid chemical isomers for $h \leq 5$ ($\Delta_{max} = 0, 1$)

| $h$ | $n_i$ | $d_s$ | Formula | Δ = 0 | | | o (non-Kekuléan) | | | | | | Total isomers |
|---|---|---|---|---|---|---|---|---|---|---|---|---|---|
| | | | | n | e | Total Kek. | Δ = 0 | 1 | 2 | 3 | 4 | Total non-Kek. | |
| 1 | 0 | −1 | $C_6H_6$ | 1[a] | | 1[b] | | | | | | | 1[b] |
| 2 | 0 | 0 | $C_{10}H_8$ | 1[a] | | 1[b] | | | | | | | 1[b] |
| 3 | 0 | 1 | $C_{14}H_{10}$ | 2[a] | | 2[b] | | | | | | | 2[b] |
| | 1 | 0 | $C_{13}H_9$ | 0 | | 0 | | 1[c] | | | | 1[b] | 1[b] |
| 4 | 0 | 2 | $C_{18}H_{12}$ | 5[a] | | 5[b] | | | | | | | 5[b] |
| | 1 | 1 | $C_{17}H_{11}$ | 0 | | 0 | | 1[c] | | | | 1[b] | 1[b] |
| | 2 | 0 | $C_{16}H_{10}$ | 1[d] | | 1[b] | 0 | | 0 | | | 0 | 1[b] |
| 5 | 0 | 3 | $C_{22}H_{14}$ | 12[e] | | 12[b] | | | | | | | 12[b] |
| | 1 | 2 | $C_{21}H_{13}$ | 0 | | 0 | | 6[c] | | | | 6[b] | 6[b] |
| | 2 | 1 | $C_{20}H_{12}$ | 2[d] | 1[d] | 3[b] | | | 0 | | | 0 | 3[b] |
| | 3 | 0 | $C_{19}H_{11}$ | 0 | 0 | 0 | | 1[c] | | | | 1[b] | 1[b] |

[a] Harary (1967) [36]; [b] Elk (1980) [23]; [c] Brunvoll, Cyvin, Cyvin and Gutman (1988) [33]; [d] Brunvoll and Cyvin (1990) [25]; [e] Balaban and Harary (1968) [32].

**Table 2.** Numbers of benzenoid chemical isomers for $h = 6, 7, 8$ ($\Delta_{max} = 2$)

| $h$ | $n_i$ | $d_s$ | Formula | Δ = 0 | | | o (non-Kekuléan) | | | | | | Total isomers |
|---|---|---|---|---|---|---|---|---|---|---|---|---|---|
| | | | | n | e | Total Kek. | Δ = 0 | 1 | 2 | 3 | 4 | Total non-Kek. | |
| 6 | 0 | 4 | $C_{26}H_{16}$ | 36[a,b] | | 36[a,b] | | | | | | | 36[a] |
| | 1 | 3 | $C_{25}H_{15}$ | 0 | | 0 | | 24[c] | | | | 24[d] | 24[a] |
| | 2 | 2 | $C_{24}H_{14}$ | 10[c] | 3[c] | 13[e] | | 0 | 1[f] | | | 1[d] | 14[a] |
| | 3 | 1 | $C_{23}H_{13}$ | 0 | 0 | 0 | | 4[c] | 0 | | | 4[d] | 4[a] |
| | 4 | 0 | $C_{22}H_{12}$ | 2[c] | 0 | 2[e] | | 0 | 1[f] | | | 1[e] | 3[e] |
| 7 | 0 | 5 | $C_{30}H_{18}$ | 118[a,(b,h)] | | 118[a] | | | | | | | 118[a] |
| | 1 | 4 | $C_{29}H_{17}$ | 0 | | 0 | | 106[c] | | | | 106[d] | 106[a] |
| | 2 | 3 | $C_{28}H_{16}$ | 40[c] | 22[c] | 62[g] | | 0 | 6[f] | | | 6[d] | 68[a] |
| | 3 | 2 | $C_{27}H_{15}$ | 0 | 0 | 0 | | 25[c] | 0 | | | 25[d] | 25[a] |
| | 4 | 1 | $C_{26}H_{14}$ | 8[c] | 1[c] | 9[e] | | 0 | 1[f] | | | 1[d] | 10[a] |
| | 5 | 0 | $C_{25}H_{13}$ | 0 | 0 | 0 | | 3[c] | 0 | | | 3[d] | 3[a] |
| | 6 | −1 | $C_{24}H_{12}$ | 1[c] | 0 | 1[e] | | 0 | 0 | | | 0 | 1[a] |
| 8 | 0 | 6 | $C_{34}H_{20}$ | 411[a] | | 411[a] | | | | | | | 411[a] |
| | 1 | 5 | $C_{33}H_{19}$ | 0 | | 0 | | 453[c] | | | | 453[d] | 453[a] |
| | 2 | 4 | $C_{32}H_{18}$ | 180[c] | 107[c] | 287[d] | | 0 | 42[f] | | | 42[d] | 329[a] |
| | 3 | 3 | $C_{31}H_{17}$ | 0 | 0 | 0 | | 144[c] | 0 | | | 144[d] | 144[a] |
| | 4 | 2 | $C_{30}H_{16}$ | 45[c] | 13[c] | 58[g] | | 0 | 9[f] | | | 9[d] | 67[a] |
| | 5 | 1 | $C_{29}H_{15}$ | 0 | 0 | 0 | | 21[c] | 0 | | | 21[d] | 21[a] |
| | 6 | 0 | $C_{28}H_{14}$ | 7[c] | 1[c] | 8[e] | | 0 | 1[f] | | | 1[g] | 9[a] |
| | 7 | −1 | $C_{27}H_{13}$ | 0 | 0 | 0 | | 1[c] | 0 | | | 1[d] | 1[a] |

[a] Knop, Szymanski, Jeričević and Trinajstić (1983) [42]; [b] Balaban and Harary (1968) [32]; [c] Brunvoll and Cyvin (1990) [25]; [d] Knop, Müller, Szymanski and Trinajstić (1985) [5]; [e] Dias (1982) [7], incorrect data therein are omitted; [f] Brunvoll, Cyvin, Cyvin and Gutman (1988) [33]; [g] Dias (1984) [12], incorrect data therein are omitted; [h] Balaban (1969) [34].

**Table 3.** Numbers of benzenoid chemical isomers for $h = 9, 10, 11$ ($\Delta_{max} = 3$)

| h | $n_i$ | $d_s$ | Formula | $\Delta = 0$ | | Total Kek. | o (non-Kekuléan) | | | | | Total non-Kek. | Total isomers |
|---|---|---|---|---|---|---|---|---|---|---|---|---|---|
| | | | | n | e | | $\Delta = 0$ | 1 | 2 | 3 | 4 | | |
| 9 | 0 | 7 | $C_{38}H_{22}$ | 1489[a] | | 1489[a] | | | | | | | 1489[a] |
| | 1 | 6 | $C_{37}H_{21}$ | | | 0 | | 1966[b] | | | | 1966[c] | 1966[a] |
| | 2 | 5 | $C_{36}H_{20}$ | 777[b] | 575[b] | 1352[c] | | | 249[b] | | | 249[c] | 1601[a] |
| | 3 | 4 | $C_{35}H_{19}$ | | | 0 | | 823[b] | | 2[d] | | 825[c] | 825[a] |
| | 4 | 3 | $C_{34}H_{18}$ | 225[b] | 108[b] | 333[e] | | | 63[b] | | | 63[c] | 396[a] |
| | 5 | 2 | $C_{33}H_{17}$ | | | 0 | | 153[b] | | 1[d] | | 154[c] | 154[a] |
| | 6 | 1 | $C_{32}H_{16}$ | 37[b] | 9[b] | 46[e] | | | 9[b] | | | 9[c] | 55[a] |
| | 7 | 0 | $C_{31}H_{15}$ | | | 0 | | 15[b] | | | | 15[c] | 15[a] |
| | 8 | -1 | $C_{30}H_{14}$ | 3[b] | | 3[f] | | | 1[b] | | | 1[f] | 4[f] |
| 10 | 0 | 8 | $C_{42}H_{24}$ | 5572[a] | | 5572[a] | | | | | | | 5572[a] |
| | 1 | 7 | $C_{41}H_{23}$ | | | 0 | | 8395[b] | | | | 8395[b] | 8395[a] |
| | 2 | 6 | $C_{40}H_{22}$ | 3403[b] | 2853[b] | 6256[b] | | | 1396[b] | | | 1396[b] | 7652[a] |
| | 3 | 5 | $C_{39}H_{21}$ | | | 0 | | 4491[b] | | 27[d] | | 4518[b] | 4518[a] |
| | 4 | 4 | $C_{38}H_{20}$ | 1132[b] | 775[b] | 1907[b] | | | 433[b] | | | 433[b] | 2340[a] |
| | 5 | 3 | $C_{37}H_{19}$ | | | 0 | | 1007[b] | | 11[d] | | 1018[b] | 1018[a] |
| | 6 | 2 | $C_{36}H_{18}$ | 236[b] | 101[b] | 337[b] | | | 79[b] | | | 79[b] | 416[a] |
| | 7 | 1 | $C_{35}H_{17}$ | | | 0 | | 123[b] | | | | 123[b] | 123[a] |
| | 8 | 0 | $C_{34}H_{16}$ | 31[b] | 3[b] | 34[b,g] | | | 8[b] | | | 8[b,g] | 42[a] |
| | 9 | -1 | $C_{33}H_{15}$ | | | 0 | | 8[b] | | 1[d] | | 9[h] | 9[a] |
| | 10 | -2 | $C_{32}H_{14}$ | 1[b] | | 1[e] | | | | | | 0 | 1[a] |
| 11 | 0 | 9 | $C_{46}H_{26}$ | 21115[i] | | 21115[i] | | | | | | | 21115[i] |
| | 1 | 8 | $C_{45}H_{25}$ | | | 0 | | 35885 | | | | 35885[j] | 35885[i] |
| | 2 | 7 | $C_{44}H_{24}$ | 14699 | 14038 | 28737[j] | | | 7372 | | | 7372[j] | 36109[i] |
| | 3 | 6 | $C_{43}H_{23}$ | | | 0 | | 23755 | | 265 | | 24020[j] | 24020[i] |
| | 4 | 5 | $C_{42}H_{22}$ | 5534 | 5018 | 10552[j] | 5[k] | | 2858 | | | 2863[j] | 13415[i] |
| | 5 | 4 | $C_{41}H_{21}$ | | | 0 | 0 | 6308 | | 97 | | 6405[j] | 6405[i] |

**Table 3.** (continued)

| h | $n_i$ | $d_s$ | Formula | $\Delta = 0$ | | | o (non-Kekuléan) | | | | | | Total isomers |
|---|---|---|---|---|---|---|---|---|---|---|---|---|---|
| | | | | n | e | Total Kek. | $\Delta = 0$ | 1 | 2 | 3 | 4 | Total non-Kek. | |
| 11 | 6 | 3 | $C_{40}H_{20}$ | 1342 | 849 | 2191[j] | 2[k] | 0 | 618 | 0 | 0 | 620[j] | 2811[i] |
| | 7 | 2 | $C_{39}H_{19}$ | 0 | 0 | 0 | 0 | 998 | 0 | 10 | 0 | 1008[j] | 1008[i] |
| | 8 | 1 | $C_{38}H_{18}$ | 211 | 53 | 264[j] | 1[k] | 0 | 68 | 0 | 0 | 69[j] | 333[i] |
| | 9 | 0 | $C_{37}H_{17}$ | 0 | 0 | 0 | 0 | 98 | 0 | 2 | 0 | 100[j] | 100[i] |
| | 10 | −1 | $C_{36}H_{16}$ | 18[b] | 2[b] | 20[l] | 0 | 0 | 6[b] | 0 | 0 | 6[b] | 26[i] |
| | 11 | −2 | $C_{35}H_{15}$ | 0 | 0 | 0 | 0 | 2[b] | 0 | 0 | 0 | 2[h] | 2[i] |

[a] Knop, Szymanski, Jeričević and Trinajstić (1983) [42]; [b] Brunvoll and Cyvin (1990) [25]; [c] Knop, Müller, Szymanski and Trinajstić (1985) [5]; [d] Brunvoll, Cyvin, Cyvin and Gutman (1988) [33]; [e] Dias (1984) [12], incorrect data therein are omitted; [f] Dias (1982) [7], incorrect data therein are omitted; [g] Cyvin (1990) [24]; [h] Dias (1986) [15], incorrect data therein are omitted; [i] Stojmenović, Tošić and Doroslovački (1986) [45]; [j] Cyvin and Brunvoll (1990) [47]; [k] Hosoya (1986) [50]; [l] Dias (1987) [19], incorrect data therein are omitted.

**Table 4.** Numbers of benzenoid chemical isomers for h = 12, 13, 14 $(\Delta_{max} = 4)$

| h | $n_i$ | $d_s$ | Formula | $\Delta = 0$ | | | o (non-Kekuléan) | | | | | | Total isomers |
|---|---|---|---|---|---|---|---|---|---|---|---|---|---|
| | | | | n | e | Total Kek. | $\Delta = 0$ | 1 | 2 | 3 | 4 | Total non-Kek. | |
| 12 | 0 | 10 | $C_{50}H_{28}$ | 81121[a,b] | | 81121[a,b] | | | | | | | 81121[a,b] |
| | 1 | 9 | $C_{49}H_{27}$ | 0 | | 0 | | 152688 | | | | 152688[c] | 152688[b] |
| | 2 | 8 | $C_{48}H_{26}$ | 63436 | 67229 | 130665 | | | 37653 | | | 37653 | 168318[d] |
| | 3 | 7 | $C_{47}H_{25}$ | 0 | | 0 | | 122060 | | 2059 | | 124119 | 124119[d] |
| | 4 | 6 | $C_{46}H_{24}$ | 26645 | 30374 | 57019 | 66[e] | | 17893 | | 7 | 17966 | 74985[d] |
| | 5 | 5 | $C_{45}H_{23}$ | 0 | | 0 | 0 | 37943 | | 784 | 0 | 38727 | 38727[d] |
| | 6 | 4 | $C_{44}H_{22}$ | 7402 | 6403 | 13805 | 29[e] | | 4466 | | 6 | 4501 | 18306[d] |

| | | | Formula | | | | | | | | | | |
|---|---|---|---|---|---|---|---|---|---|---|---|---|---|
| | 7 | 3 | C43H21 | 0 | 0 | 0 | 0 | 7243 | 0 | 132 | 0 | 7375[c] | 7375[d] |
| | 8 | 2 | C42H20 | 1410 | 675 | 2085[c] | 3[e] | 0 | 624 | 0 | 1 | 628[c] | 2713[d] |
| | 9 | 1 | C41H19 | 0 | 0 | 0 | 0 | 864 | 0 | 14 | 0 | 878[c] | 878[d] |
| | 10 | 0 | C40H18 | 181 | 32 | 213[c] | 0 | 0 | 66 | 0 | 0 | 66[c] | 279[d] |
| | 11 | −1 | C39H17 | 0 | 0 | 0 | 0 | 60 | 0 | 1 | 0 | 61[c] | 61[d] |
| | 12 | −2 | C38H16 | 10[f] | 0 | 108[g,h] | 0 | 0 | 3[f] | 0 | 0 | 3[i] | 13[d] |
| | 13 | −3 | C37H15 | 0 | 0 | 0 | 0 | 1[f] | 0 | 0 | 0 | 1[i] | 1[d] |
| 13 | 0 | 11 | C54H30 | 314075[c] | 314075[c] | 314075[c] | | 648632 | 186749 | 14233 | 121 | 648632 | 314075[c] |
| | 1 | 10 | C53H29 | 0 | 0 | 0 | 752 | 0 | 0 | 0 | 0 | 186749 | 648632 |
| | 2 | 9 | C52H28 | 272114 | 317589 | 589703 | 0 | 613746 | 107079 | 5966 | 83 | 627979 | 776452 |
| | 3 | 8 | C51H27 | 0 | 0 | 0 | 299 | 0 | 0 | 0 | 0 | 107952 | 627979 |
| | 4 | 7 | C50H26 | 126133 | 174700 | 300833 | 0 | 220871 | 30568 | 1297 | 6 | 226837 | 408785[d] |
| | 5 | 6 | C49H25 | 0 | 0 | 0 | 43 | 0 | 0 | 0 | 0 | 30950 | 226837[d] |
| | 6 | 5 | C48H24 | 39192 | 44184 | 83376 | 0 | 49108 | 5174 | 163 | 1 | 50405 | 114326[d] |
| | 7 | 4 | C47H23 | 0 | 0 | 0 | 3 | 0 | 0 | 0 | 0 | 5223 | 50405[d] |
| | 8 | 3 | C46H22 | 8617 | 6279 | 14896 | 0 | 7090 | 619 | 15 | 0 | 7253[c] | 20119[d] |
| | 9 | 2 | C45H21 | 0 | 0 | 0 | 0 | 0 | 0 | 0 | 1 | 623[c] | 7253[d] |
| | 10 | 1 | C44H20 | 1340 | 496 | 1836[c] | 0 | 693 | 48 | 0 | 0 | 708[c] | 2459[d] |
| | 11 | 0 | C43H19 | 0 | 0 | 0 | 0 | 0 | 0 | 1[f] | 0 | 708[c] | 708[d] |
| | 12 | −1 | C42H18 | 125 | 14 | 139[c] | 0 | 34[f] | 48 | 15 | 0 | 48[c] | 187[d] |
| | 13 | −2 | C41H17 | 0 | 0 | 0 | 0 | 0 | 0 | 0 | 0 | 35[f] | 35[d] |
| | 14 | −3 | C40H16 | 3[f] | 0 | 3[g,h] | 0 | 0 | 1[f] | 0 | 0 | 1[i] | 4[d] |
| 14 | 0 | 12 | C58H32 | 1224528[c] | 1224528[c] | 1224528[c] | 0 | 2749719 | 907390 | 89678 | 1561 | 2749719 | 1224528[c] |
| | 1 | 11 | C57H31 | 0 | 0 | 0 | 6585 | 0 | 0 | 0 | 0 | 907390 | 2749719 |
| | 2 | 10 | C56H30 | 1164197 | 1479635 | 2643832 | 0 | 3029961 | 616950 | 42870 | 893 | 3119639 | 3551222 |
| | 3 | 9 | C55H29 | 0 | 0 | 0 | 2696 | 0 | 0 | 0 | 0 | 625096 | 3119639 |
| | 4 | 8 | C54H28 | 589990 | 967179 | 1557169 | 0 | 1250831 | 199432 | 11143 | 113 | 1293701 | 2182265 |
| | 5 | 7 | C53H27 | 0 | 0 | 0 | 478 | 0 | 0 | 0 | 0 | 203021 | 1293701 |
| | 6 | 6 | C52H26 | 202251 | 286661 | 488912 | 0 | 317449 | 39730 | 1632 | 21 | 328592 | 691933[d] |
| | 7 | 5 | C51H25 | 0 | 0 | 0 | 44 | 0 | 0 | 0 | 0 | 40321 | 328592[d] |
| | 8 | 4 | C50H24 | 50132 | 50815 | 100947 | 0 | 53492 | 5563 | 0 | 0 | 55124 | 141268[d] |
| | 9 | 3 | C49H23 | 0 | 0 | 0 | 0 | 0 | 0 | 0 | 0 | 5628 | 55124[d] |
| | 10 | 2 | C48H22 | 9222 | 5471 | 14693 | 0 | 5563 | 0 | 0 | 0 | | 20321[d] |

191

**Table 4.** (continued)

| $h$ | $n_i$ | $d_s$ | Formula | $\Delta = 0$ | | | o (non-Kekuléan) | | | | | | Total isomers |
|---|---|---|---|---|---|---|---|---|---|---|---|---|---|
| | | | | n | e | Total Kek. | $\Delta = 0$ | 1 | 2 | 3 | 4 | Total non-Kek. | |
| 14 | 11 | 1 | $C_{47}H_{21}$ | 0 | 0 | 0 | 0 | 6426 | 0 | 168 | 0 | 6594 | 6594[d] |
| | 12 | 0 | $C_{46}H_{20}$ | 1182 | 289 | 1471 | 1 | 0 | 529 | 0 | 0 | 530 | 2001[d] |
| | 13 | −1 | $C_{45}H_{19}$ | 0 | 0 | 0 | 0 | 517 | 0 | 16 | 0 | 533 | 533[d] |
| | 14 | −2 | $C_{44}H_{18}$ | 81 | 8 | 89 | 0 | 0 | 31 | 0 | 0 | 31 | 120[d] |
| | 15 | −3 | $C_{43}H_{17}$ | 0 | 0 | 0 | 0 | 15[f] | 0 | 1[f] | 0 | 16[f] | 16[d] |
| | 16 | −4 | $C_{42}H_{16}$ | 1[f] | 0 | 1[g,h] | 0 | 0 | 0 | 0 | 0 | 0 | 1[d] |

[a] Balaban, Brunvoll, Cyvin and Cyvin (1988) [46]; [b] He, He, Wang, Brunvoll and Cyvin (1988) [31]; [c] Cyvin and Brunvoll (1990) [47]; [d] Stojmenović, Tošić and Doroslovački (1968) [45]; [e] He, He, Cyvin, Cyvin and Brunvoll (1988) [51]; [f] Brunvoll and Cyvin (1990) [25]; [g] Dias (1984) [12], incorrect data therein are omitted; [h] Dias (1984) [10], incorrect data therein are omitted; [i] Dias (1986) [15], incorrect data therein are omitted.

## 5.2 Classification According to the Number of Internal Vertices

An enumeration and classification of the benzenoids according to $h$ and $n_i$ up to $h = 10$, executed by the Düsseldorf-Zagreb group [5, 42], gives precise information about numbers of $C_nH_s$ isomers. Knop et al. [5, 44] were aware of this fact when they compared some Dias numbers to their own. For the pertinent relations between the pairs of invariants $(h, n_i)$ and $(n, s)$, see Eq. (1).

## 5.3 Classification According to the Perimeter Length

Doroslovački and Tošić [48] in their characterization of benzenoid systems used the perimeter length $(n_e)$ as the leading parameter, i.e. they enumerated benzenoids with given (increasing) $n_e$ values. These data are also reproduced by Tošić et al. [49]. In a later work Stojmenović et al. [45] supplemented the data in question substantially and classified the set of systems with a given $n_e$ according to the number of hexagons, $h$. This material again gives precise information about the numbers of $C_nH_s$ isomers. The relations between the pairs of invariants $(h, n_e)$ and $(n, s)$ are given in Eq. (2). The extensive material of Stojmenović et al. [45] ranges up to $n_e = 46$. All the numbers for $h \leq 10$ therein coincide with the corresponding numbers of Knop et al. [5, 42].

# 6 Complete Data for Some Benzenoid Isomers

## 6.1 Arrangement of Tables

Here we refer to the enumeration data of benzenoid isomers as complete if, for a given $h$, all the numbers of $C_nH_s$ isomers are given at least for the Kekuléan and non-Kekuléan systems separately. Such data are known for $h$ values up to 14; cf. Tables 1–4. In addition, we have specified the numbers of normal (n) and essentially disconnected (e) benzenoids among the Kekuléan systems in the tables. The smallest essentially disconnected benzenoid, viz. perylene, occurs at $h = 5$. Furthermore, the non-Kekuléan systems are classified according to the $\Delta$ values (Tables 1–4).

Table 1 accounts for the systems with $\Delta_{max} = 0$ or 1 ($h \leq 5$). The following tables should, according to our arrangement of them, comprise three and three $h$ values as $\Delta_{max}$ is increased stepwise. Thus Table 2 pertains to $\Delta_{max} = 2$ ($h = 6, 7, 8$), Table 3 to $\Delta_{max} = 3$ ($h = 9, 10, 11$), and Table 4 to $\Delta_{max} = 4$ ($h = 12, 13, 14$).

With respect to the documentation of the data it was especially difficult to choose proper references to the smallest values. Benzenoids with $h$ up to 4 or more have certainly been generated independently by many investigators. We have chosen to give Harary [36] credit for the catacondensed ($n_i = 0$) systems with $h \leq 4$ because his paper is the first place where we have located that these benzenoids are depicted. Furthermore, we have given much credit to Elk [23] as to the isomers with $h \leq 5$. To our best knowledge, this researcher speaks for the first time explicitly about the numbers of $C_nH_s$ isomers in the mathematical-chemical context, and he has also characterized these systems in a way that immediately identifies them with Kekuléans and non-Kekuléans.

Much information can be extracted from Knop et al. [5], where all benzenoids with $h \leq 9$ are depicted. These computer-generated pictures are ordered according to the numbers of internal vertices ($n_i$) within each $h$ value. The Kekulé structure counts are indicated ($K > 0$ for Kekuléan and $K = 0$ for non-Kekuléan systems). In Tables 2 and 3 this reference is quoted in appropriate places for some total Kekuléan and total non-Kekuléan systems. We have not taken into account the corresponding mammoth listing for $h = 10$, on which it was informed by Knop et al. [44]. It was stated that a very limited number of copies were available for distribution in 1984. We are not in the possession of any of these copies.

For a documentation pertaining to the normal pericondensed and essentially disconnected benzenoids, as well as the classification according to $\Delta$ values, references are made to Brunvoll et al. [25, 33]. For $h \leq 9$ the information on $\Delta$ values could, of course, be extracted from Knop et al. [5] by studying all the forms of the benzenoids depicted therein.

Concealed non-Kekuléan benzenoids occur for the first time at $h = 11$. In Table 3 we are giving Hosoya [50] credit for the enumeration of such systems; he depicted for the first time all eight of them as a group. Those for $h = 12$ (Table 4) were depicted by He et al. [51].

## 6.2 Preliminary Account on Extremal Benzenoids

It is of interest to know the maximum value of $n_i$ for a given $h$ when setting up tables like those under consideration here. In this connection we shall refer to some benzenoids as *extremal*, i.e. those who have $n_i = (n_i)_{\text{max}}$, the maximum number of internal vertices for a given number of hexagons. In other words, these benzenoids are "extremely pericondensed". The upper bound of $n_i$ as a function of $h$ is readily obtained from the known results of Harary and Harborth [52]. One has [40]

$$0 \leq n_i \leq 2h + 1 - \lceil (12h - 3)^{1/2} \rceil \tag{12}$$

when $\lceil x \rceil$ is used to denote the smallest integer larger than or equal to $x$. Similarly one finds for the upper and lower bound of the Dias parameter:

$$\lceil (12h - 3)^{1/2} \rceil - h - 3 \leq d_s \leq h - 2 \tag{13}$$

where the minimum values (lower bound) pertain to the extremal benzenoids.

For the coefficients of $C_nH_s$ the following inequalities are valid. Firstly, for the number of carbon atoms ($n$) [52]

$$2h + 1 + \lceil (12h - 3)^{1/2} \rceil \leq n \leq 4h + 2 . \tag{14}$$

Secondly, for the number of hydrogens ($s$) [29]

$$3 + \lceil (12h - 3)^{1/2} \rceil \leq s \leq 2h + 4 . \tag{15}$$

In conclusion, for a given $h$: $(n_i)_{\text{min}} = 0$, $(d_s)_{\text{max}}$, $n_{\text{max}}$ and $s_{\text{max}}$ pertain to the catacondensed benzenoids, while $(n_i)_{\text{max}}$, $(d_s)_{\text{min}}$, $n_{\text{min}}$ and $s_{\text{min}}$ pertain to the extremal benzenoids. One has also $(n_e)_{max}$ in the case of catacondensed and $(n_e)_{min}$ in the case of extremal benzenoids. The latter bounds are strongly related to those of $s$; cf. Eq. (3).

A catacondensed benzenoid has the formula $C_{4h+2}H_{2h+4}$; cf. the right-hand sides of Eqs. (14) and (15).

# 7 Periodic Table for Benzenoid Hydrocarbons

## 7.1 Description of the Table

In the periodic table for benzenoid hydrocarbons [7] the formulas $C_nH_s$ are arranged in an array with coordinates $(d_s, n_i)$. The Dias parameters $(d_s)$ are found on a horizontal axis (increasing from left to right), while the numbers of internal vertices $(n_i)$ are on a vertical axis (increasing downwards). The table extends infinitely to the right and downwards. To the left the formulas form a line in the shape of an uneven staircase, which shall be referred to as the staircase-like boundary.

The periodic table for benzenoid hydrocarbons has been reproduced (to different extents) many times [5, 7–15, 19–22, 25]. Usually it is given only for even-numbered

Table 5. Periodic table for benzenoid hydrocarbons

| $d_s$ | | | | | | | | | |
|---|---|---|---|---|---|---|---|---|---|
| −5 | −4 | −3 | −2 | −1 | 0 | 1 | 2 | | $n_i$ |
| | | | | | $C_{10}H_8$ | $C_{14}H_{10}$ | $C_{18}H_{12}$ | | 0 |
| | | | | | $C_{13}H_9$ | $C_{17}H_{11}$ | $C_{21}H_{13}$ | | 1 |
| | | | | $h = 4$ | $C_{16}H_{10}$ | $C_{20}H_{12}$ | $C_{24}H_{14}$ | | 2 |
| | | | | | $C_{19}H_{11}$ | $C_{23}H_{13}$ | $C_{27}H_{15}$ | | 3 |
| | | | | | $C_{22}H_{12}$ | $C_{26}H_{14}$ | $C_{30}H_{16}$ | | 4 |
| | | | | $C_{25}H_{13}$ | $C_{29}H_{15}$ | | $C_{33}H_{17}$ | | 5 |
| | | | $h = 7$ | $C_{24}H_{12}$ | $C_{28}H_{14}$ | $C_{32}H_{16}$ | $C_{36}H_{18}$ | | 6 |
| | | | | $C_{27}H_{13}$ | $C_{31}H_{15}$ | $C_{35}H_{17}$ | $C_{39}H_{19}$ | | 7 |
| | | | | $C_{30}H_{14}$ | $C_{34}H_{16}$ | $C_{38}H_{18}$ | $C_{42}H_{20}$ | | 8 |
| | | | | $C_{33}H_{15}$ | $C_{37}H_{17}$ | $C_{41}H_{19}$ | $C_{45}H_{21}$ | | 9 |
| | | $h = 10$ | $C_{32}H_{14}$ | $C_{36}H_{16}$ | $C_{40}H_{18}$ | $C_{44}H_{20}$ | $C_{48}H_{22}$ | | 10 |
| | | | $C_{35}H_{15}$ | $C_{39}H_{17}$ | $C_{43}H_{19}$ | $C_{47}H_{21}$ | $C_{51}H_{23}$ | | 11 |
| | | | $C_{38}H_{16}$ | $C_{42}H_{18}$ | $C_{46}H_{20}$ | $C_{50}H_{22}$ | $C_{54}H_{24}$ | | 12 |
| | | $C_{37}H_{15}$ | $C_{41}H_{17}$ | $C_{45}H_{19}$ | $C_{49}H_{21}$ | $C_{53}H_{23}$ | $C_{57}H_{25}$ | | 13 |
| | $h = 13$ | $C_{40}H_{16}$ | $C_{44}H_{18}$ | $C_{48}H_{20}$ | $C_{52}H_{22}$ | $C_{56}H_{24}$ | $C_{60}H_{26}$ | | 14 |
| | | $C_{43}H_{17}$ | $C_{47}H_{19}$ | $C_{51}H_{21}$ | $C_{55}H_{23}$ | $C_{59}H_{25}$ | $C_{63}H_{27}$ | | 15 |
| | $C_{42}H_{16}$ | $C_{46}H_{18}$ | $C_{50}H_{20}$ | $C_{54}H_{22}$ | $C_{58}H_{24}$ | $C_{62}H_{26}$ | $C_{66}H_{28}$ | | 16 |
| | $C_{45}H_{17}$ | $C_{49}H_{19}$ | $C_{53}H_{21}$ | $C_{57}H_{23}$ | $C_{61}H_{25}$ | $C_{65}H_{27}$ | $C_{69}H_{29}$ | | 17 |
| | $C_{48}H_{18}$ | $C_{52}H_{20}$ | $C_{56}H_{22}$ | $C_{60}H_{24}$ | $C_{64}H_{26}$ | $C_{68}H_{28}$ | $C_{72}H_{30}$ | | 18 |
| $C_{47}H_{17}$ | $C_{51}H_{19}$ | $C_{55}H_{21}$ | $C_{59}H_{23}$ | $C_{63}H_{25}$ | $C_{67}H_{27}$ | $C_{71}H_{29}$ | $C_{75}H_{31}$ | | 19 |
| $h = 16$ | | | $h = 19$ | | | $h = 22$ | | | |

195

Jon Brunvoll, Björg N. Cyvin, and Sven J. Cyvin

carbon atoms ($n$ even). In Table 5 we show the periodic table for both even- and odd-numbered carbon atoms. This table is a fusion of the two periodic tables given separately in one of the Dias publications [15]. Only a small portion of this version of the table has been displayed previously [25].

When setting up the periodic table for benzenoid hydrocarbons one observes easily the regularities as to the steps of the coefficients of $C_nH_s$. They are obtained recursively by the scheme:

$$C_nH_s(d_s, n_i) \rightarrow C_{n+4}H_{s+2}(d_{s+1}, n_i)$$
$$\downarrow$$
$$C_{n+3}H_{s+1}(d_s, n_i + 1).$$

As the initial condition one has $C_{10}H_8(0, 0)$, the formula for naphthalene, at the upper-left corner.

An acute problem arises, however, concerning the limitation at the staircase-like boundary. In other words, where to stop writing up the formulas to the left? This question is answered implicitly in Sect. 6.2; cf. Eqs. (12)–(15). In the subsequent sections a more detailed description of the staircase-like boundary is provided.

**Table 5.** (continued)

| $d_s$ | | | | | | | | |
|---|---|---|---|---|---|---|---|---|
| $-12$ | $-11$ | $-10$ | $-9$ | $-8$ | $-7$ | $-6$ | $-5$ | $n_i$ |
| | | | | | | | $C_{50}H_{18}$ | 20 |
| | | | | | | | $C_{53}H_{19}$ | 21 |
| | | | | | | $C_{52}H_{18}$ | $C_{56}H_{20}$ | 22 |
| | | | | | | $C_{55}H_{19}$ $\cdots$ | $C_{59}H_{21}$ | 23 |
| | | | | $h = 19$ $\cdots$ | $C_{54}H_{18}$ $\cdots$ | $C_{58}H_{20}$ | $C_{62}H_{22}$ | 24 |
| | | | | | $C_{57}H_{19}$ | $C_{61}H_{21}$ | $C_{65}H_{23}$ | 25 |
| | | | | | $C_{60}H_{20}$ | $C_{64}H_{22}$ $\cdots$ | $C_{68}H_{24}$ | 26 |
| | | | | $C_{59}H_{19}$ $\cdots$ | $C_{63}H_{21}$ $\cdots$ | $C_{67}H_{23}$ | $C_{71}H_{25}$ | 27 |
| | | | $h = 22$ $\cdots$ | $C_{62}H_{20}$ $\cdots$ | $C_{66}H_{22}$ | $C_{70}H_{24}$ | $C_{74}H_{26}$ | 28 |
| | | | | $C_{65}H_{21}$ | $C_{69}H_{23}$ | $C_{73}H_{25}$ $\cdots$ | $C_{77}H_{27}$ | 29 |
| | | | $C_{64}H_{20}$ | $C_{68}H_{22}$ $\cdots$ | $C_{72}H_{24}$ $\cdots$ | $C_{76}H_{26}$ | $C_{80}H_{28}$ | 30 |
| | | | $C_{67}H_{21}$ | $C_{71}H_{23}$ | $C_{75}H_{25}$ | $C_{79}H_{27}$ | $C_{83}H_{29}$ | 31 |
| | | $C_{66}H_{20}$ | $C_{70}H_{22}$ $\cdots$ | $C_{74}H_{24}$ | $C_{78}H_{26}$ | $C_{82}H_{28}$ $\cdots$ | $C_{86}H_{30}$ | 32 |
| | $h = 25$ $\cdots$ | $C_{69}H_{21}$ $\cdots$ | $C_{73}H_{23}$ | $C_{77}H_{25}$ $\cdots$ | $C_{81}H_{27}$ $\cdots$ | $C_{85}H_{29}$ | $C_{89}H_{31}$ | 33 |
| | | $C_{72}H_{22}$ | $C_{76}H_{24}$ $\cdots$ | $C_{80}H_{26}$ | $C_{84}H_{28}$ | $C_{88}H_{30}$ | $C_{92}H_{32}$ | 34 |
| | $C_{71}H_{21}$ | $C_{75}H_{23}$ $\cdots$ | $C_{79}H_{25}$ | $C_{83}H_{27}$ | $C_{87}H_{29}$ | $C_{91}H_{31}$ $\cdots$ | $C_{95}H_{33}$ | 35 |
| | $C_{74}H_{22}$ $\cdots$ | $C_{78}H_{24}$ | $C_{82}H_{26}$ | $C_{86}H_{28}$ | $C_{90}H_{30}$ | $C_{94}H_{32}$ | $C_{98}H_{34}$ | 36 |
| $C_{73}H_{21}$ | $C_{77}H_{23}$ $\cdots$ | $C_{81}H_{25}$ | $C_{85}H_{27}$ | $C_{89}H_{29}$ | $C_{93}H_{31}$ | $C_{97}H_{33}$ | $C_{101}H_{35}$ | 37 |
| $C_{76}H_{22}$ $\cdots$ | $C_{80}H_{24}$ | $C_{84}H_{26}$ | $C_{88}H_{28}$ | $C_{92}H_{30}$ | $C_{96}H_{32}$ | $C_{100}H_{34}$ $\cdots$ | $C_{104}H_{36}$ | 38 |
| $C_{79}H_{23}$ | $C_{83}H_{25}$ | $C_{87}H_{27}$ $\cdots$ | $C_{91}H_{29}$ | $C_{95}H_{31}$ | $C_{99}H_{33}$ | $C_{103}H_{35}$ | $C_{107}H_{37}$ | 39 |
| $C_{82}H_{24}$ | $C_{86}H_{26}$ $\cdots$ | $C_{90}H_{28}$ | $C_{94}H_{30}$ | $C_{98}H_{32}$ | $C_{102}H_{34}$ | $C_{106}H_{36}$ | $C_{110}H_{38}$ | 40 |
| | $\vdots$ | | | $\vdots$ | | | | |
| | $h = 31$ | | | $h = 34$ | | | | |

## 7.2 How to Find the Place of a Formula in the Periodic Table?

Assume a formula $C_nH_s$ which corresponds to a benzenoid hydrocarbon. Then the coordinates of the periodic table for benzenoid hydrocarbons are readily obtained from Eqs. (1) and (6) as

$$d_s = (1/2)(3s - n) - 7, \qquad n_i = n - 2s + 6. \tag{16}$$

It may be advantageous to keep track of the number of hexagons ($h$) also during the studies of benzenoid isomers. On eliminating $n_i$ from (1) the following relation emerges, which is of interest in this connection.

$$h = (1/2)(n - s) + 1. \tag{17}$$

The formulas $C_nH_s$ associated with the same $h$ value are found along diagonals in the periodic table. Some of these diagonals are indicated in Table 5.

It was assumed that $C_nH_s$ is compatible with a benzenoid hydrocarbon in order to have a place in the periodic table at all. This can be decided most directly in the following way. Firstly, $n$ and $s$ have the same parity; it means that either both $n$ and $s$ are even or both of them are odd. Then, if $n$ is given, the possible values of $s$, which all are realized, are found within the range [29, 52]

$$2\lceil (1/2)(n + 6^{1/2}n^{1/2}) \rceil - n \le s \le n + 2 - 2\lceil (1/4)(n - 2) \rceil. \tag{18}$$

The possible values of $n$ are 6, 10, 13, 14 and all integers $n \ge 16$. Example: for $n = 60$, $20 \le s \le 32$; hence the following formulas exist for the $C_{60}$ benzenoid hydrocarbons – $C_{60}H_{20}$, $C_{60}H_{22}$, $C_{60}H_{24}$, $C_{60}H_{26}$, $C_{60}H_{28}$, $C_{60}H_{30}$, $C_{60}H_{32}$ [18, 24].

Conversely, if $s$ is given, the always realized possible values of $n$ (under the restriction of same parity) are

$$s - 6 + 2\lceil s/2 \rceil \le n \le + 2\lfloor (1/12)(s^2 - 6s) \rfloor. \tag{19}$$

The possible values of $s$ are 6 and all integers $s \ge 8$. Example: for $s = 20$, $34 \le n \le 64$; hence the following formulas exist for the $H_{20}$ benzenoid hydrocarbons – $C_{34}H_{20}$, $C_{36}H_{20}$, $C_{38}H_{20}$, ..., $C_{60}H_{20}$, $C_{62}H_{20}$, $C_{64}H_{20}$.

## 7.3 The Position of Benzene

On extrapolating the above scheme the formula for benzene, $C_6H_6$, should be placed in the periodic table for benzenoids in the first row (among catacondensed benzenoids) just to the left of $C_{10}H_8$ (naphthalene) [25]. There are serious arguments against this position, which implies $d_s = -1$, $n_i = 0$. Firstly, $C_6H_6$ in this position would be the only formula not having any formula in the same column just below it. Secondly, the staircase-like boundary would be obscured at the top of the table. It is recalled that $d_s = -1$ for benzene, although it fits into Eq. (6), can not be interpreted in terms of the tree disconnections and tree connections. It seems safest to keep $C_6H_6$ outside the periodic table for benzenoids, although benzene is a benzenoid (with $h = 1$) by definition. By the way, some authors do not reckon

Jon Brunvoll, Björg N. Cyvin, and Sven J. Cyvin

benzene among benzenoid hydrocarbons, but start this class with naphthalene ($h = 2$).

Dias [16] has stretched the analogy between the periodic table for benzenoid hydrocarbons and the Mendeleev periodic table (for elements) rather far. It can be stretched still farther by comparing the position of benzene to the unique position of hydrogen in the Mendeleev table.

## 7.4 Shape of the Staircase-Like Boundary

The left-hand side boundary of the periodic table starts with one extremely high step (six formulas), followed by a four-formula step. When measured in the same way the following steps, as can be proved rigorously, always hold either three or two formulas. The first ("low") two-formula step starts with $C_{52}H_{18}$ ($h = 18$); cf. Table 5. The last column in Table 5 (starting with $C_{73}H_{21}$ at the top) contains a three-formula ("high") step, which is followed by a low step:

$$
\begin{array}{ccc}
& & C_{73}H_{21} \ \ (h = 27) \\
& & C_{76}H_{22} \\
& & C_{79}H_{23} \\
& C_{78}H_{22} & \cdot \\
& C_{81}H_{23} & \cdot \\
C_{80}H_{22} & \cdot & \cdot \\
\cdot & \cdot & \cdot
\end{array}
$$

# 8 Detailed Analysis of the Formulas

## 8.1 Notation and Circumscribing

Let a formula $C_nH_s$ be denoted alternatively as

$$C_nH_s \equiv (n; s).$$

The generation of a (larger) benzenoid by circumscribing another (smaller) benzenoid is an important process in the studies of benzenoid isomers [7, 15, 19]. Let B be a benzenoid which can be circumscribed and has the formula $C_nH_s$. Further, let $B' \equiv$ circum-B have the formula $C_{n'}H_{s'}$. Then [25]

$$(n'; s') \equiv (n + 2s + 6; s + 6). \tag{20}$$

Repeated application of (20) is covered by the following explicit formula.

$k$-circum-B: $(n_k; s_k) = (6k^2 + 2sk + n; 6k + s). \tag{21}$

## 8.2 More Classes of Benzenoids

It is easily found that the extremal benzenoids, as defined in Sect. 6.2, say A, have the formulas given by

A:  $\quad (2h + 1 + \lceil(12h - 3)^{1/2}\rceil; 3 + \lceil(12h - 3)^{1/2}\rceil)$

for $h = 1, 2, 3, 4, \ldots$. Notice that benzene ($C_6H_6$) is reckoned among the extremal benzenoids. Any formula for extremal benzenoids, except benzene, is obviously situated on the staircase-like boundary so that it has no formula in the same row to the left of it.

The *protrusive* benzenoids form a subclass of the extremal benzenoids. By definition a protrusive benzenoid has a formula with no other formula in the same column above it, and no formula in the same row to the left of it in the periodic table. All pericondensed protrusive benzenoids are generated by circumscribing the extremal benzenoids. Consequently they have the formulas as given below [53].

circum-A: $(2h + 13 + 3\lceil(12h - 3)^{1/2}\rceil; 9 + \lceil(12h - 3)^{1/2}\rceil)$.

In addition comes naphthalene ($C_{10}H_8$), which also by definition is a protrusive benzenoid, the only catacondensed system of this class.

The *extreme-left* benzenoids are defined by formulas on the staircase-like boundary so that, in each case, there is no formula in the same row to the left. Hence the extremal benzenoids without benzene form a subclass of the extreme-left benzenoids. But there exist extreme-left benzenoids, say x, which are not extremal. Their formulas are found one step up and one step to the right from every formula for the pericondensed protrusive benzenoids. Hence one obtains readily the following expression.

x:  $\quad (2h + 14 + 3\lceil(12h - 3)^{1/2}\rceil; 10 + \lceil(12h - 3)^{1/2}\rceil)$.

The smallest extreme-left benzenoids which are not extremal, have the formula $C_{25}H_{13}$ ($h = 7$); cf. Table 5.

The *circular* benzenoids are defined by having $h = h_{max}$ for a given $s$. These systems have, loosely speaking, the largest area in relation to the circumference; hence the term "circular". More precisely, they have the largest number of hexagons for a given perimeter length. In this connection we give the always realized upper and lower bounds for $h$ when $s$ is given:

$$\lceil s/2\rceil - 2 \le h \le \lfloor(1/12)(s^2 - 6s + 12)\rfloor. \tag{22}$$

Here the upper bound ($h_{max}$) is a deduction from a formula for primitive coronoids [54]. For the number of internal vertices one obtains

$$2\lceil s/2\rceil - s \le n_i \le 2\lfloor(1/12)(s^2 - 6s + 12)\rfloor - s + 4 \tag{23}$$

where the upper and lower bounds are realized simultaneously with those of Eq. (22). When $s$ is even, the lower bound represents the catacondensed benzenoids ($n_i = 0$); when $s$ is odd it corresponds to $n_i = 1$. But here we

199

are most interested in the upper bound. It gives the general formula for a circular benzenoid, say O.

O: $\qquad (s + 2\lfloor(1/12)(s^2 - 6s)\rfloor); s)$

where $s = 6, 8, 9, 10, 11, \ldots$ as specified above. All circular benzenoids are extremal and therefore also extreme-left if benzene is excluded. The first (smallest) circular benzenoids have the formulas $C_6H_6$, $C_{10}H_8$, $C_{13}H_9$, $C_{16}H_{10}$, $C_{19}H_{11}$, $C_{24}H_{12}$ and $C_{27}H_{13}$, of which $C_{24}H_{12}$ pertains to a protrusive benzenoid (coronene). All the higher circular benzenoids are protrusive.

# 9 Strictly Pericondensed Benzenoid and Excised Internal Structure

The pericondensed extreme-left benzenoids constitute a subclass of the *strictly pericondensed* benzenoids in the sense of Dias [12, 15, 19, 21, 55–57]; they are defined by having all their internal vertices connected and no catacondensed appendages. Phenalene, $C_{13}H_9$, which has only one internal vertex, is reckoned among the strictly pericondensed benzenoids. An equivalent definition in a most succint form reads:

A strictly pericondensed benzenoid is a benzenoid with $h > 2$ and all its internal edges connected.

The formulas at the extreme left in the periodic table for benzenoid hydrocarbons, except $C_{10}H_8$ (naphthalene), represent exclusively strictly pericondensed benzenoids [12, 16, 21], viz. the pericondensed extreme-left benzenoids. Formulas for $d_s \le 0$ and not at the extreme left represent both strictly pericondensed and non-strictly pericondensed benzenoids. For $d_s > 0$ there are no strictly pericondensed benzenoids [56].

It should be clear that strictly pericondensed benzenoids occur for formulas at unlimited distances from the staircase-like boundary. Consider, for instance, the homolog series of hydrocarbons as shown in Fig. 2. The systems have in general (for $h \ge 2$, $s \ge 8$) the formulas:

$$(3h + 4; h + 6) = (3s - 14; s). \qquad (24)$$

The Dias parameter is constantly zero ($d_s = 0$), which means that the formulas are found in the same column, where $C_{10}H_8$ (naphthalene) is at the top. There is no limitation as to how far down one can get in this way, moving steadily away from the staircase-like boundary.

The extremal benzenoids have no coves and no fjords, but this property is not valid for strictly pericondensed benzenoids in general. Figure 3 shows some counterexamples.

The *excised internal structure* [10, 12, 19, 21, 22, 55] is defined in connection with strictly pericondensed benzenoids. It is the set of internal vertices and the edges

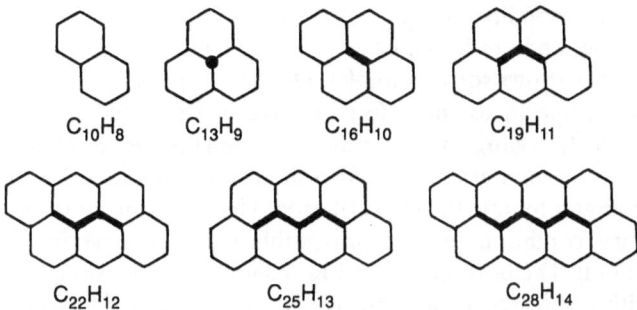

$C_{10}H_8$ $\qquad$ $C_{13}H_9$ $\qquad$ $C_{16}H_{10}$ $\qquad$ $C_{19}H_{11}$

$C_{22}H_{12}$ $\qquad$ $C_{25}H_{13}$ $\qquad$ $C_{28}H_{14}$

**Fig. 2.** A homolog series of hydrocarbons generated successively by two-contact additions (attachments of $C_3H$). The excised internal structures are indicated by *heavy lines* (and the *dot* in $C_{13}H_9$)

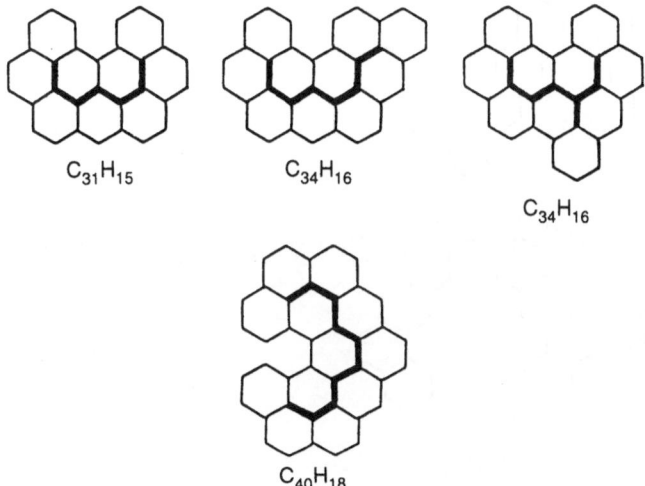

$C_{31}H_{15}$ $\qquad$ $C_{34}H_{16}$

$C_{34}H_{16}$

$C_{40}H_{18}$

**Fig. 3.** The smallest strictly pericondensed benzenoid with a cove, $C_{31}H_{15}$, which is non-Kekuléan ($\Delta = 1$). The two smallest Kekuléan strictly pericondensed benzenoids with a cove, $C_{34}H_{16}$, which are normal. The smallest strictly pericondensed benzenoid with a fjord, $C_{40}H_{18}$, which is normal (Kekuléan)

connecting them. Thus the excised internal structure emerges by deleting the external vertices and their incident edges ("excising" the benzenoid).

If A is a benzenoid which can be circumscribed, then it is the excised internal structure of circum-A. But an excised internal structure is not necessarily a benzenoid. In Fig. 2, for instance, one finds one vertex (corresponding to $CH_3$) as the excised internal structure of phenalene ($C_{13}H_9$). Further, two connected vertices (corresponding to ethene, $C_2H_4$) is the excised internal structure of pyrene ($C_{16}H_{10}$).

The circumscribing is obviously an important process in connection with excised internal structures. It is clear that a benzenoid with a cove or a fjord (or

Jon Brunvoll, Björg N. Cyvin, and Sven J. Cyvin

non-benzenoids with the corresponding formations) can not be circumscribed. But also benzenoids without coves and fjords can be constructed so that they can not be circumscribed [25]. Figure 4 shows some examples. Dias [12, 15] has formulated the two-carbon atom gap criterion to this effect. A cove is associated with a two-edge gap (or one-carbon atom gap), on the perimeter, while a fjord is associated with a one-edge gap (or zero-carbon atom gap). Also the third example of Fig. 4 ($C_{46}H_{26}$), having a two-edge gap, follows the Dias criterion. This criterion, however, fails to give the necessary condition for the impossibility to circumscribe a benzenoid (or non-benzenoid). The bottom row in Fig. 4 shows counterexamples: a benzenoid ($C_{42}H_{24}$) with a three-edge gap and one ($C_{38}H_{22}$) with a four-edge gap; yet none of them can be circumscribed.

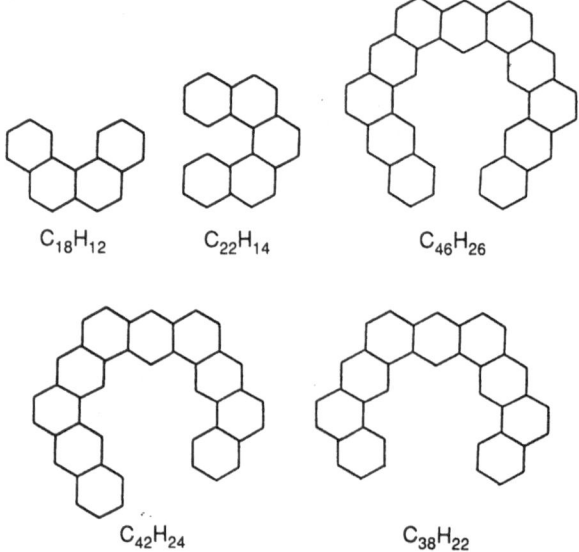

$C_{18}H_{12}$ $\quad\quad C_{22}H_{14}$ $\quad\quad\quad C_{46}H_{26}$

$C_{42}H_{24}$ $\quad\quad\quad\quad C_{38}H_{22}$

**Fig. 4.** The smallest benzenoid with a cove ($C_{18}H_{12}$), the smallest benzenoid with a fjord ($C_{22}H_{14}$), and other benzenoids which can not be circumscribed

## 10 Incomplete Data for Some Benzenoid Isomers

Table 6 is a compilation of data for numbers of benzenoid isomers with $h \geq 15$. They are incomplete in the sense that, for each $h$, the list of $n_i$ values is not complete. Furthermore, in many cases the subdivisions into normal and essentially disconnected benzenoids are unknown.

The table starts with the number of $h = 15$ catacondensed benzenoids, which has not been published previously (but see the preceding chapter). In general most of the numbers, especially for the higher $h$ values, pertain to extreme-left benzenoids. This is the case for all entries at $h \geq 39$.

**Table 6.** Numbers of benzenoid isomers for $h \geqq 15$

| $h$ | $n_i$ | $d_s$ | Formula | $\Delta = 0$ | | | o (non-Kekuléan) | | | | | | Total isomers |
|---|---|---|---|---|---|---|---|---|---|---|---|---|---|
| | | | | n | e | Total Kek. | $\Delta = 0$ | 1 | 2 | 3 | 4 | Total non-Kek. | |
| 15 | 0 | 13 | $C_{62}H_{34}$ | 4799205 | 0 | 4799205 | 0 | 0 | 0 | 0 | 0 | 0 | 4799205 |
| | 13 | 0 | $C_{49}H_{21}$ | 0 | 0 | 0 | 0 | 5438 | 0 | 174 | 0 | 5612[a] | 5612[b] |
| | 14 | −1 | $C_{48}H_{20}$ | 943 | 181 | 1124[a] | 0 | 0 | 445 | 0 | 1 | 446[a] | 1570[b] |
| | 15 | −2 | $C_{47}H_{19}$ | 0 | 0 | 0 | 0 | 336 | 0 | 11 | 0 | 347[a] | 347[b] |
| | 16 | −3 | $C_{46}H_{18}$ | 47[c] | 2[c] | 49[c] | 0 | 0 | 20[c] | 0 | 1[c] | 21[c] | 70[b] |
| | 17 | −4 | $C_{45}H_{17}$ | 0 | 0 | 0 | 0 | 4[c] | 0 | 0 | 0 | 4[c-f] | 4[b] |
| 16 | 15 | −1 | $C_{51}H_{21}$ | 0 | 0 | 0 | 0 | 4341 | 0 | 160 | 0 | 4501[2] | 4501[b] |
| | 16 | −2 | $C_{50}H_{20}$ | 702 | 88 | 790[a] | 1 | 0 | 329 | 0 | 1 | 331[a] | 1121[b] |
| | 17 | −3 | $C_{49}H_{19}$ | 0 | 0 | 0 | 0 | 213 | 0 | 10 | 0 | 223[a] | 223[b] |
| | 18 | −4 | $C_{48}H_{18}$ | 22[c] | 0 | 22[g,h] | 0 | 0 | 8[i] | 0 | 0 | 8[i] | 30[b] |
| | 19 | −5 | $C_{47}H_{17}$ | 0 | 0 | 0 | 0 | 1[c] | 0 | 0 | 0 | 1[i] | 1[b] |
| 17 | 17 | −2 | $C_{53}H_{21}$ | 0 | 0 | 0 | 0 | 3272 | 0 | 142 | 0 | 3414[a] | 3414[b] |
| | 18 | −3 | $C_{52}H_{20}$ | 478 | 47 | 525[a] | 0 | 0 | 236 | 0 | 2 | 238[a] | 763[b] |
| | 19 | −4 | $C_{51}H_{19}$ | 0 | 0 | 0 | 0 | 111 | 0 | 6 | 0 | 117[a] | 117[b] |
| | 20 | −5 | $C_{50}H_{18}$ | 7[c] | 0 | 7[g,h] | 0 | 0 | 2[c] | 0 | 0 | 2[i] | 9[b] |
| 18 | 19 | −3 | $C_{55}H_{21}$ | 0 | 0 | 0 | 0 | 2322 | 0 | 115 | 0 | 2437[a] | 2437[b] |
| | 20 | −4 | $C_{54}H_{20}$ | 304 | 17 | 321[a] | 0 | 0 | 149 | 0 | 1 | 150[a] | 471[b] |
| | 21 | −5 | $C_{53}H_{19}$ | 0 | 0 | 0 | 0 | 50 | 0 | 3 | 0 | 53[a] | 53[b] |
| | 22 | −6 | $C_{52}H_{18}$ | 2[c] | 0 | 2[g,h] | 0 | 0 | 1[c] | 0 | 0 | 1[g] | 3[b] |
| 19 | 21 | −4 | $C_{57}H_{21}$ | 0 | 0 | 0 | 0 | 1558 | 0 | 89 | 0 | 1647[a] | 1647[b] |
| | 22 | −5 | $C_{56}H_{20}$ | 164 | 7 | 171[a] | 0 | 0 | 84 | 0 | 1 | 85[a] | 256[b] |
| | 23 | −6 | $C_{55}H_{19}$ | 0 | 0 | 0 | 0 | 17[c] | 0 | 1[c] | 0 | 18[c] | 18[b] |
| | 24 | −7 | $C_{54}H_{18}$ | 1[c] | 0 | 1[j] | 0 | 0 | 0 | 0 | 0 | 0 | 1[b] |

**Table 6.** (continued)

| $h$ | $n_i$ | $d_s$ | Formula | $\Delta = 0$ | | Total Kek. | o (non-Kekuléan) | | | | | Total non-Kek. | Total isomers |
|---|---|---|---|---|---|---|---|---|---|---|---|---|---|
| | | | | n | e | | $\Delta = 1$ | 2 | 3 | 4 | 5 | | |
| 20 | 23 | −5 | $C_{59}H_{21}$ | 0 | 0 | 0 | 950 | 0 | 59 | 0 | 0 | 1009[a] | 1009[b] |
| | 24 | −6 | $C_{58}H_{20}$ | 87 | 1 | 88[a] | 0 | 40 | 0 | 1 | 0 | 41[a] | 129[b] |
| | 25 | −7 | $C_{57}H_{19}$ | 0 | 0 | 0 | 4[c] | 0 | 0 | 0 | 0 | 4[i] | 4[b] |
| 21 | 25 | −6 | $C_{61}H_{21}$ | 0 | 0 | 0 | 548 | 0 | 38 | 0 | 1 | 587[a] | 587[b] |
| | 26 | −7 | $C_{60}H_{20}$ | 32[c] | 0 | 32[c,k] | 1[c] | 15[c] | 0 | 0 | 0 | 15[c] | 47[b] |
| | 27 | −8 | $C_{59}H_{19}$ | 0 | 0 | 0 | 0 | 0 | 0 | 0 | 0 | 1[i] | 1[b] |
| 22 | 27 | −7 | $C_{63}H_{21}$ | 0 | 0 | 0 | 270 | 0 | 20 | 0 | 0 | 290[a] | 290[b] |
| | 28 | −8 | $C_{62}H_{20}$ | 12[c] | 0 | 12[g] | 0 | 4[c] | 0 | 0 | 0 | 4[c−f] | 16[b] |
| 23 | 29 | −8 | $C_{65}H_{21}$ | 0 | 0 | 0 | 118[a] | 0 | 8 | 0 | 0 | 126[a] | 126[b] |
| | 30 | −9 | $C_{64}H_{20}$ | 3[c] | 0 | 3[g] | 0 | 1[c] | 0 | 0 | 0 | 1[g] | 4[b] |
| 24 | 30 | −8 | $C_{68}H_{22}$ | † | † | 503 | 0 | 280 | 0 | 6 | 0 | 286[a] | 789[b] |
| | 31 | −9 | $C_{67}H_{21}$ | 0 | 0 | 0 | 41 | 0 | 2 | 0 | 0 | 43[a] | 43[b] |
| | 32 | −10 | $C_{66}H_{20}$ | 1[c] | 0 | 1[g,h] | 0 | 0 | 0 | 0 | 0 | 0 | 1[b] |
| 25 | 32 | −9 | $C_{70}H_{22}$ | † | † | 241 | 0 | 133 | 0 | 3 | 0 | 136 | 377[b] |
| | 33 | −10 | $C_{69}H_{21}$ | 0 | 0 | 0 | 12[c] | 0 | 1[c] | 0 | 0 | 13[c−f] | 13[b] |
| 26 | 34 | −10 | $C_{72}H_{22}$ | 103 | 0 | 103 | 0 | 52 | 0 | 1 | 0 | 53 | 156[b] |
| | 35 | −11 | $C_{71}H_{21}$ | 0 | 0 | 0 | 2[c] | 0 | 0 | 0 | 0 | 2[c−f] | 2[b] |
| 27 | 35 | −10 | $C_{75}H_{23}$ | 0 | 0 | 0 | 1013 | 0 | 91 | 0 | 1 | 1105 | 1105[b] |
| | 36 | −11 | $C_{74}H_{22}$ | 36 | 0 | 36 | 0 | 17 | 0 | 0 | 0 | 17 | 53[b] |
| | 37 | −12 | $C_{73}H_{21}$ | 0 | 0 | 0 | 1[c] | 0 | 0 | 0 | 0 | 1[i] | 1[b] |
| 28 | 37 | −11 | $C_{77}H_{23}$ | 0 | 0 | 0 | 469 | 0 | 40 | 0 | 0 | 509 | 509[b] |
| | 38 | −12 | $C_{76}H_{22}$ | 12[c] | 0 | 12[c−f,k] | 0 | 4[c] | 0 | 0 | 0 | 4[c−f] | 16[b] |

| # | pair | h | Formula | | | | | | | | | | Total |
|---|---|---|---|---|---|---|---|---|---|---|---|---|---|
| 29 | 39 | −12 | $C_{79}H_{23}$ | 0 | 0 | 0 | 192 | 0 | 15 | 0 | 0 | 207 | 207[b] |
|  | 40 | −13 | $C_{78}H_{22}$ | 3[c] | 0 | 3[g] | 0 | 1[c] | 0 | 0 | 0 | 1[g] | 4[b] |
| 30 | 40 | −12 | $C_{82}H_{24}$ | + | + | 1092 | 0 | 687 | 0 | 20 | 0 | 707 | 1799[b] |
|  | 41 | −13 | $C_{81}H_{23}$ | 0 | 0 | 0 | 64 | 0 | 4 | 0 | 0 | 4 | 68[b] |
|  | 42 | −14 | $C_{80}H_{22}$ | 1[c] | 0 | 1[g,h] | 0 | 0 | 0 | 0 | 0 | 0 | 1[b] |
| 31 | 42 | −13 | $C_{84}H_{24}$ | + | + | 508 | 0 | 309 | 0 | 8 | 0 | 317 | 825[b] |
|  | 43 | −14 | $C_{83}H_{23}$ | 0 | 0 | 0 | 19[c] | 0 | 1[c] | 0 | 0 | 20[c-f] | 20[b] |
| 32 | 44 | −14 | $C_{86}H_{24}$ | + | + | 217 | 0 | 125 | 0 | 2 | 0 | 127 | 344[b] |
|  | 45 | −15 | $C_{85}H_{23}$ | 0 | 0 | 0 | 4[c] | 0 | 0 | 0 | 0 | 4[c-f] | 4[b] |
| 33 | 46 | −15 | $C_{88}H_{24}$ | 79 | 0 | 79 | 0 | 41 | 0 | 1 | 0 | 42 | 121[b] |
|  | 47 | −16 | $C_{87}H_{23}$ | 0 | 0 | 0 | 1[c] | 0 | 0 | 0 | 0 | 1[i] | 1[b] |
| 34 | 47 | −15 | $C_{91}H_{25}$ | 0 | 0 | 0 | 1324 | 0 | 136 | 0 | 1 | 1461 | 1461[b] |
|  | 48 | −16 | $C_{90}H_{24}$ | 27[c] | 0 | 27[c-f] | 0 | 12[c] | 0 | 0 | 0 | 12[c-f] | 39[b] |
| 35 | 49 | −16 | $C_{93}H_{25}$ | 0 | 0 | 0 | 564 | 0 | 52 | 0 | 0 | 616 | 616[b] |
|  | 50 | −17 | $C_{92}H_{24}$ | 7[c] | 0 | 7[g] | 0 | 2[c] | 0 | 0 | 0 | 2[g] | 9[b] |
| 36 | 51 | −17 | $C_{95}H_{25}$ | 0 | 0 | 0 | 211 | 0 | 17 | 0 | 0 | 228 | 228[b] |
|  | 52 | −18 | $C_{94}H_{24}$ | 2[c] | 0 | 2[g] | 0 | 1[c] | 0 | 0 | 0 | 1[g] | 3[b] |
| 37 | 52 | −17 | $C_{98}H_{26}$ | + | + | 1748 | 0 | 1191 | 0 | 40 | 0 | 1231 | 2979[b] |
|  | 53 | −18 | $C_{97}H_{25}$ | 0 | 0 | 0 | 68 | 0 | 4 | 0 | 0 | 72 | 72[b] |
|  | 54 | −19 | $C_{96}H_{24}$ | 1[c] | 0 | 1[j] | 0 | 0 | 0 | 0 | 0 | 0 | 1[b] |
| 38 | 54 | −18 | $C_{100}H_{26}$ | + | + | 798 | 0 | 511 | 0 | 15 | 0 | 525 | 1324[b] |
|  | 55 | −19 | $C_{99}H_{25}$ | 0 | 0 | 0 | 0 | 19[c] | 0 | 1[c] | 0 | 20[c-f] | 20[b] |
| 39 | 56 | −19 | $C_{102}H_{26}$ | + | + | 322 | 0 | 199 | 0 | 4 | 0 | 203 | 525[b] |
|  | 57 | −20 | $C_{101}H_{25}$ | 0 | 0 | 0 | 4[c] | 0 | 0 | 0 | 0 | 4[i] | 4[b] |
| 40 | 58 | −20 | $C_{104}H_{26}$ | 121 | 0 | 121 | 0 | 64 | 0 | 1 | 0 | 65 | 186[b] |
|  | 59 | −21 | $C_{103}H_{25}$ | 0 | 0 | 0 | 1[c] | 0 | 0 | 0 | 0 | 1[i] | 1[b] |

Table 6. (continued)

| h | $n_i$ | $d_s$ | Formula | $\Delta = 0$ | | | o (non-Kekuléan) | | | | | | | Total isomers |
|---|---|---|---|---|---|---|---|---|---|---|---|---|---|---|
| | | | | n | e | Total Kek. | $\Delta = 1$ | 2 | 3 | 4 | 5 | Total non-Kek. | |
| 41 | 60 | −21 | $C_{106}H_{26}$ | 38[c] | 0 | 38[c-f,k] | 0 | 19[c] | 0 | 0 | 0 | 19[c-f] | 57[b] |
| 42 | 61 | −21 | $C_{109}H_{27}$ | 0[c] | 0 | 0 | 1172 | 0 | 125 | 0 | 1 | 1298 | 1298 |
| | 62 | −22 | $C_{108}H_{26}$ | 12[c] | 0 | 12[c-f] | 0 | 4[c] | 0 | 0 | 0 | 4[c-f] | 16[b] |
| 43 | 63 | −22 | $C_{111}H_{27}$ | 0[c] | 0 | 0 | 449 | 0 | 41 | 0 | 0 | 490 | 490 |
| | 64 | −23 | $C_{110}H_{26}$ | 3[c] | 0 | 3[g] | 0 | 1[c] | 0 | 0 | 0 | 1[g] | 4[b] |
| 44 | 65 | −23 | $C_{113}H_{27}$ | 0 | 0 | 0 | 155 | 0 | 12 | 0 | 0 | 167 | 167 |
| | 66 | −24 | $C_{112}H_{26}$ | 1[c] | 0 | 1[g,h] | 0 | 0 | 0 | 0 | 0 | 0 | 1[b] |
| 45 | 67 | −24 | $C_{115}H_{27}$ | 0 | 0 | 0 | 46 | 0 | 2 | 0 | 0 | 48[d-f] | 48[d-f] |
| 46 | 68 | −24 | $C_{118}H_{28}$ | † | † | 889 | 0 | 591 | 0 | 17 | 0 | 608 | 1497 |
| | 69 | −25 | $C_{117}H_{27}$ | 0 | 0 | 0 | 12 | 0 | 1 | 0 | 0 | 13[d-f] | 13[d-f] |
| 47 | 70 | −25 | $C_{120}H_{28}$ | † | † | 343 | 0 | 214 | 0 | 4 | 0 | 218 | 561 |
| | 71 | −26 | $C_{119}H_{27}$ | 0 | 0 | 0 | 2[c] | 0 | 0 | 0 | 0 | 2[c-f] | 2[c-f] |
| 48 | 72 | −26 | $C_{122}H_{28}$ | † | † | 125 | 0 | 68 | 0 | 1 | 0 | 69 | 194[e] |
| | 73 | −27 | $C_{121}H_{27}$ | 0 | 0 | 0 | 1[c] | 0 | 0 | 0 | 0 | 1[i] | 1[i] |
| 49 | 74 | −27 | $C_{124}H_{28}$ | 38 | 0 | 38[e,f] | 0 | 19 | 0 | 0 | 0 | 19[e,f] | 57[e,f] |
| 50 | 75 | −27 | $C_{127}H_{29}$ | 0 | 0 | 0 | 1744 | 0 | 199 | 0 | 1 | 1944 | 1944 |
| | 76 | −28 | $C_{126}H_{28}$ | 12 | 0 | 12[d-f] | 0 | 4 | 0 | 0 | 0 | 4[d-f] | 16[d-f] |
| 51 | 77 | −28 | $C_{129}H_{29}$ | 0 | 0 | 0 | 666 | 0 | 64 | 0 | 0 | 730 | 730 |
| | 78 | −29 | $C_{128}H_{28}$ | 3[c] | 0 | 3[c-f] | 0 | 1[c] | 0 | 0 | 0 | 1[c-f] | 4[c-f] |
| 52 | 79 | −29 | $C_{131}H_{29}$ | 0 | 0 | 0 | 232 | 0 | 19 | 0 | 0 | 251 | 251 |
| | 80 | −30 | $C_{130}H_{28}$ | 1[c] | 0 | 1[g,h] | 0 | 0 | 0 | 0 | 0 | 0 | 1[g,h] |

| No. | | Formula | | | | | | | | | | |
|---|---|---|---|---|---|---|---|---|---|---|---|---|
| 53 | −30 | $C_{133}H_{29}$ | 0 | 0 | 0 | 70 | 0 | 4 | 0 | 0 | $74^{d-f}$ | $74^{d-f}$ |
| 54 | −30 | $C_{136}H_{30}$ | † | † | 1730 | 0 | 1220 | 0 | 41 | 0 | 1261 | 2991 |
|  | −31 | $C_{135}H_{29}$ | 0 | 0 | 0 | 19 | 0 | 1 | 0 | 0 | $20^{d-f}$ | $20^{d-f}$ |
| 55 | −31 | $C_{138}H_{30}$ | † | † | 691 | 0 | 457 | 0 | 12 | 0 | 469 | 1160 |
|  | −32 | $C_{137}H_{29}$ | 0 | 0 | 0 | $4^c$ | 0 | 0 | 0 | 0 | $4^{c-f}$ | $4^{c-f}$ |
| 56 | −32 | $C_{140}H_{30}$ | † | † | 256 | 0 | 155 | 0 | 2 | 0 | 157 | 413 |
|  | −33 | $C_{139}H_{29}$ | 0 | 0 | 0 | $1^c$ | 0 | 0 | 0 | 0 | $1^i$ | $1^i$ |
| 57 | −33 | $C_{142}H_{30}$ | 86 | 86 | $86^f$ | 0 | 46 | 0 | 1 | 0 | $47^f$ | $133^{e,f}$ |
| 58 | −34 | $C_{144}H_{30}$ | 27 | 27 | $27^{d-f}$ | 0 | 12 | 0 | 0 | 0 | $12^{d-f}$ | $39^{d-f}$ |
| 59 | −34 | $C_{147}H_{31}$ | 0 | 0 | 0 | 1859 | 0 | 214 | 1 | 1 | 2074 | 2074 |
|  | −35 | $C_{146}H_{30}$ | $7^c$ | $7^c$ | $7^{c-f}$ | 0 | $2^c$ | 0 | 0 | 0 | $2^{c-f}$ | $9^{c-f}$ |
| 60 | −35 | $C_{149}H_{31}$ | 0 | 0 | 0 | 693 | 0 | 68 | 0 | 0 | 761 | 761 |
|  | −36 | $C_{148}H_{30}$ | $2^c$ | $2^c$ | $2^{c-f}$ | 0 | $1^c$ | 0 | 0 | 0 | $1^{c-f}$ | $3^{c-f}$ |
| 61 | −36 | $C_{151}H_{31}$ | 0 | 0 | 0 | 236 | 0 | 19 | 0 | 0 | 255 | 255 |
|  | −37 | $C_{150}H_{30}$ | $1^c$ | $1^c$ | $1^j$ | 0 | 0 | 0 | 0 | 0 | 0 | $1^j$ |
| 62 | −37 | $C_{153}H_{31}$ | 0 | 0 | 0 | 70 | 0 | 4 | 0 | 0 | $74^{d-f}$ | $74^{d-f}$ |
| 63 | −37 | $C_{156}H_{32}$ | † | † | 2454 | 0 | 1795 | 0 | 64 | 0 | 1859 | 4313 |
|  | −38 | $C_{155}H_{31}$ | 0 | 0 | 0 | 19 | 0 | 1 | 0 | 0 | $20^{d-f}$ | $20^{d-f}$ |
| 64 | −38 | $C_{158}H_{32}$ | † | † | 994 | 0 | 675 | 0 | 19 | 0 | 694 | 1688 |
|  | −39 | $C_{157}H_{31}$ | 0 | 0 | 0 | 4 | 0 | 0 | 0 | 0 | $4^{d-f}$ | $4^{d-f}$ |
| 65 | −39 | $C_{160}H_{32}$ | † | † | 364 | 0 | 232 | 0 | 4 | 0 | 236 | 600 |
|  | −40 | $C_{159}H_{31}$ | 0 | 0 | 0 | $1^c$ | 0 | 0 | 0 | 0 | $1^i$ | $1^i$ |
| 66 | −40 | $C_{162}H_{32}$ | 128 | 128 | $128^f$ | 0 | 70 | 0 | 1 | 0 | $71^f$ | $199^{e,f}$ |
| 67 | −41 | $C_{164}H_{32}$ | 38 | 38 | $38^{d-f}$ | 0 | 19 | 0 | 0 | 0 | $19^{d-f}$ | $57^{d-f}$ |

**Table 6.** (continued)

| $h$ | $n_i$ | $d_s$ | Formula | $\Delta = 0$ | | | o (non-Kekuléan) | | | | | | Total non-Kek. | Total isomers |
|---|---|---|---|---|---|---|---|---|---|---|---|---|---|---|
| | | | | $n$ | $e$ | Total Kek. | $\Delta = 1$ | 2 | 3 | 4 | 5 | | | |
| 68 | 108 | $-42$ | $C_{166}H_{32}$ | 12 | 0 | $12^{d-f}$ | 0 | 4 | 0 | 0 | 0 | $4^{d-f}$ | $16^{d-f}$ |
| 69 | 109 | $-42$ | $C_{169}H_{33}$ | 0 | 0 | 0 | 1403 | 0 | 155 | 0 | 1 | 1559 | 1559 |
| | 110 | $-43$ | $C_{168}H_{32}$ | $3^c$ | 0 | $3^{c-f}$ | 0 | $1^c$ | 0 | 0 | 0 | $1^{c-f}$ | $4^{c-f}$ |
| 70 | 111 | $-43$ | $C_{171}H_{33}$ | 0 | 0 | 0 | 496 | 0 | 46 | 0 | 0 | 542 | 542 |
| | 112 | $-44$ | $C_{170}H_{32}$ | $1^c$ | 0 | $1^{g,h}$ | 0 | 0 | 0 | 0 | 0 | 0 | $1^{g,h}$ |
| 71 | 113 | $-44$ | $C_{173}H_{33}$ | 0 | 0 | 0 | 162 | 0 | 12 | 0 | 0 | $174^{e,f}$ | $174^{e,f}$ |
| 72 | 115 | $-45$ | $C_{175}H_{33}$ | 0 | 0 | 0 | 46 | 0 | 2 | 0 | 0 | $48^{d-f}$ | $48^{d-f}$ |
| 73 | 117 | $-46$ | $C_{177}H_{33}$ | 0 | 0 | 0 | 12 | 0 | 1 | 0 | 0 | $13^{d-f}$ | $13^{d-f}$ |
| 74 | 118 | $-46$ | $C_{180}H_{34}$ | † | † | 1018 | 0 | 696 | 0 | 19 | 0 | 715 | $1733^{d-f}$ |
| | 119 | $-47$ | $C_{179}H_{33}$ | 0 | 0 | 0 | 2 | 0 | 0 | 0 | 0 | $2^{d-f}$ | $2^{d-f}$ |
| 75 | 120 | $-47$ | $C_{182}H_{34}$ | † | † | 370 | 0 | 236 | 0 | 4 | 0 | 240 | 610 |
| | 121 | $-48$ | $C_{181}H_{33}$ | 0 | 0 | 0 | 1 | 0 | 0 | 0 | 0 | 1 | 1 |
| 76 | 122 | $-48$ | $C_{184}H_{34}$ | 128 | 0 | $128^f$ | 0 | 70 | 0 | 1 | 0 | $71^f$ | $199^{e,f}$ |
| 77 | 124 | $-49$ | $C_{186}H_{34}$ | 38 | 0 | $38^{e,f}$ | 0 | 19 | 0 | 0 | 0 | $19^{e,f}$ | $57^{e,f}$ |
| 78 | 126 | $-50$ | $C_{188}H_{34}$ | 12 | 0 | $12^{d-f}$ | 0 | 4 | 0 | 0 | 0 | $4^{d-f}$ | $16^{d-f}$ |
| 79 | 127 | $-50$ | $C_{191}H_{35}$ | 0 | 0 | 0 | 1991 | 0 | 232 | 0 | 1 | 2224 | 2224 |
| | 128 | $-51$ | $C_{190}H_{34}$ | 3 | 0 | 3 | 0 | 1 | 0 | 0 | 0 | 1 | 4 |
| 80 | 129 | $-51$ | $C_{193}H_{35}$ | 0 | 0 | 0 | 717 | 0 | 70 | 0 | 0 | 787 | 787 |
| | 130 | $-52$ | $C_{192}H_{34}$ | 1 | 0 | 1 | 0 | 0 | 0 | 0 | 0 | 0 | 1 |

| | | | | | | | | | | | |
|---|---|---|---|---|---|---|---|---|---|---|---|
| 81 | 131 | −52 | $C_{195}H_{35}$ | 0 | 0 | 239 | 0 | 19 | 0 | 0 | 258[e,f] | 258[e,f] |
| 82 | 133 | −53 | $C_{197}H_{35}$ | 0 | 0 | 70 | 0 | 4 | 0 | 0 | 74[d-f] | 74[d-f] |
| 83 | 135 | −54 | $C_{199}H_{35}$ | 0 | 0 | 19 | 0 | 1 | 0 | 0 | 20[d-f] | 20[d-f] |
| 84 | 136 | −54 | $C_{202}H_{36}$ | † | 1954 | 0 | 1413 | 0 | 46 | 0 | 1459 | 3413 |
| | 137 | −55 | $C_{201}H_{35}$ | 0 | 0 | 4 | 0 | 0 | 0 | 0 | 4[d-f] | 4[d-f] |
| 85 | 138 | −55 | $C_{204}H_{36}$ | † | 739 | 0 | 496 | 0 | 12 | 0 | 508 | 1247 |
| | 139 | −56 | $C_{203}H_{35}$ | 0 | 0 | 1 | 0 | 0 | 0 | 0 | 1 | 1 |
| 86 | 140 | −56 | $C_{206}H_{36}$ | † | 264 | 0 | 162 | 0 | 2 | 0 | 164 | 428[e,f] |
| 87 | 142 | −57 | $C_{208}H_{36}$ | 86 | 86[f] | 0 | 46 | 0 | 1 | 0 | 47[f] | 133[e,f] |
| 88 | 144 | −58 | $C_{210}H_{36}$ | 27 | 27[d-f] | 0 | 12 | 0 | 0 | 0 | 12[d-f] | 39[d-f] |
| 89 | 146 | −59 | $C_{212}H_{36}$ | 7 | 7 | 0 | 2 | 0 | 0 | 0 | 2 | 9 |
| 90 | 147 | −59 | $C_{215}H_{37}$ | 0 | 0 | 2021 | 0 | 236 | 0 | 1 | 2258 | 2258 |
| | 148 | −60 | $C_{214}H_{36}$ | 2 | 2 | 0 | 1 | 0 | 0 | 0 | 1 | 3 |
| 91 | 149 | −60 | $C_{217}H_{37}$ | 0 | 0 | 723 | 0 | 70 | 0 | 0 | 793 | 793 |
| | 150 | −61 | $C_{216}H_{36}$ | 1 | 1 | 0 | 0 | 0 | 0 | 0 | 0 | 1 |
| 92 | 151 | −61 | $C_{219}H_{37}$ | 0 | 0 | 239 | 0 | 19 | 0 | 0 | 258[e,f] | 258[e,f] |
| 93 | 153 | −62 | $C_{221}H_{37}$ | 0 | 0 | 70 | 0 | 4 | 0 | 0 | 74[d-f] | 74[d-f] |
| 94 | 155 | −63 | $C_{223}H_{37}$ | 0 | 0 | 19 | 0 | 1 | 0 | 0 | 20[d-f] | 20[d-f] |
| 95 | 156 | −63 | $C_{226}H_{38}$ | † | 2693 | 0 | 2001 | 0 | 70 | 0 | 2071 | 4764 |
| | 157 | −64 | $C_{225}H_{37}$ | 0 | 0 | 4 | 0 | 0 | 0 | 0 | 4[d-f] | 4[d-f] |

ᵃ Cyvin, Brunvoll and Cyvin (1991) [53]; ᵇ Stojmenović, Tošić and Doroslovački (1986) [45]; ᶜ Brunvoll and Cyvin (1990) [25]; ᵈ Dias (1990) [56]; ᵉ Dias (1990) [58]; ᶠ Dias (1990) [57]; ᵍ Dias (1984) [12], incorrect data therein are omitted; ʰ Dias (1984) [10], incorrect data therein are omitted; ⁱ Dias (1986) [15], incorrect data therein are omitted; ʲ Dias (1982) [7], incorrect data therein are omitted; ᵏ Dias (1990) [21]; † Unknown.

## 11 Benzenoid Isomers and Number of Edges

The number of edges (C–C bonds), $m$, is an invariant, being the same for all isomers with a given formula $C_nH_s$ [16, 21]. Explicitly one has:

$$m = (1/2)(3n-s).\tag{25}$$

This does not mean, however, that two benzenoids with the same number of edges ($m$) must be isomers in general. The coefficients of the formula ($n$, $s$) are in general determined by two independent invariants, e.g. ($m$, $n_i$). In this case the transformation reads

$$n = (1/5)(4m - n_i + 6), \qquad s = (1/5)(2m - 3n_i + 18).\tag{26}$$

The $m$ values are limited between upper and lower bounds as functions of other invariants. Gutman [40] deduced from the work of Harary and Harborth [52]:

$$3h + \lceil(12h - 3)^{1/2}\rceil \le m \le 5h + 1.\tag{27}$$

In the original work [52] the following relation is given.

$$n - 1 + \lceil(1/4)(n - 2)\rceil \le m \le 2n - \lceil(1/2)(n + 6^{1/2}n^{1/2})\rceil.\tag{28}$$

As a supplement we give:

$$s + 3\lceil s/2\rceil - 9 \le m \le s + \lfloor(1/12)(s^2 - 6s)\rfloor - 2.\tag{29}$$

It was stated that a value of $m$ does not determine the pair of coefficients ($n$, $s$) of a formula for benzenoid isomers in general. Nevertheless, for the smallest values of $m$ the $C_nH_s$ formula for benzenoids with that number of edges is determined uniquely. This feature can be analysed in terms of upper and lower bounds as functions of $m$. Harary and Harborth [52] have given the relation

$$\lceil(1/5)(m - 1)\rceil \le h \le m - \lceil(1/3)[2m - 2 + (4m + 1)^{1/2}]\rceil.\tag{30}$$

We can supplement it by

$$5\lceil(1/5)(m - 1)\rceil - m + 1 \le n_i \le 4m + 1 - 5\lceil(1/3)[2m - 2 + (4m + 1)^{1/2}]\rceil.\tag{31}$$

The possible values of $m$ are 6, 11, 15, 16, 19, 20, 21 and all integers $m \ge 23$.

It is found that for all $m \le 29$ and $m = 32, 33, 37$ the upper and lower bounds are coincident so that the relations (30) and (31) give a precise determination of the invariants in question ($h$, $n_i$). Example: for $m = 20, 4 \le h \le 4$ and $1 \le n_i \le 1$. In Table 7 all the invariant pairs ($h$, $n_i$) and the corresponding formulas $C_n H_s$ are listed which, for benzenoid isomers, are determined by the $m$ values.

The analysis becomes more oriented towards the $C_nH_s$ isomers by using the pertinent coefficients ($n$, $s$) in the inequalities of the considered type. Again according to Harary and Harborth [52]:

$$1 + \lceil(1/3)[2m - 2 + (4m + 1)^{1/2}]\rceil \le n \le m + 1 - \lceil(1/5)(m - 1)\rceil.\tag{32}$$

**Table 7.** All formulas for benzenoid isomers which are determined by the number of edges ($m$)

| $m$ | $h$ | $n_i$ | Formula |
|---|---|---|---|
| 6 | 1 | 0 | $C_6H_6$ |
| 11 | 2 | 0 | $C_{10}H_8$ |
| 15 | 3 | 1 | $C_{13}H_9$ |
| 16 | 3 | 0 | $C_{14}H_{10}$ |
| 19 | 4 | 2 | $C_{16}H_{10}$ |
| 20 | 4 | 1 | $C_{17}H_{11}$ |
| 21 | 4 | 0 | $C_{18}H_{12}$ |
| 23 | 5 | 3 | $C_{19}H_{11}$ |
| 24 | 5 | 2 | $C_{20}H_{12}$ |
| 25 | 5 | 1 | $C_{21}H_{13}$ |
| 26 | 5 | 0 | $C_{22}H_{14}$ |
| 27 | 6 | 4 | $C_{22}H_{12}$ |
| 28 | 6 | 3 | $C_{23}H_{13}$ |
| 29 | 6 | 2 | $C_{24}H_{14}$ |
| 32 | 7 | 4 | $C_{26}H_{14}$ |
| 33 | 7 | 3 | $C_{27}H_{15}$ |
| 37 | 8 | 4 | $C_{30}H_{16}$ |

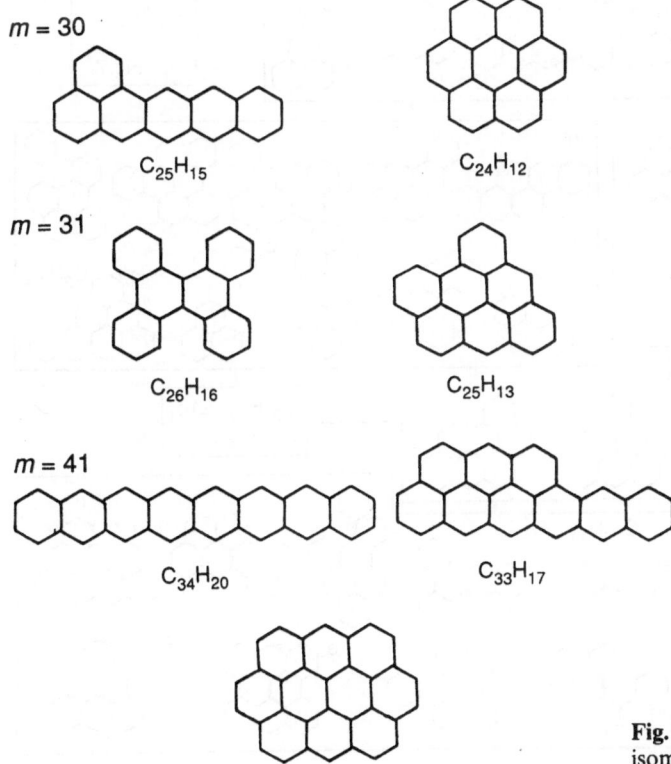

$m = 30$

$C_{25}H_{15}$

$C_{24}H_{12}$

$m = 31$

$C_{26}H_{16}$

$C_{25}H_{13}$

$m = 41$

$C_{34}H_{20}$

$C_{33}H_{17}$

$C_{32}H_{14}$

**Fig. 5.** Different benzenoid isomers with the same number of edges ($m$)

211

Jon Brunvoll, Björg N. Cyvin, and Sven J. Cyvin

We give as a supplement:

$$3 - 2m + 3\lceil (1/3) [2m - 2 + (4m + 1)^{1/2}]\rceil$$
$$\leq s \leq m + 3 - 3\lceil (1/5) (m - 1)\rceil. \tag{33}$$

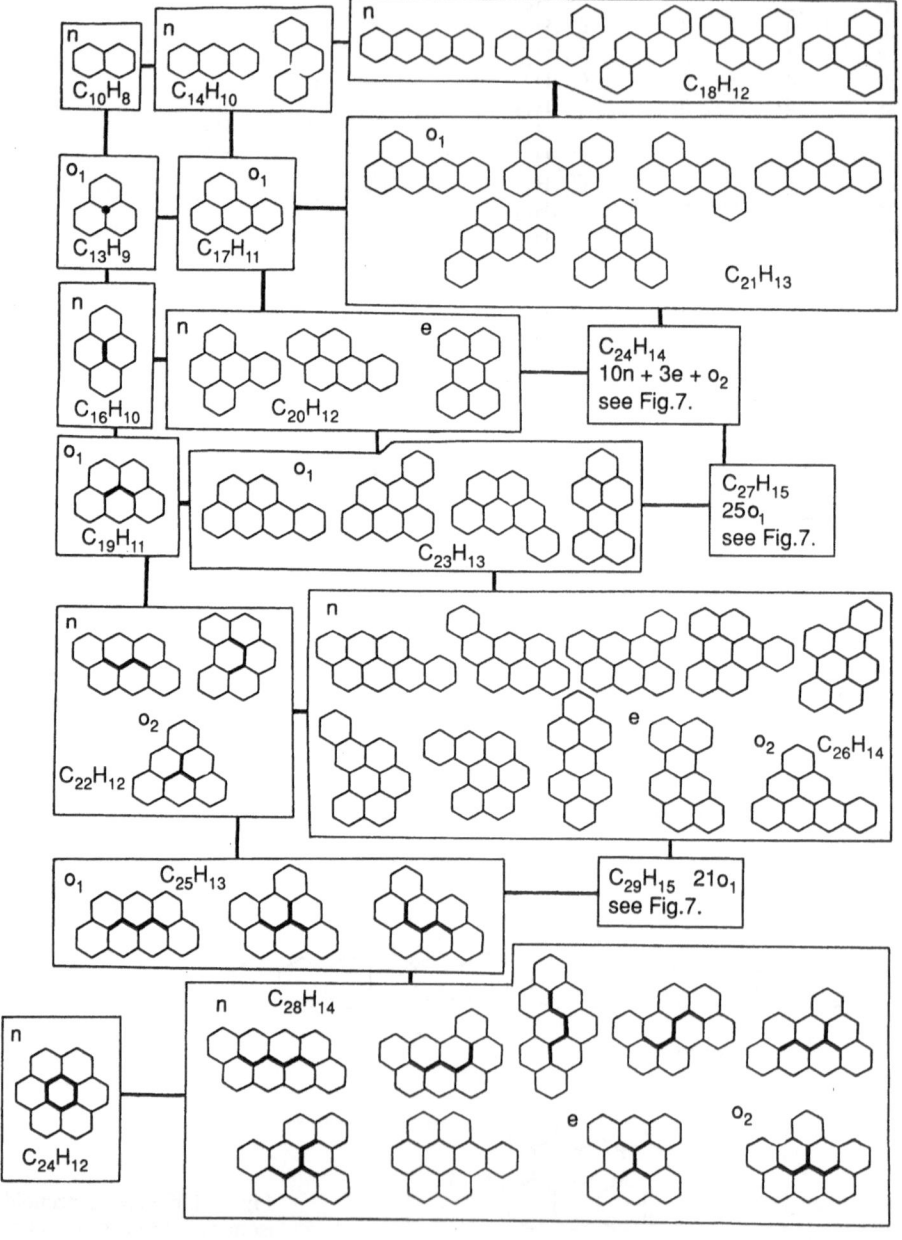

**Fig. 6.** Benzenoid isomers in the upper-left part of the periodic table (cf. Table 5)

212

The smallest different benzenoid isomers with the same number of edges occur for $m = 30$: $C_{25}H_{15}$ ($h = 6$, $n_i = 1$; 24 isomers) and $C_{24}H_{12}$ ($h = 7$, $n_i = 6$; coronene); see Fig. 5. The next example pertains to $m = 31$: $C_{26}H_{16}$ ($h = 6, n_i = 0$; 36 isomers) and $C_{25}H_{13}$ ($h = 7$, $n_i = 5$; 3 isomers). Similarly, for $m = 34, 35, 36$, 38, 39 and 40 there are in each case two formulas, one even-carbon and one odd-carbon. For $m = 41$ the situation occurs for the first time that three different formulas are compatible with the same $m$ value: $C_{34}H_{20}$ ($h = 8$, $n_i = 0$; 411 isomers), $C_{33}H_{17}$ ($h = 9$, $n_i = 5$; 154 isomers) and $C_{32}H_{14}$ ($h = 10$, $n_i = 10$; ovalene); cf. Fig. 5.

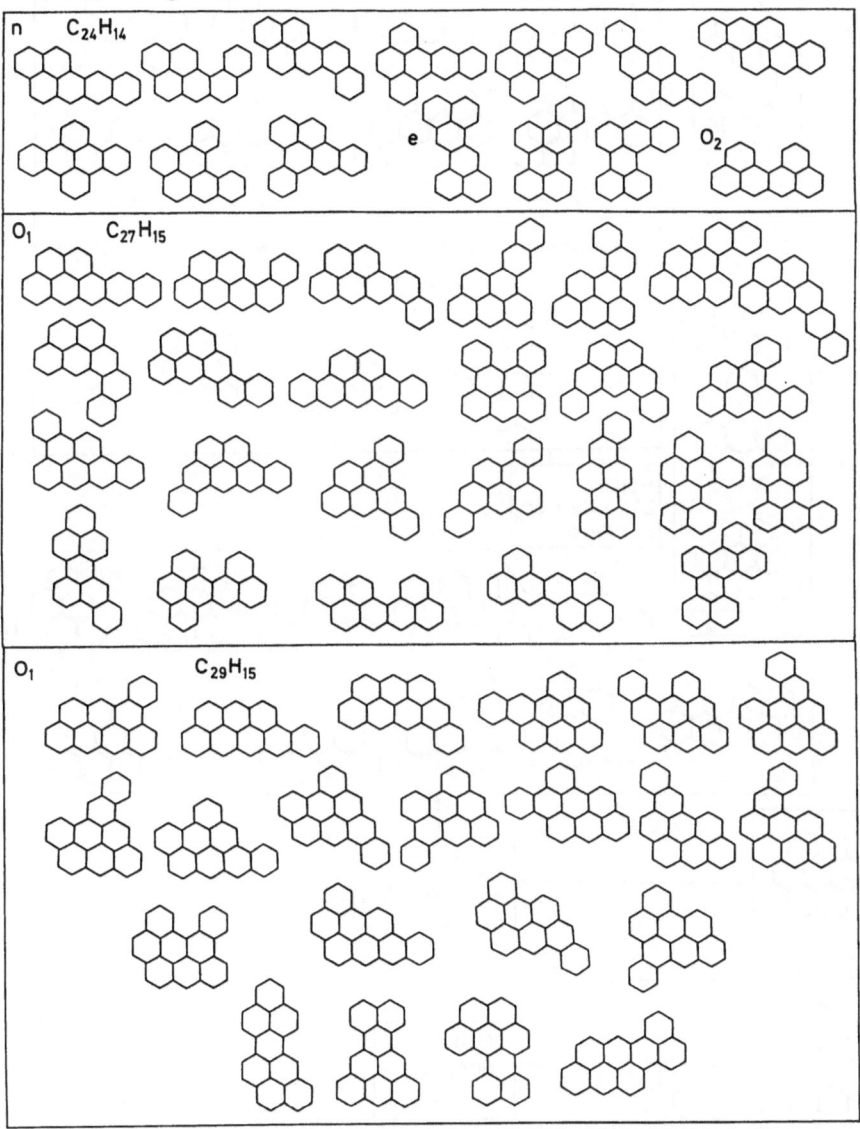

**Fig. 7.** The $C_{24}H_{14}$ ($h = 6$), $C_{27}H_{15}$ ($h = 7$) and $C_{29}H_{15}$ ($h = 8$) benzenoid isomers; see Fig. 6 for the positions in the periodic table

Jon Brunvoll, Björg N. Cyvin, and Sven J. Cyvin

## 12 Forms of Some Benzenoid Isomers

In Fig. 6–13, a number of forms of different benzenoid isomers are displayed. The contours of excised internal structures for the strictly pericondensed systems are indicated in bold lines. The depictions of this type have proved to be very useful for the studies of benzenoid isomers in general, and especially for the extremal

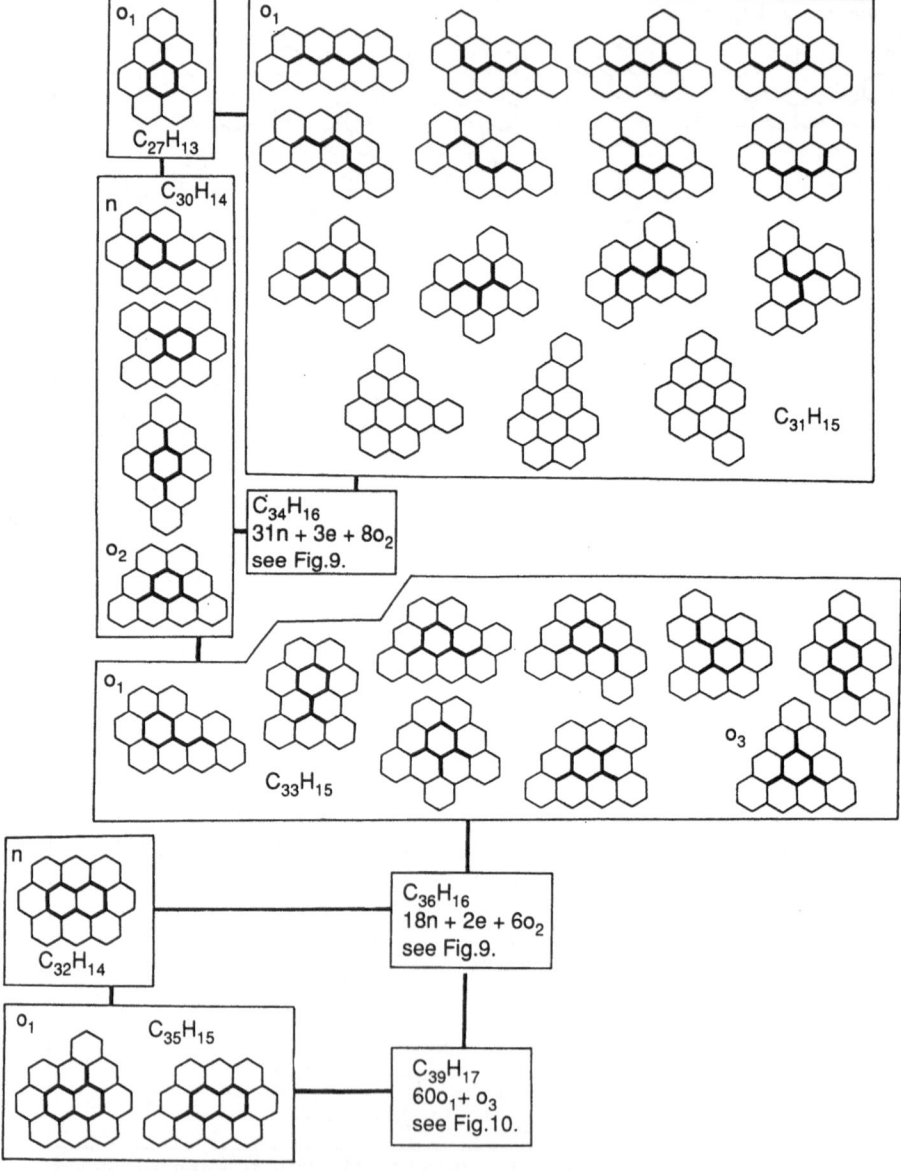

**Fig. 8** Next portion of benzenoid isomers with formulas below those of Fig. 6

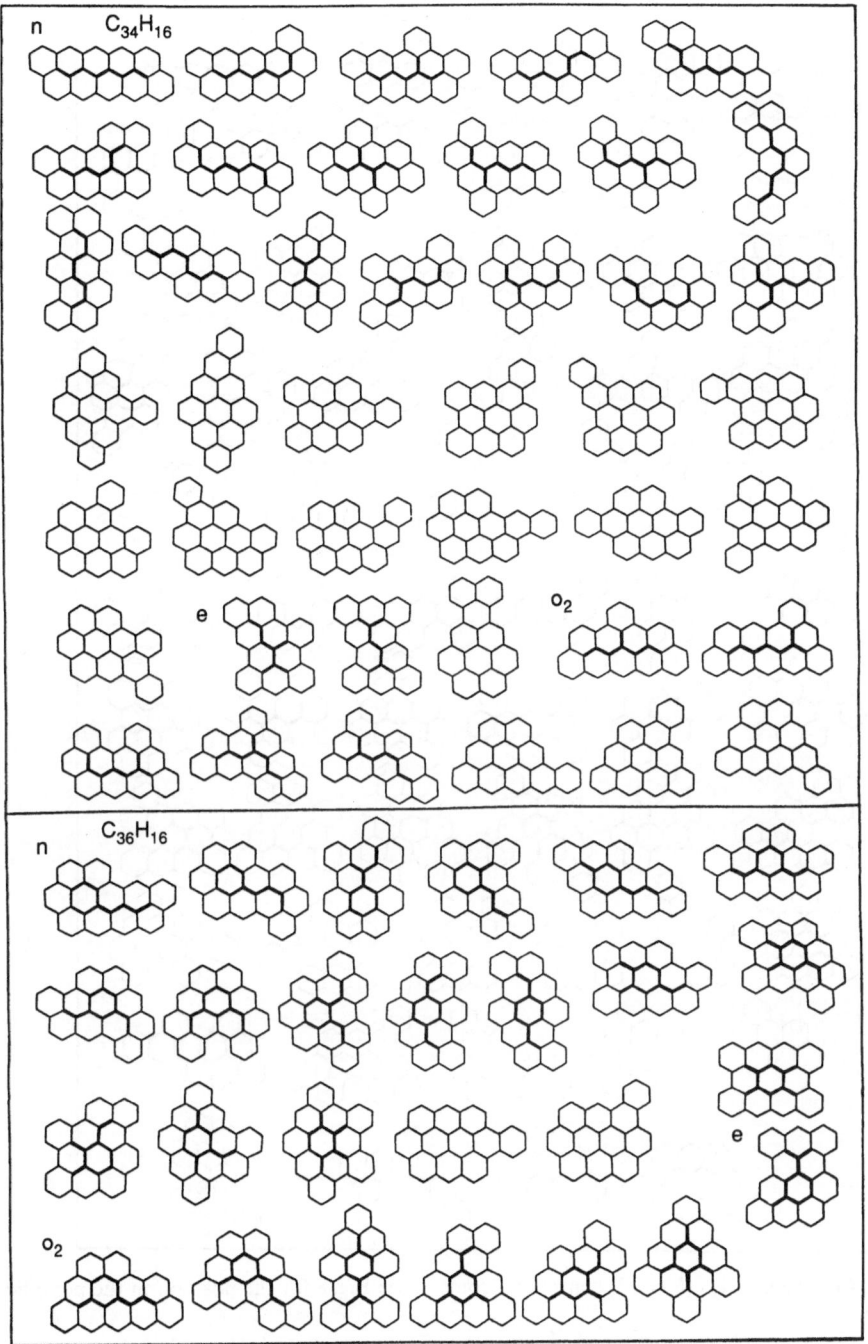

**Fig. 9.** The $C_{34}H_{16}$ ($h = 10$) and $C_{36}H_{16}$ ($h = 11$) benzenoid isomers; see Fig. 8 for the positions in the periodic table

**Fig. 10.** The $C_{39}H_{17}$ ($h = 12$) benzenoid isomers; see Fig. 8 for the position in the periodic table

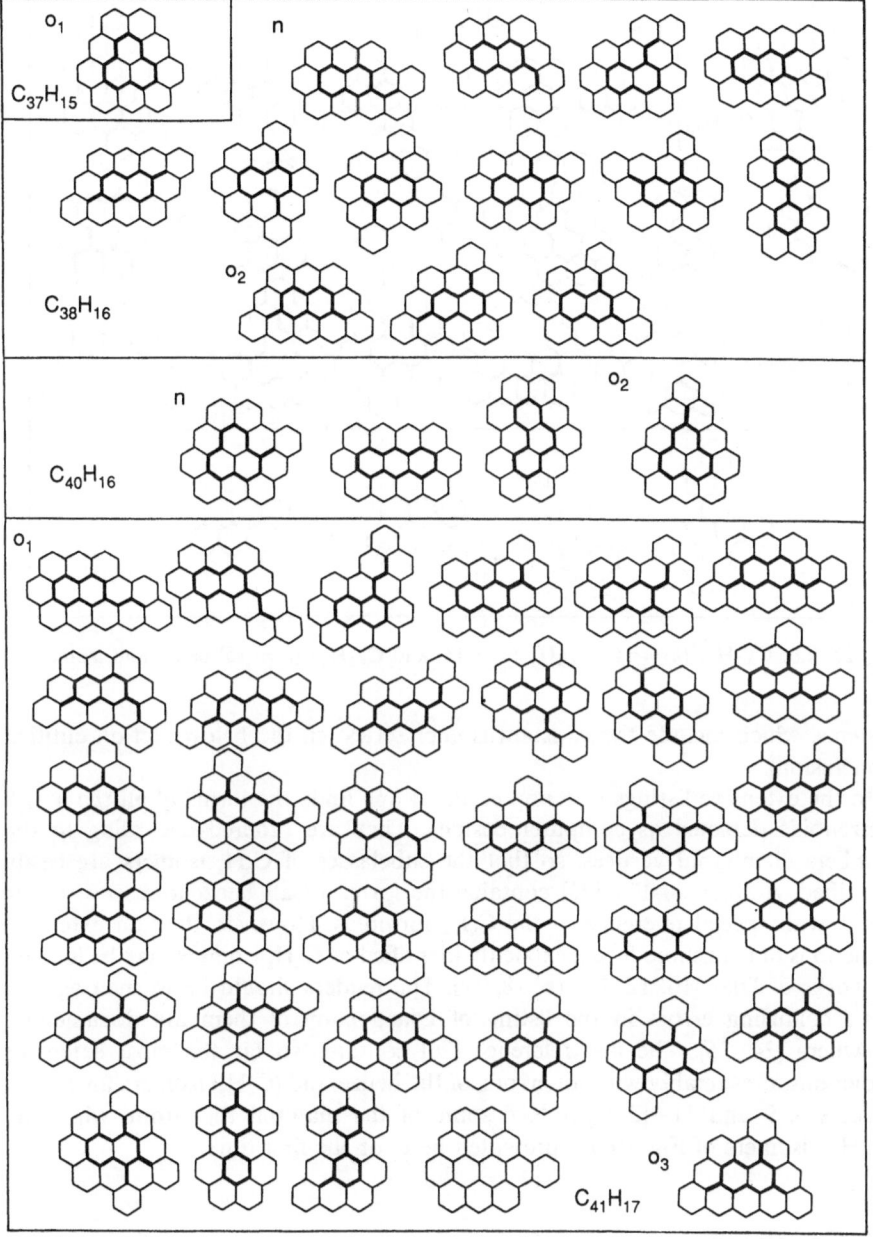

**Fig. 11.** The $C_{37}H_{15}$ ($h = 12$), $C_{38}H_{16}$ ($h = 12$), $C_{40}H_{16}$ ($h = 13$) and $C_{41}H_{17}$ ($h = 13$) benzenoid isomers

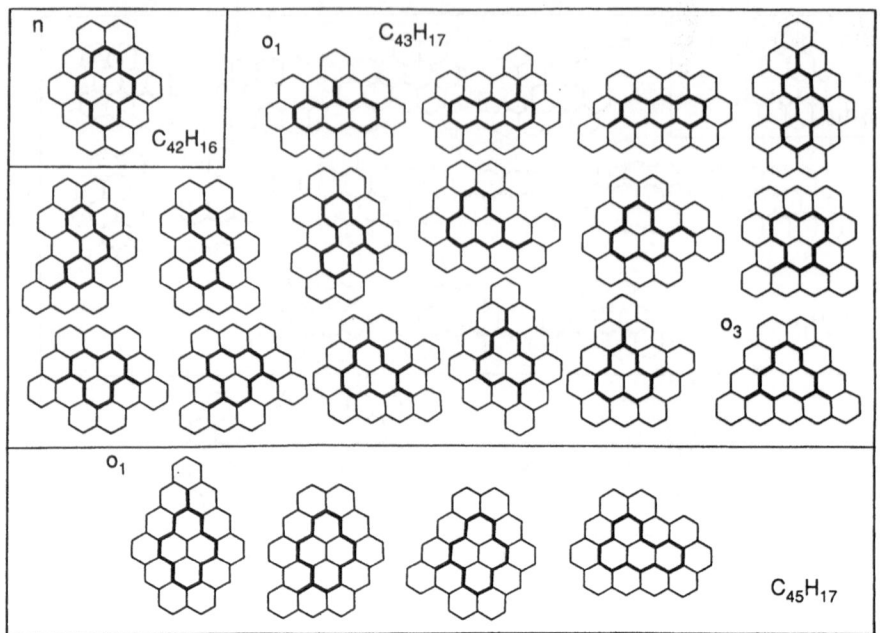

**Fig. 12.** The $C_{42}H_{16}$ ($h = 14$), $C_{43}H_{17}$($h = 14$) and $C_{45}H_{17}$ ($h = 15$) benzenoid isomers

systems, which include the constant-isomer series (cf. the below section entitled Conclusion).

In the extensive listings of Knop et al. [5] one finds the forms of all the $h \leq 9$ benzenoids depicted as computer designs. They are ordered according to the numbers of internal vertices, so that the subclasses of $C_nH_s$ isomers are easily identified. A work of Elk [23] contains the forms of all benzenoid isomers for $h \leq 5$ with special reference to the $C_nH_s$ formulas. However, the first extensive depictions of benzenoid forms explicitly identified as $C_nH_s$ isomers may be located to works of Dias [10, 12, 15, 16, 18, 19]. The readers should be warned against some confusing errors in the listing of Dias; many of them are documented elsewhere [24, 26]. The last reference [26] contains so far the most extensive depictions, presumably without errors, of the benzenoid ($C_nH_s$) isomers for $h > 9$. Figures 8, 9 and 11–13 reproduce some of the material therefrom, while the $C_{39}H_{17}$ isomers of Fig. 10 are presented here for the first time.

## 13 Conclusion

In the present chapter the studies of benzenoid isomers, or benzenoid systems compatible with a formula $C_nH_s$, are reviewed. On one hand the emphasis is laid on precise definitions and relations. Some of the relations, especially for certain upper and lower bounds, have not been published before. On the other hand a comprehensive collection of enumeration data with documentations is presented.

**Fig. 13.** The $C_{46}H_{18}$ ($h = 15$) benzenoid isomers

Jon Brunvoll, Björg N. Cyvin, and Sven J. Cyvin

Many of the numbers are new. However, many of the unclassified data from Stojmenović et al. [45] are not quoted here. All of them are found in a review by Brunvoll and Cyvin [25], who linked them explicitly to the appropriate $C_nH_s$ formulas.

Another topic of considerable interest concerns the *constant-isomer series* of benzenoids. Different problems in this connection have recently been taken up by Dias [56–58], who simultaneously managed to extend the data of Stojmenović et al. [45]. Also in our laboratory some work with these problems is in progress, but the topic does not seem to be mature enough for a review at present.

*Acknowledgement:* Financial support to BNC from The Norwegian Research Council for Science and the Humanities is gratefully acknowledged.

# 14 References

1. Cayley A (1875) Ber Dtsch Chem Ges 8: 172
2. Herrmann F (1880) Ber Dtsch Chem Ges 13: 792
3. Henze HR, Blair CM (1931) J Am Chem Soc 53: 3077
4. Davis CC, Cross K, Ebel M (1971) J Chem Educ 48: 675
5. Knop JV, Müller WR, Szymanski K, Trinajstić N (1985) Computer generation of certain classes of molecules, SKTH/Kemija u industriji (Association of Chemists and Technologists of Croatia), Zagreb
6. Trinajstić N (1983) Chemical graph theory, Vols I, II, CRC Press, Boca Raton, FL
7. Dias JR (1982) J Chem Inf Comput Sci 22: 15
8. Dias JR (1982) J Chem Inf Comput Sci 22: 139
9. Dias JR (1983) Match 14: 83
10. Dias JR (1984) J Chem Inf Comput Sci 24: 124
11. Dias JR (1985) J Macromol Sci-Chem A 22: 335
12. Dias JR (1984) Can J Chem 62: 2914
13. Dias JR (1985) Nouv J Chim 9: 125
14. Dias JR (1985) Theor Chim Acta 68: 107
15. Dias JR (1986) J Mol Struct (Theochem) 137: 9
16. Dias JR (1987) J Mol Struct (Theochem) 149: 213
17. Dias JR (1987) Thermochim Acta 122: 313
18. Dias JR (1989) J Mol Struct (Theochem) 185: 57; erratum ibid 207: 141
19. Dias JR (1987) Handbook of polycyclic hydrocarbons − Part A − Benzenoid hydrocarbons, Elsevier, Amsterdam
20. Dias JR (1985) Accounts Chem Res 18: 241
21. Dias JR (1990) In: Gutman I, Cyvin SJ (eds) Advances in the theory of benzenoid hydrocarbons (Topics in Current Chemistry 153), Springer, Berlin Heidelberg New York, p 123
22. Dias JR (1990) J Math Chem 4: 17
23. Elk SB (1980) Match 8: 121
24. Cyvin SJ (1990) J Mol Struct (Theochem) 208: 173
25. Brunvoll J, Cyvin SJ (1990) Z Naturforsch 45a: 69
26. Cyvin SJ, Brunvoll J, Cyvin BN (1991) Match 26: 27
27. Gutman I, Polansky OE (1986) Mathematical concepts in organic chemistry, Springer, Berlin Heidelberg New York
28. Cyvin SJ, Gutman I (1988) Kekulé structures in benzenoid hydrocarbons (Lecture Notes in Chemistry 46), Springer, Berlin Heidelberg New York

29. Gutman I, Cyvin SJ (1989) Introduction to the theory of benzenoid hydrocarbons, Springer, Berlin Heidelberg New York
30. Balaban AT, Brunvoll J, Cioslowski J, Cyvin BN, Cyvin SJ, Gutman I, He WC, He WJ, Knop JV, Kovačević M, Müller WR, Szymanski K, Tošić R, Trinajstić N (1987) Z Naturforsch 42a: 863
31. He WJ, He WC, Wang QX, Brunvoll J, Cyvin SJ (1988) Z Naturforsch 43a: 693
32. Balaban AT, Harary F (1968) Tetrahedron 24: 2505
33. Brunvoll J, Cyvin BN, Cyvin SJ, Gutman I (1988) Z Naturforsch 43a: 889
34. Balaban AT (1969) Tetrahedron 25: 2949
35. Harary F, Read RC, (1970) Proc Edinburgh Math Soc 17 (ser. II): 1
36. Harary F (1967) In: Harary F (ed) Graph theory and theoretical physics, Academic, London, p 1
37. Yamaguchi T, Suzuki M, Hosoya H (1975) Natural Science Report, Ochanomizu University 26: 39
38. Balasubramanian K, Kaufman JJ, Koski WS, Balaban AT (1980) J Comput Chem 1: 149
39. Bonchev D, Balaban AT (1981) J Chem Inf Comput Sci 21: 223
40. Gutman I (1982) Bull Soc Chim Beograd 47: 453
41. Džonova-Jerman-Blažič B, Trinajstić N (1982) Croat Chem Acta 55: 347
42. Knop JV, Szymanski K, Jeričević Ž, Trinajstić N (1983) J Comput Chem 4: 23
43. Trinajstić N, Jeričević Ž, Knop JV, Müller WR, Szymanski K (1983) Pure Appl Chem 55: 379
44. Knop JV, Szymanski K, Jeričević Ž, Trinajstić N (1984) Match 16: 119
45. Stojmenović I, Tošić R, Doroslovački R (1986) In: Tošić R, Acketa D, Petrović V (eds) Graph theory, Proceedings of the sixth Yugoslav seminar on graph theory, Dubrovnik, April 18–19, 1985, University of Novi Sad, p 189
46. Balaban AT, Brunvoll J, Cyvin BN, Cyvin SJ (1988) Tetrahedron 44: 221
47. Cyvin SJ, Brunvoll J (1990) Chem Phys Letters 170: 364
48. Doroslovački R, Tošić R (1984) Review of Research Fac Sci Univ Novi Sad, Mathematics Series 14: 201
49. Tošić R, Doroslovački R, Gutman I (1986) Match 19: 219
50. Hosoya H (1986) Croat Chem Acta 59: 583
51. He WC, He WJ, Cyvin BN, Cyvin SJ, Brunvoll J (1988) Match 23: 201
52. Harary F, Harborth H (1976) J Combinat Inf System Sci 1: 1
53. Cyvin SJ, Brunvoll J, Cyvin BN (1991) J Math Chem 8: 63
54. Brunvoll J, Cyvin BN, Cyvin SJ, Gutman I, Tošić R, Kovačević M (1989) J Mol Struct (Theochem) 184: 165
55. Dias JR (1989) Z Naturforsch 44a: 765
56. Dias JR (1990) J Chem Inf Comput Sci 30: 61
57. Dias JR (1990) Theor Chim Acta 77: 143
58. Dias JR (1990) J Chem Inf Comput Sci 30: 251

# Author Index Volumes 151–162

Author Index Vols. 26–50 see Vol. 50
Author Index Vols. 50–100 see Vol. 100
Author Index Vols. 101–150 see Vol. 150

*The volume numbers are printed in italics*

Allamandola, L. J.: Benzenoid Hydrocarbons in Space: The Evidence and Implications *153*, 1–26 (1990).
Astruc, D.: The Use of π-Organoiron Sandwiches in Aromatic Chemistry. *160*, 47–96 (1991)

Balzani, V., Barigelletti, F., De Cola, L.: Metal Complexes as Light Absorption and Light Emission Sensitizers. *158*, 31–71 (1990).
Barigelletti, F., see Balzani, V.: *158*, 31–71 (1990).
Bignozzi, C. A., see Scandola, F.: *158*, 73–149 (1990).
Billing, R., Rehorek, D., Hennig, H.: Photoinduced Electron Transfer in Ion Pairs. *158*, 151–199 (1990).
Brunvoll, J., see Chen, R. S.: *153*, 227–254 (1990).
Brunvoll, J., Cyvin, B. N., and Cyvin, S. J.: Benzenoid Chemical Isomers and Their Enumeration. *162*, 181–221 (1992).
Brunvoll, J., see Cyvin, B. N.: *162*, 65–180 (1992).
Bundle, D. R.: Synthesis of Oligosaccharides Related to Bacterial O-Antigens. *154*, 1–37 (1990).
Burrell, A. K., see Sessler, J. L.: *161*, 177–274 (1991).

Caffrey, M.: Structural, Mesomorphic and Time-Resolved Studies of Biological Liquid Crystals and Lipid Membranes Using Synchrotron X-Radiation. *151*, 75–109 (1989).
Chen, R. S., Cyvin, S. J., Cyvin, B. N., Brunvoll, J., and Klein, D. J.: Methods of Enumerating Kekulé Structures, Exemplified by Applified by Applications to Rectangle-Shaped Benzenoids. *153*, 227–254 (1990).
Chen, R. S., see Zhang, F. J.: *153*, 181–194 (1990).
Chiorboli, C., see Scandola, F.: *158*, 73–149 (1990).
Ciolowski, J.: Scaling Properties of Topological Invariants. *153*, 85–100 (1990).
Cooper, D. L., Gerratt, J., and Raimondi, M.: The Spin-Coupled Valence Bond Description of Benzenoid Aromatic Molecules. *153*, 41–56 (1990).
Cyvin, B. N., see Chen, R. S.: *153*, 227–254 (1990).
Cyvin, S. J., see Chen, R. S.: *153*, 227–254 (1990).
Cyvin, B. N., Brunvoll, J. and Cyvin, S. J.: Enumeration of Benzenoid Systems and Other Polyhexes. *162*, 65–180 (1992).
Cyvin, S. J., see Cyvin, B. N.: *162*, 65–180 (1992).

Dartyge, E., see Fontaine, A.: *151*, 179–203 (1989).
De Cola, L., see Balzani, V.: *158*, 31–71 (1990).
Descotes, G.: Synthetic Saccharide Photochemistry. *154*, 39–76 (1990).
Dias, J. R.: A Periodic Table for Benzenoid Hydrocarbons. *153*, 123–144 (1990).
Dohm, J., Vögtle, F.: Synthesis of (Strained) Macrocycles by Sulfone Pyrolysis *161*, 69–106 (1991).

Eaton, D. F.: Electron Transfer Processes in Imaging. *156*, 199–226 (1990).

El-Basil, S.: Caterpillar (Gutman) Trees in Chemical Graph Theory. *153*, 273–290 (1990)

Fontaine, A., Dartyge, E., Itie, J. P., Juchs, A., Polian, A., Tolentino, H. and Tourillon, G.: Time-Resolved X-Ray Absorption Spectroscopy Using an Energy Dispensive Optics: Strengths and Limitations. *151*, 179–203 (1989).

Fox, M. A.: Photoinduced Electron Transfer in Arranged Media. *159*, 67–102 (1991).

Fuller, W., see Greenall, R.: *151*, 31–59 (1989).

Gehrke, R.: Research on Synthetic Polymers by Means of Experimental Techniques Employing Synchrotron Radiation. *151*, 111–159 (1989).

Gerratt, J., see Cooper, D. L.: *153*, 41–56 (1990).

Gigg, J., and Gigg, R.: Synthesis of Glycolipids. *154*, 77–139 (1990).

Gislason, E. A.: see Guyon, P.-M.: *151*, 161–178 (1989).

Greenall, R., Fuller, W.: High Angle Fibre Diffraction Studies on Conformational Transitions DNA Using Synchrotron Radiation. *151*, 31–59 (1989).

Guo, X. F., see Zhang, F. J.: *153*, 181–194 (1990).

Gust, D., and Moore, T. A.: Photosynthetic Model Systems. *159*, 103–152 (1991).

Gutman, I.: Topological Properties of Benzenoid Systems. *162*, 1–28 (1992).

Gutman, I.: Total *u*-Electron Energy of Benzenoid Hydrocarbons. *162*, 29–64 (1992).

Guyon, P.-M., Gislason, E. A.: Use of Synchrotron Radiation to Study State-Selected Ion-Molecule Reactions. *151*, 161–178 (1989).

Harbottle, G.: Neutron Activation Analysis in Archaeological Chemistry. *157*, 57–92 (1990).

He, W. C. and He, W. J.: Peak-Valley Path Method on Benzenoid and Coronoid Systems. *153*, 195–210 (1990).

He, W. J., see He, W. C.: *153*, 195–210 (1990).

Heinze, J.: Electronically Conducting Polymers. *152*, 1–19 (1989).

Helliwell, J., see Moffat, J. K.: *151*, 61–74 (1989).

Hennig, H., see Billing, R.: *158*, 151–199 (1990).

Hesse, M., see Meng, Q.: *161*, 107–176 (1991).

Hiberty, P. C.: The Distortive Tendencies of Delocalized π Electronic Systems. Benzene, Cyclobutadiene and Related Heteroannulenes. *153*, 27–40 (1990).

Ho, T. L.: Trough-Bond Modulation of Reaction Centers by Remote Substituents. *155*, 81–158 (1990).

Holmes, K. C.: Synchrotron Radiation as a Source for X-Ray Diffraction – The Beginning. *151*, 1–7 (1989).

Hopf, H., see Kostikov, R. R.: *155*, 41–80 (1990).

Indelli, M. T., see Scandola, F.: *158*, 73–149 (1990).

Itie, J. P., see Fontaine, A.: *151*, 179–203 (1989).

Ito, Y.: Chemical Reactions Induced and Probed by Positive Muons. *157*, 93–128 (1990).

John, P. and Sachs, H.: Calculating the Numbers of Perfect Matchings and of Spanning Tress, Pauling's Bond Orders, the Characteristic Polynomial, and the Eigenvectors of a Benzenoid System. *153*, 145–180 (1990).

Jucha, A., see Fontaine, A.: *151*, 179–203 (1989).

Kavarnos, G. J.: Fundamental Concepts of Photoinduced Electron Transfer. *156*, 21–58 (1990).

Kim, J. I., Stumpe, R., and Klenze, R.: Laser-induced Photoacoustic Spectroscopy for the Speciation of Transuranic Elements in Natural Aquatic Systems. *157*, 129–180 (1990).

Klaffke, W. see Thiem, J.: *154*, 285–332 (1990).

Klein, D. J.: Semiempirical Valence Bond Views for Benzenoid Hydrocarbons. *153*, 57–84 (1990).

Klein, D. J., see Chen, R. S.: *153*, 227–254 (1990).

Klenze, R., see Kim, J. I.: *157*, 129–180 (1990).

Knops, P., Sendhoff, N., Mekelburger, H.-B., Vögtle, F.: High Dilution Reactions – New Synthetic Applications *161*, 1–36 (1991).

Koepp, E., see Ostrowicky, A.: *161*, 37–68 (1991).

Kostikov, R. R., Molchanov, A. P., and Hopf, H.: Gem-Dihalocyclopropanos in Organic Synthesis. *155*, 41–80 (1990).

Krogh, E., and Wan, P.: Photoinduced Electron Transfer of Carbanions and Carbacations. *156*, 93–116 (1990).

Kunkeley, H., see Vogler, A.: *158*, 1–30 (1990).

Kuwajima, I. and Nakamura, E.: Metal Homoenolates from Siloxycyclopropanes. *155*, 1–39 (1990).

Lange, F., see Mandelkow, E.: *151*, 9–29 (1989).

Lopez, L.: Photoinduced Electron Transfer Oxygenations. *156*, 117–166 (1990).

Lymar, S. V., Parmon, V. N., and Zamarev, K. I.: Photoinduced Electron Transfer Across Membranes. *159*, 1–66 (1991).

Mandelkow, E., Lange, G., Mandelkow, E.-M.: Applications of Synchrotron Radiation to the Study of Biopolymers in Solution: Time-Resolved X-Ray Scattering of Microtubule Self-Assembly and Oscillations. *151*, 9–29 (1989).

Mandelkow, E.-M., see Mandelkow, E.: *151*, 9–29 (1989).

Mattay, J., and Vondenhof, M.: Contact and Solvent-Separated Radical Ion Pairs in Organic Photochemistry. *159*, 219–255 (1991).

Mekelburger, H.-B., see Knops, P.: *161*, 1–36 (1991).

Meng, Q., Hesse, M.: Ring Closure Methods in the Synthesis of Macrocyclic Natural Products *161*, 107–176 (1991).

Merz, A.: Chemically Modified Electrodes. *152*, 49–90 (1989).

Meyer, B.: Conformational Aspects of Oligosaccharides. *154*, 141–208 (1990).

Moffat, J. K., Helliwell, J.: The Laue Method and its Use in Time-Resolved Crystallography. *151*, 61–74 (1989).

Molchanov, A. P., see Kostikov, R. R.: *155*, 41–80 (1990).

Moore, T. A., see Gust, D.: *159*, 103–152 (1991).

Nakamura, E., see Kuwajima, I.: *155*, 1–39 (1990).

Okuda, J.: Transition Metal Complexes of Sterically Demanding Cyclopentadienyl Ligands. *160*, 97–146 (1991).

Ostrowicky, A., Koepp, E., Vögtle, F.: The "Vesium Effect": Syntheses of Medio- and Macrocyclic Compounds *161*, 37–68 (1991).

Parmon, V. N., see Lymar, S. V.: *159*, 1–66 (1991).

Polian, A., see Fontaine, A.: *151*, 179–203 (1989).

Raimondi, M., see Copper, D. L.: *153*, 41–56 (1990).

Riekel, C.: Experimental Possibilities in Small Angle Scattering at the European Synchrotron Radiation Facility. *151*, 205–229 (1989).

Roth, H. D.: A Brief History of Photoinduced Electron Transfer and Related Reactions. *156*, 1–20 (1990).

Sachs, H., see John, P.: *153*, 145–180 (1990).

Saeva, F. D.: Photoinduced Electron Transfer (PET) Bond Cleavage Reactions. *156*, 59–92 (1990).

Sendhoff, N., see Knops, P.: *161*, 1–36 (1991).

Sessler, J. L., Burrell, A. K.: Expanded Porphyrins *161*, 177–274 (1991).

Sheng, R.: Rapid Ways to Recognize Kekuléan Benzenoid Systems. *153*, 211–226 (1990).

Schäfer, H.-J.: Recent Contributions of Kolbe Electrolysis to Organic Synthesis. *152*, 91–151 (1989).
Stanek, Jr., J.: Preparation of Selectively Alkylated Saccharides as Synthetic Intermediates. *154*, 209–256 (1990).
Stumpe, R., see Kim. J. I.: *157*, 129–180 (1990).
Suami, T.: Chemistry of Pseudo-sugars. *154*, 257–283 (1990).
Suzuki, N.: Radiometric Determination of Trace Elements. *157*, 35–56 (1990).

Thiem, J., and Klaffke, W.: Syntheses of Deoxy Oligosaccharides. *154*, 285–332 (1990).
Timpe, H.-J.: Photoinduced Electron Transfer Polymerization. *156*, 167–198 (1990).
Tolentino, H., see Fontaine, A.: *151*, 179–203 (1989).
Tourillon, G., see Fontaine, A.: *151*, 179–203 (1989).

Vögtle, F., see Dohm, J.: *161*, 69–106 (1991).
Vögtle, F., see Knops, P.: *161*, 1–36 (1991).
Vögtle, F., see Ostrowicky, A.: *161*, 37–68 (1991).
Vogler, A., Kunkeley, H.: Photochemistry of Transition Metal Complexes Induced by Outer-Sphere Charge Transfer Excitation. *158*, 1–30 (1990).
Vondenhof, M., see Mattay, J.: *159*, 219–255 (1991).

Wan, P., see Krogh, E.: *156*, 93–116 (1990).
Willner, I., and Willner, B.: Artifical Photosynthetic Model Systems Using Light-Induced Electron Transfer Reactions in Catalytic and Biocatalytic Assemblies. *159*, 153–218 (1991).

Yoshihara, K.: Chemical Nuclear Probes Using Photon Intensity Ratios. *157*, 1–34 (1990).

Zamarev, K. I., see Lymar, S. V.: *159*, 1–66 (1991).
Zander, M: Molecular Topology and Chemical Reactivity of Polynuclear Benzenoid Hydrocarbons. *153*, 101–122 (1990).
Zhang, F. J., Guo, X. F., and Chen, R. S.: The Existence of Kekulé Structures in a Benzenoid System. *153*, 181–194 (1990).
Zybill, Ch.: The Coordination Chemistry of Low Valent Silicon. *160*, 1–46 (1991).